AF386607

Diffusion of Neutrons in Nuclear Reactors

Hubert Grard

Diffusion of Neutrons in Nuclear Reactors

An Introductory Textbook for Neutronics

 Springer

Hubert Grard (Deceased)
INSTN
CEA Paris-Saclay
Gif-sur-Yvette, France

ISBN 978-3-032-05087-8   ISBN 978-3-032-05088-5   (eBook)
https://doi.org/10.1007/978-3-032-05088-5

This Springer imprint is published by the registered company Springer Nature Switzerland AG
The registered company address is: Gewerbestrasse 11, 6330 Cham, Switzerland

If disposing of this product, please recycle the paper.

Foreword

It is with great sadness that I find myself writing this foreword, as the author, Hubert Grard[†] (1967–2025), passed away suddenly mid-November 2025, at the age of just 58, shortly after finalizing the proofs for this book and, very sadly, without having seen it published. I knew Hubert very well at the beginning of his career, as he joined around 2002 the Reactor Physics Department at *Electricité De France*/R&D as an intern, where I had been working as an engineer since 1987. Hubert was then hired by EDF at the Chooz-B nuclear power plant in northern France as an operations engineer. Personal circumstances then led him to resign and join the French Atomic Energy Commission (CEA–Saclay), and his love of teaching naturally led him to the National Institute of Nuclear Science and Technology (INSTN–Saclay), the main nuclear training center in France. It was there that his skills enabled him to write a highly accomplished first little book in French on neutron kinetics, which helped train a generation of engineering students. I followed Hubert's career from the beginning, and we always stayed in touch. Our friendship grew even stronger when we worked together for years as teachers at the Institute for Technology Transfer (ITECH), a partnership between EDF and INSTN. As he embarked on a more ambitious project, writing this book: *"Diffusion of Neutrons in Nuclear Reactors: An Introductory Textbook for Neutronics"*, he kindly asked me to proofread it. Given the importance of this work on neutron diffusion, which Hubert humbly describes as an introduction, I immediately suggested that he contact Springer Verlag, the publisher of my own books, to whom I warmly recommended Hubert, supporting the project for publication in English with international reach, rather than a book in French, which would inevitably have less impact. And so, Hubert embarked on the adventure of what would become his masterpiece. One cannot help but appreciate this book for its author's highly didactic approach, richly illustrated with relevant diagrams. Hubert never hesitated to develop valuable teaching software, which was highly educational and removed the pitfalls in transmitting knowledge of a complex subject. His work on the operation of PWRs is well known in the French nuclear community. I know that Hubert cherished the idea of writing an even more ambitious book on the operation and safety of PWRs, a field in which he excelled thanks to his early career on power plant, where he was able to see all aspects of a reactor in operation. But fate decided otherwise. I hope that this book will find its audience, as it addresses concrete issues in a clear and rigorous manner. I won't hide that I am very emotional as I finish these few lines.

February 2026   Serge Marguet

Expert at the International Institute of Nuclear Energy (I2EN)

Public Investment Bank (BPI-France)

European Community (severe accidents)

Reactor physics specialist at EDF

Author of numerous books on nuclear energy

But more than that, ...friend!

Hubert Grard (1967–2025), always smiling
Rest in peace, my friend.

Competing Interests The author has no competing interests to declare that are relevant to the content of this manuscript.

Introduction

This book is intended as an introductory course in nuclear reactor theory, for the use of Master students or engineers who are newcomers into the field of nuclear engineering, as example core operation or design.

Hence, as the title also suggests, it does not cover the entire field of modern reactor physics. It aims at introducing the fundamental concepts of nuclear fission chain reaction, in a steady state or during transients. Diffusion theory is a common thread running through most of the book. The advanced numerical methods to solve the Boltzmann equation for neutron transport are a more advanced stage, which this book does not address, but prepares for. Some problems and concepts of academic or historic interest, such as the four-factor formula and the Fermi age, are emphasized because of their great didactic value, even if they may seem outdated. In order to keep this book to a reasonable size, nuclear physics has been excluded from the scope. We have assumed that the reader is already familiar with such topics as the main nuclear reactions, the fission process, and neutron cross sections.

Although it is a young science, reactor physics is a subject that has been covered in numerous works, some of which are renowned classics. The author has taken pleasure in including numerous quotations, shedding light on difficult concepts. Page references will encourage the reader to discover these classics. Each book has its own particularities, provides an original perspective and conceals its own treasures.

The book makes every effort to develop the physical sense. Numerical calculations[1] are numerous, and they are intended to establish comparisons and to acquire knowledge of orders of magnitude. All numerical calculations are based on a coherent set of data that covers the entire book. Input data is summarized in tables.

Some Pressurized Water Reactor operation modes, like subcritical reactor in hot shutdown, zero power physics testing, or critical reactor in power operation, are described with simplified models based on diffusion theory, point kinetics, with numerical calculations: theory sheds light on practice.

---

[1] Nota bene: numerous numerical values derived from calculations have been included in the LaTex manuscript, where they are automatically formatted. The number of digits shown should not be interpreted as the number of significant digits.

A grounding in nuclear physics is necessary and is also recommended in mathematics and physics. Synthetic presentations of the mathematical and physical concepts used in the book are provided in the different appendices.

Contents

List of Figures

List of Tables

Chapter 1
A Few Calculations with Monokinetic Neutrons

Abstract The aim of this chapter is to introduce the notions of macroscopic cross-section of a material, mean length of path, mean lifetime and diffusion time. This is an opportunity to offer a first approach of the migration area. These concepts are handled thanks to a few simple examples.

1.1 Macroscopic Cross Section

1.1.1 Microscopic Cross Section

The cross section for a particular nuclear reaction (scattering, capture, fission for example), which applies to a single nucleus, is called the microscopic cross section σ. It has the dimension of a surface, often expressed in barn ($1\,\mathrm{b} = 1 \times 10^{-24}\,\mathrm{cm}^2$). Microscopic cross sections depend on neutron energy. Some reactions have an energy threshold and only occur above a certain value of neutron energy. For nuclei with medium and large atomic weights, cross sections may exhibit a resonant region: at certain energy values, the cross-section increases sharply and also drops sharply.

It is interesting to note that scattering cross sections are in most of cases constant over a wide energy range. For example, the scattering cross section of $^1_1\mathrm{H}$ (Fig. 4.18), the main moderator isotope in water reactors, exhibits an extremely flat plateau of around 20 b between 0.1 eV and 10 keV.

And of course, if the neutrons are assumed to be monokinetic, then the microscopic cross section is a constant.

1.1.2 Macroscopic Cross Section

The definition of the macroscopic cross section of a material for a specific reaction is: $\Sigma = N\sigma$ (cm^{-1}). Where N is the number of nuclei per cm (Glasstone and Edlund 1952, p. 45). It characterizes the frequency of occurrence for this particular nuclear reaction per unit distance of neutron travel and has the dimension of a reciprocal length (Bell and Glasstone 1970, p. 7).

1.1.3   Macroscopic Cross Sections of a Material

If the considered material consists in an homogeneous mixture of different nuclides, then the macroscopic cross section for a given nuclear reaction is (Glasstone and Edlund 1952, p. 46):

$$\Sigma = N_1\sigma_1 + N_2\sigma_2 + \cdots + N_i\sigma_i + \cdots$$

Where N_i is the number of nuclei per cm^3 of the ith kind present and σ_i is the microscopic cross section for the given reaction.

Within a reactor, materials composed of several isotopes are found. If we ignore the effects of binding of molecules, the microscopic cross sections combine by simple addition. For example:

$$\Sigma_{H_2O} = \Sigma_H + \Sigma_O = N_H{\cdot}\sigma_H + \frac{1}{2}N_H{\cdot}\sigma_O$$

1.1.4   Total Collision Cross Section

The quantity Σ_t is defined as the total collision (or interaction) cross section of a neutron at position r having energy E (in the laboratory frame). The probability of interaction with matter over a distance dx is defined by the dimensionless value Σdx. The interaction may be for example a fission (with the associated value for the cross section, commonly written Σ_f, a scattering Σ_s, an absorption Σ_a. This probability is logically proportional to the path length, and since it is dimensionless, Σ represents the dimension of the inverse of length. The probability of interaction of a neutron coming from origin $x = 0$ with matter over the distance dx is given by $e^{-\Sigma x}\Sigma dx$. The integral $\int_0^{+\infty} e^{-\Sigma x}\Sigma dx = 1$ proves the normalization of this probability. As the probability of reaching $x + dx$ without interacting is $e^{-\Sigma(x+dx)}$, this means that the probability of no interaction of a neutron with matter across dx is given by $e^{-\Sigma dx} \leq 1$. Thus, $1 - e^{-\Sigma dx}$ is logically the probability of interaction over the distance dx, due to the normalization of the probability. A first-order Taylor development leads to Σdx for the probability of interaction and $1 - \Sigma dx$ for the probability of no-interaction.

1.2 Mean Free Path and Mean Length of Path

1.2.1 Radioactive Decay and Interactions Through Matter Have the Same Maths Formalism

λ is the probability of decay by second. The variation of the number of decaying nuclei during dt is:

$$dN = -N\lambda dt$$

We can calculate the probability that an nucleus has a lifetime t: it is the probability of survival during t times the probability of decay between t and $t + dt$.

$$e^{-\lambda t} \cdot \lambda dt$$

An integration enables to calculate the mean lifetime:

$$\tau = \lim_{x \to \infty} \int_0^x t \cdot e^{-\lambda t} \cdot \lambda dt$$

$$\int_0^x t \cdot e^{-\lambda t} \cdot \lambda dt = \lambda \left[t \left(\frac{-1}{\lambda} \right) e^{-\lambda t} \right]_0^x - \lambda \int_0^x \frac{-1}{\lambda} \cdot e^{-\lambda t} \cdot dt = \frac{1}{\lambda} \left[e^{-\lambda x}(-\lambda x - 1) + 1 \right]$$

$$\lim_{x \to \infty} \frac{1}{\lambda} \left[e^{-\lambda x}(-\lambda x - 1) + 1 \right] = \frac{1}{\lambda}$$

$$\tau = \lim_{x \to \infty} \int_0^x t \cdot e^{-\lambda t} \cdot \lambda dt = \frac{1}{\lambda}$$

Instead of decay, we now consider interactions of a neutron through matter, and instead of λdt, we write $\Sigma_t dx$ where Σ_t is the total cross section, all interacting events considered (these interactions will also be coined shocks, or collisions).

The probability that the neutron ranges x without interacting with a nucleus is $e^{-\Sigma_t x}$.

$\tau = \frac{1}{\lambda}$ is known as the mean time period between two decays. Similarly, the mean length between two shocks is $\frac{1}{\Sigma_t}$. The mean length between two shocks is coined mean free path.

1.3 Distance to Next Collision

The probability that the distance to the next collision is between x and $x + dx$ is:

$$\Sigma_t e^{-\Sigma_t x} dx$$

The probability that the shock occurs in the interval $[0, x]$ is $1 - e^{-\Sigma_t x}$.

It is possible to sample the distance x at which the shock occurs by sampling ϵ uniformly distributed in the interval $[0, 1)$: $1 - e^{-\Sigma_t x} = \epsilon$

$$x = \frac{\ln(1 - \epsilon)}{-\Sigma_t}$$

which is equivalent to sampling ϵ uniformly distributed in the interval $[0, 1)$ and calculating x:

$$x = \frac{\ln \epsilon}{-\Sigma_t}$$

This is by this way that Monte Carlo codes calculate the distance to next collision.

1.4   Mean Length of Path in a Absorbing and Scattering Medium

We consider monokinetic neutrons within an infinite and homogeneous medium, characterized by the following cross sections: a total cross section Σ_t, a scattering cross section Σ_s and an absorption Σ_a, with $\Sigma_t = \Sigma_s + \Sigma_a$. It is interesting to point out that since the probability of capture is given as $\Sigma_a dx$ in the path length dx, the probability that neutrons are captured at a certain distance from the source is not related to the crow flight distance but instead depends on the length of the zigzag trajectory.

The macroscopic cross section for absorption is Σ_a which is the probability of absorption per unit length. The probability of absorption during a shock is $\dfrac{\Sigma_a}{\Sigma_t}$. In the previous section, we established that the mean length between two scattering events is $\dfrac{1}{\Sigma_t}$.

Thus we can similarly write that the mean length of the path before an absorption is $\dfrac{1}{\Sigma_a}$.

A more complicated approach is possible. The probability that absorption occurs at the nth shock is $\left(1 - \frac{\Sigma_a}{\Sigma_t}\right)^{n-1} \cdot \frac{\Sigma_a}{\Sigma_t}$

Thus we can calculate the mean length of the neutron's path:

$$\frac{1}{\Sigma_t}\left(1 - \frac{\Sigma_a}{\Sigma_t}\right)^0 \left(\frac{\Sigma_a}{\Sigma_t}\right) + \frac{2}{\Sigma_t}\left(1 - \frac{\Sigma_a}{\Sigma_t}\right)^1 \left(\frac{\Sigma_a}{\Sigma_t}\right) + \frac{3}{\Sigma_t}\left(1 - \frac{\Sigma_a}{\Sigma_t}\right)^2 \left(\frac{\Sigma_a}{\Sigma_t}\right) + (\ldots)$$

$$= \sum_{n=1}^{\infty} \frac{n}{\Sigma_t}\left(1 - \frac{\Sigma_a}{\Sigma_t}\right)^{n-1} \left(\frac{\Sigma_a}{\Sigma_t}\right)$$

We write $p = 1 - \dfrac{\Sigma_a}{\Sigma_t}$.

So we need to calculate $\displaystyle\sum_{n=1}^{\infty} np^{n-1}$.

Let's consider the function $f(x) = \left(\dfrac{1}{1-x}\right)^2$ f is infinitely derivable on 0 which enables to write f as a Mac-laurin series:

$$f(x) = 1 + \frac{1}{1!}f^{(1)}(0)x + \frac{1}{2!}f^{(2)}(0)x^2 + \cdots + \frac{1}{n!}f^{(n)}(0)x^n + \cdots$$

$$= 1 + \frac{2}{1!}x^1 + \frac{2 \times 3}{2!}x^2 + \frac{2 \times 3 \times 4}{3!}x^3 + \cdots$$

$$= \sum_{n=1}^{\infty} nx^{n-1} = \left(\frac{1}{1-x}\right)^2$$

Thus $\displaystyle\sum_{n=1}^{\infty} np^{n-1} = \left(\dfrac{1}{1-p}\right)^2$.

The mean length of the neutron's path is then:

$$= \sum_{n=1}^{\infty} \frac{1}{\Sigma_t} np^{n-1} \left(\frac{\Sigma_a}{\Sigma_t}\right)$$

$$= \frac{1}{\Sigma_t} \left(\frac{1}{1-p}\right)^2 \left(\frac{\Sigma_a}{\Sigma_t}\right)$$

$$= \frac{1}{\Sigma_t} \left(\frac{1}{1-p}\right)$$

$$= \frac{1}{\Sigma_t} \frac{\Sigma_t}{\Sigma_a} = \frac{1}{\Sigma_a}$$

1.5 Example of Light Water

We consider light water at $20.4\,°C$. At this temperature, the energy of neutrons is $0.0253\,eV$ and $\sigma_s(^{1}_{1}H) = 30.14\,b$, $\sigma_s(^{16}_{8}O) = 3.9733\,b$.
The density of water is $0.9982\,g/cm^3$.

The scattering macroscopic cross section can be calculated, for monokinetic neutrons:

$$\Sigma_s = 0.9982 \times 30.14 \times 10^{-23} \times 2 \times \frac{N_A}{18}$$

$$+ 0.9982 \times 3.9733 \times 10^{-24} \times 1 \times \frac{N_A}{18}$$

$$= 2.1458\,cm^{-1}$$

And the absorbing macroscopic cross section:

$$\Sigma_a = 0.9982 \times 0.3321 \times 10^{-25} \times 2 \times \frac{N_A}{18}$$

$$= 0.022\,18\,\text{cm}^{-1}$$

The mean path of neutrons is:

$$\frac{1}{\Sigma_a} = 45.08\,\text{cm}$$

And the mean free path:

$$\frac{1}{\Sigma_a + \Sigma_s} = 0.4613\,\text{cm}$$

1.6 Rigorous Migration Area for Monokinetic Neutrons in an Infinite Homogeneous Medium

1.6.1 *If Scattering is Assumed to be Isotropic*

Chapter 4 deals with the neutron scattering angle and in which conditions the scattering is isotropic in the laboratory frame.

Liu et al. (2020) establish the following result in the case of an isotropic scattering in the laboratory:

$$\frac{1}{6}\overline{r^2} = \frac{1}{3\Sigma_t \Sigma_a} \tag{1.1}$$

$\overline{r^2}$ is the mean squared crow flight distance a neutron travels from the position where it is born to the position where it is absorbed.

1.6.2 *If Scattering is Not Isotropic*

In this case, anisotropy is characterized by the average cosine of the neutron scattering angle in the medium $\overline{\mu}$. The rigorous migration area for anisotropic scattering neutron transport (Liu et al. 2020) is:

$$\frac{1}{6}\overline{r^2} = \frac{1}{3\,\Sigma_t \Sigma_a\,(1 - \overline{\mu})} - \frac{\overline{\mu}\Sigma_a}{3\,\Sigma_t^2\,(1 - \overline{\mu})^2}\left(\frac{1}{\Sigma_s} - \frac{\overline{\mu}}{\Sigma_t - \overline{\mu}\Sigma_s}\right) \tag{1.2}$$

1.7 Lifetime of Monokinetic Neutrons

The neutrons are supposed to have a velocity v_0, which is supposed to remain constant in the event of scattering. During the duration dt, in an infinite medium, the length of the path is $dx = v_0\,dt$.

In a absorbing medium, the probability of absorption during this path is $\Sigma_a dx = \Sigma_a v_0 dt$.

The variation of the population of neutrons is $dn = -n\Sigma_a v_0 dt$. It can be concluded that the lifetime of neutrons is:

$$l = \frac{1}{\Sigma_a v_0} \tag{1.3}$$

The higher the absorption is, the lower the lifetime is.

1.8 Thermal Diffusion Time

Section 3.2 will show us that thermal neutrons have various speeds, statistically distributed according a Maxwellian speed spectrum. In addition, we admit that the absorption cross section $\Sigma_a(E)$ is a function of $1/v$, as as illustrated in Fig. 4.18.

The mean lifetime of neutrons at energy E is:

$$t(E) = \frac{1}{\Sigma_a(E)v(E)}$$

With $1/v$ absorption:

$$\Sigma_a(E) = \Sigma_a(E_0)\frac{v_0}{v}$$

Where E_0 is the energy at the most probable velocity v_0 (see the Chap. 3, Sect. 3.2.4). Then:

$$t(E) = \frac{1}{\Sigma_a(E_0)\frac{v_0}{v}v} = \frac{1}{\Sigma_a(E_0)v_0}$$

This result shows that for a $1/v$ absorbing medium the mean lifetime of neutrons is independent of energy E. Therefore, the mean lifetime of neutrons over a thermal distribution is called thermal diffusion time t_d (Lamarsh 1966, p. 263).

$$t_d = \frac{1}{\Sigma_a(E_0)v_0} = \frac{1}{\Sigma_a(\overline{v})\overline{v}}$$

Where $\bar{v}$ is the average velocity in Maxwellian distribution (see Sects. 3.2.2 and 3.2.4).

t_d is defined as the average time that a thermal neutron spends in an infinite scattering and absorbing medium before it is captured.

1.9   Thermal Diffusion Time in Heavy Water

We consider heavy water at 30 °C, 2 bar. The density of water is $1.1033\,\text{g·cm}^{-3}$. At $T = 30\,°\text{C}$, we calculate that $v_0 = 2235.6\,\text{m·s}^{-1}$ and $E_0 = 0.026123\,\text{eV}$.

The D_2O molecular concentration is $\frac{1.1033\,\text{g·cm}^{-3}}{20\,\text{g·mol}^{-1}} = 0.055165\,\text{mol·cm}^{-3}$, and the deuterium atomic concentration is $N_D = 6.6442 \times 10^{22}\,\text{atoms/cm}^3$.

Since $\sigma_{a,D}(E_0) = 0.49110\,\text{mb}$, we can calculate $\Sigma_{a,D}(E_0) = \sigma_{a,D}(E_0)N_D$: $\Sigma_{a,D}(E_0) = 3.2630 \times 10^{-5}\,\text{cm}^{-1}$.

An then, the lifetime of thermal neutrons in heavy water at 30 °C is $t_d = 0.13709\,\text{s}$.

1.10   Thermal Diffusion Time in Light Water

In the example of light water at 20.4 °C:

$$t_d = \frac{1}{0.02218 \times 2200 \times 100}$$
$$= 204.9 \times 10^{-4}\,\text{s}$$

This is 10 times higher than the lifetime at criticality in a PWR, which is about $1.5 \times 10^{-5}\,\text{s}$. The difference can be explained by the much higher absorption of neutrons in the fuel than in the water, and secondarily by boron absorption (boric acid in the solution, in the primary coolant) and neutron leakage.

References

G.I. Bell, S. Glasstone, *Nuclear Reactor Theory* (Van Nostrand Reinhold Company, 1970). https://books.google.fr/books?id=RNQmAQAAMAAJ

S. Glasstone, M.C. Edlund, *The Elements of Nuclear Reactor Theory* (D. Van Nostrand Company, 1952). ISBN 9780598464224. https://books.google.fr/books?id=MSRRAAAAMAAJ

J.R. Lamarsh, *Introduction to Nuclear Reactor Theory*. Addison-Wesley Series in Nuclear Engineering (Addison-Wesley Publishing Company, 1966). ISBN 9780201041262. https://books.google.fr/books?id=by5RAAAAMAAJ

Z. Liu, K. Smith, B. Forget, Calculation of multi-group migration areas in deterministic transport simulations. Ann. Nucl. Energy *140*, 107110 (2020). ISSN 0306-4549. https://doi.org/10.1016/j.anucene.2019.107110. https://www.sciencedirect.com/science/article/pii/S0306454919306206

Chapter 2
Neutron Density, Flux and Current

Abstract In transport theory, a neutron is considered to be a point particle in the sense that it can be described completely by its position and velocity vector. That makes possible to investigate the process of neutron transport, that is, the motion of the neutrons as they stream about the reactor core, frequently scattering off of atomic nuclei and being absorbed, sometimes leaking out of the reactor. In this chapter, we recall the classic definitions of fundamental neutronic quantities in transport theory: neutron angular density, angle dependant flux and angular neutron current. The time rate of change of neutron angular density is described by the transport equation, introduced in the Chap. 7. The reader can refer to Bell and Glasstone (1970, pp. 3–7).

2.1 Neutron Density

2.1.1 Neutron Angular Density

In the most general case, neutrons may have privileged directions, and they also have different energies covering a wide range.

To specify the state of an individual neutron, its direction of motion must be indicated using the unit vector $\boldsymbol{\Omega}$ which is defined as $\dfrac{v}{v}$.

The energy and angular neutron density is represented by $\nu(\boldsymbol{r}, \boldsymbol{\Omega}, E, t)$ which is the number of neutrons at time t, at the position $\boldsymbol{r}$, in the direction $\boldsymbol{\Omega}$, at the energy E per unit solid angle, per unit energy and per unit volume, e.g., per cm^3, per sr (steradian), per MeV.

Then $\nu(\boldsymbol{r}, \boldsymbol{\Omega}, E, t)dV d\Omega dE$ is the number of neutrons at time t in the element space, that is the element volume dV about $\boldsymbol{r}$, restricted to neutrons having directions within $d\Omega$ about $\boldsymbol{\Omega}$ and energies in dE about E.

H. Grard, *Diffusion of Neutrons in Nuclear Reactors*,
https://doi.org/10.1007/978-3-032-05088-5_2

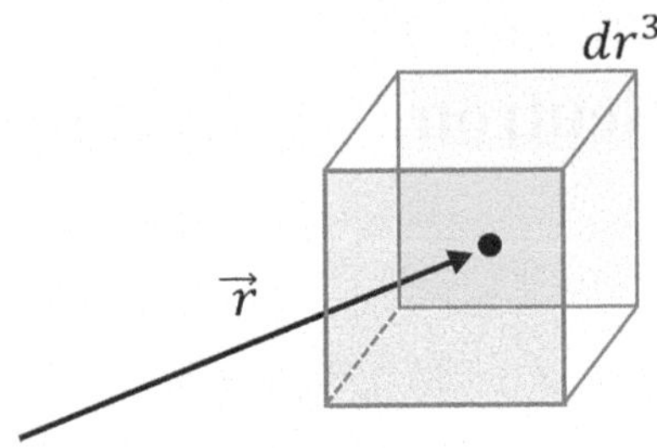

Fig. 2.1 Neutrons at the position *r*

2.1.2 Neutron Energy Density

Neutron energy density is angle integrated. If angular neutron density is integrated over 4π steradians, this defines the neutron energy density $n(r, E, t)$, which is the number of neutrons at time t, at the position *r*, at the energy E, per unit energy and per unit volume.

$$n(r, E, t) = \int_{4\pi} \nu(r, \Omega, E, t)d\Omega$$

Then $n(r, E, t)dV dE$ is the number of neutrons in the element volume dV about *r*, having energies in dE about E, at time t.

2.1.3 Neutron Angle and Energy Integrated Density

The neutron angle and energy integrated density $N(r, t)$ is the number of neutrons at the position *r* at time t (n/cm^3), regardless of their velocity vector.

$$N(r, t) = \int_0^\infty n(r, E, t)dE$$

The number of neutrons in the element volume dr^3, as indicated in Fig. 2.1 , is $N(r, t) \cdot dr^3$.

Let's make an estimation of this neutron density in the case of a PWR. We consider the French N4 (New four loops) reactor: 1450 MWe and 4350 MWth.

We assume that the energy released by fissions is 200 MeV, the lifetime of neutrons in the reactor is 1.8×10^{-5} s (see the Chap. 14, Sect. 14.3.2) and the volume of the core is 40 m^3.

The calculations steps are the following:

1. The number of fissions necessary to obtain 1 J :

$$\frac{1}{1.6022 \times 10^{-19} \times 1 \times 10^6 \times 200} = 3.1207 \times 10^{10}.$$

2. The fission rate in the N4 reactor core:
   $4.350 \times 10^9 \times 3.1207 \times 10^{10} = 1.3575 \times 10^{20}$ fission/s.
3. The average number of neutrons emitted by second:
   $2.4355 \times 1.3575 \times 10^{20} = 3.3063 \times 10^{20}$ s^{-1}.
4. The total population of neutrons in the core:
   $3.3063 \times 10^{20} \times 1.8 \times 10^{-5} = 5.9513 \times 10^{15}$.
5. The neutron density:
   $$\frac{5.9513 \times 10^{15}}{40 \times 1 \times 10^6} = 1.4878 \times 10^8 \text{ neutron/cm}^3.$$
6. The corrected thermal neutron density, taking into account the antitrap factor:
   $1.4878 \times 10^8 \times 0.72 = 1.0712 \times 10^8$
   where 0.72 is a value of the resonance escape probability p, or antitrap factor, calculated in Sect. 13.1.6. The multiplication by p gives the number of neutrons that escape to resonant absorption, complete the slowing down process and become thermal neutrons. A major part of neutron life, about 1.8×10^5 s, is spent slowing down.

This neutron density can be compared with the atomic concentration of hydrogen (atoms/cm^3) in water (155 bar, 300 °C) : 4.86×10^{22} atom/cm^3.

There are 1×10^{14} more hydrogen atoms in one cm^3 of coolant than neutrons. Contrary to common belief, very few neutrons are present in a reactor, as compared to matter.

2.2 Neutron Flux

2.2.1 Angle Dependent Flux

The angle dependent flux, or angular neutron flux, is defined by:

$$\varphi(\boldsymbol{r}, \boldsymbol{\Omega}, E, t) = v(E)\nu(\boldsymbol{r}, \boldsymbol{\Omega}, E, t) \tag{2.1}$$

2.2.2 Angle Integrated Flux

If the angular distribution of the neutron velocity vectors is isotropic, or nearly isotropic, the expression of the conservation of neutrons no longer includes the direction of neutron velocity vectors as a variable. In this case, the conservation of neutrons can be expressed by the diffusion equation (Glasstone and Edlund 1952, p. 91).

The diffusion equation is presented in the Chaps. 10 and 12.

The integral of the angular flux over all directions is the angle integrated flux, or flux spectrum. It is also the product of the neutron energy density with the speed of these neutrons :

$$\psi(\boldsymbol{r}, E, t) = v(E)n(\boldsymbol{r}, E, t) = \int_{4\pi} \underbrace{v(E)\nu(\boldsymbol{r}, \boldsymbol{\Omega}, E, t)}_{\varphi(\boldsymbol{r},\boldsymbol{\Omega},E,t)} \, d\Omega \qquad (2.2)$$

Some renewed authors call it total flux (Bell and Glasstone 1970, p. 5), but this may create a confusion with the angle and energy integrated flux.

2.2.3 Angle and Energy Integrated Flux

The angle and energy integrated flux, or neutron flux, is defined by:

$$\begin{aligned}
\Phi(\boldsymbol{r}, t) &= \int_0^\infty v(E)n(\boldsymbol{r}, E, t)dE \\
&= \int_0^\infty \psi(\boldsymbol{r}, E, t)dE \\
&= \int_0^\infty v(E) \left(\int_{4\pi} \nu(\boldsymbol{r}, \boldsymbol{\Omega}, E, t)d\Omega \right) dE
\end{aligned}$$

Glasstone and Edlund (1952, p. 48) points out that, in the case of monokinetic neutrons, the product nv is the sum of the distances traveled by all the neutrons in one cubic centimeter in one second. Similarly, $\Phi(\boldsymbol{r}, t)$ is the total path of all neutrons, by per unit volume per unit time. The unit for Φ is $n\,cm/cm^3/s$, usually written $n/cm^2/s$.

2.3 Neutron Current

2.3.1 Angular Neutron Current

The angular neutron current is a vector quantity that depends on space, speed (or energy), direction and time.

$$i(\boldsymbol{r}, \boldsymbol{\Omega}, E, t) = \boldsymbol{\Omega} \cdot \varphi(\boldsymbol{r}, \boldsymbol{\Omega}, E, t)$$

2.3.2 Angle Integrated Current

The angle integrated current, at energy E, informs about the direction and magnitude of the net neutrons movements.

$$j(r, E, t) = \int_{4\pi} i(r, \boldsymbol{\Omega}, E, t)\,d\Omega$$

2.3.3 Angle and Energy Integrated Current

The angle and energy integrated current, neutron current, is defined by:

$$J(r, t) = \int_0^{\infty} j(r, E, t)\,dE$$

The unit of the norm of J is n/cm^2/s.

Caution: the norm of the neutron current and the neutron flux have both the same unit n/cm^2/s, which stems a certain amount of confusion. These two physical quantities do not have the same physical meaning. Neutron flux is a scalar, not a vector, and is not analogous to such things as heat flux, light flux, or magnetic flux that are encountered in other branches of engineering and physics.

2.4 Reaction Rate

Reaction rates are quantities of primary importance and are defined as the number of reactions per unit volume and unit time.

For example, the fission reaction rate is: $\Sigma_f \Phi$, where Φ is the flux in the element volume dV about r.

$$R_f = \Sigma_f \Phi \tag{2.3}$$

Similarly, scattering and absorption reaction rates can be defined. By definition, reaction rates depend on angle integrated flux.

References

G.I. Bell, S. Glasstone, *Nuclear Reactor Theory* (Van Nostrand Reinhold Company, 1970). https://books.google.fr/books?id=RNQmAQAAMAAJ

S. Glasstone, M.C. Edlund, *The Elements of Nuclear Reactor Theory* (D. Van Nostrand Company, 1952). ISBN 9780598464224. https://books.google.fr/books?id=MSRRAAAAMAAJ

Chapter 3
Fission Spectrum and Thermal Spectrum

Abstract In this chapter we shall study some properties of an idealized thermal neutron spectrum in which the neutrons have a true equilibrium distribution with the moderator. Then we provide an overview of the production of neutrons by neutron induced fission. The number of neutrons emitted by fission, the share between prompt and delayed neutrons, and the prompt-neutron spectrum which is the continuous distribution of energies of prompt neutrons are presented.

3.1   Introduction

> The neutron energy spectrum is, for convenience, often divided into three energy regions: the thermal region, consisting of neutrons in thermal equilibrium with the moderator, the fast or fission region in which the neutrons from fission are produced, and the intermediate or slowing-down region which joins these two. It must be emphasized that this division is purely arbitrary and the neutron energy spectrum in a reactor is a continuous function of energy with no clearly defined boundaries.
>
> IAEA (1970)

In nuclear reactors, neutrons are born at million electron-volts (MeV) energies, and they slow down (are moderated) by elastic and inelastic collisions with the materials in the reactor until they reach the thermal range below a few electron-volts (eV). There exists therefore, in a thermal reactor, a spectrum of neutron energies covering a range of more than eight decades (IAEA 1970, p. 7). In this range, in addition to losing energy in collisions, they can also gain energy by collisions with atoms and molecules in thermal motion. After some time, the distribution of the neutrons will come into equilibrium with the thermal motion of the atoms or molecules of the material and show a Maxwellian-like shape.

Theory predicts that in the absence of absorption, thermal neutrons are distributed according to a Maxwell flux distribution and slowing-down (epithermal) neutrons obey a flux that is inversely proportional to energy.

© The Author(s), under exclusive license to Springer Nature Switzerland AG 2026

H. Grard, *Diffusion of Neutrons in Nuclear Reactors*,

https://doi.org/10.1007/978-3-032-05088-5_3

3.2 The Concept of Thermal Neutrons

As the number of neutrons in a reactor is very small, collision between neutrons may be disregarded. The nuclear instability of the neutron can also be neglected, since the life time of a neutron in a reactor is very small compared to the half-life of the neutron. Hence, if the medium is considered as non absorbing, the population of neutrons maybe assimilated to a mono atomic perfect gas.

As the neutron slows down, the kinetic energy of target nuclides due to thermal agitation at temperature T for the non-absorbing moderating medium cannot be neglected anymore. The population of neutrons can be considered as a mono atomic gas tending towards the thermodynamic equilibrium with the medium. Neutrons will thus ultimately have the same speed spectrum as the medium described by the statistical mechanics Boltzmann law, established in 1859.

Although the Maxwellian spectrum is a good approximation for many positions in a thermal reactor, neutrons only reach equilibrium with the moderator in regions where neutron absorption is small such as graphite or heavy-water thermal columns (IAEA 1970, p. 9).

This spectrum is commonly expressed in several different forms which must be clearly distinguished. Either the neutron flux or the neutron density is quoted and either may be given as a function of neutron velocity or of neutron energy (IAEA 1970, p. 7).

3.2.1 Density as a Function of Velocity or Energy Distributions

The neutron density as a function of velocity $n(v)$ is given in Eq. 3.1, and the density as a function of energy $n(E)$ in Eq. 3.2. Both of these equations are normalized to unit area.

$$n(v)dv = \frac{4}{\sqrt{\pi}} n \cdot \sqrt{\frac{m_n}{2kT}} \cdot \frac{\frac{1}{2}m_n v^2}{kT} \cdot \exp\left(-\frac{\frac{1}{2}m_n v^2}{kT}\right) dv \qquad (3.1)$$

Where: k is the Boltzmann constant, $k = 1.3806 \times 10^{-23}$ J $\cdot$ K^{-1} ; T is the moderator temperature in K ; m_n is the neutron mass in kg ; v is the neutron speed in m $\cdot$ s^{-1}.

n is the density of neutrons, n/cm^3.

$n(v)$ is the number of neutrons at velocity v per unit volume and velocity.

$\frac{n(v)dv}{n}$ is the probability that a neutron has a velocity v, within in an interval of width dv. In theory, this distribution is valid only for a non-absorbing medium. For slightly absorbing moderators, the thermal spectrum is shifted towards higher energies since absorption increases at low energy. This effect is often neglected in light water reactors.

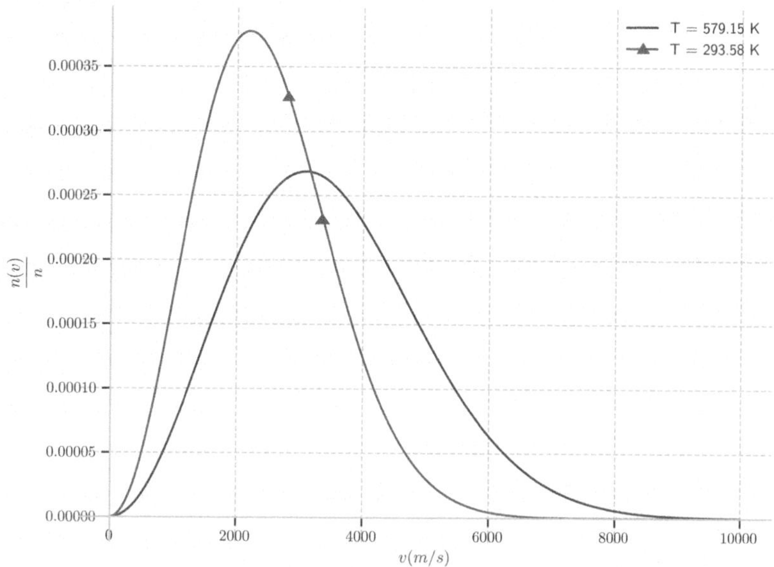

Fig. 3.1 Maxwellian distribution of velocities for two temperatures, 20.43 and 306 °C

Figure 3.1 represents $\dfrac{n(v)}{\int_0^\infty n(v)dv}$, which is written on the figure as $\dfrac{n(v)}{n}$.
The Maxwellian speed distribution can be turned into an energy distribution:

$$n(E)dE = \frac{2}{\sqrt{\pi}}n \cdot \sqrt{\frac{E}{kT}} \cdot \exp\left(\frac{-E}{kT}\right) \cdot \frac{dE}{kT} \tag{3.2}$$

It is interesting to note that:
$$\int_0^\infty n(v)dv = n, \text{ and also } \int_0^\infty n(E)dE = n.$$
And $n(v)dv = n(E)dE$.

Figure 3.2 represents $\dfrac{n(E)}{\int_0^\infty n(E)dE}$, which is written on the figure as $\dfrac{n(E)}{n}$.

From Eq. 3.1 the most probable velocity v_0 can be derived as $\frac{1}{2}m_n v_0^2 = kT$.

We write that $E_0 = kT$, and $v_0 = \sqrt{\dfrac{2E_0}{m_n}}$. Then another expression of the velocity

distribution can be established:

$$n(v)dv = \frac{4}{\sqrt{\pi}}n \cdot \left(\frac{v}{v_0}\right)^2 \cdot \exp\left(\frac{-v}{v_0}\right)^2 \frac{dv}{v_0} \tag{3.3}$$

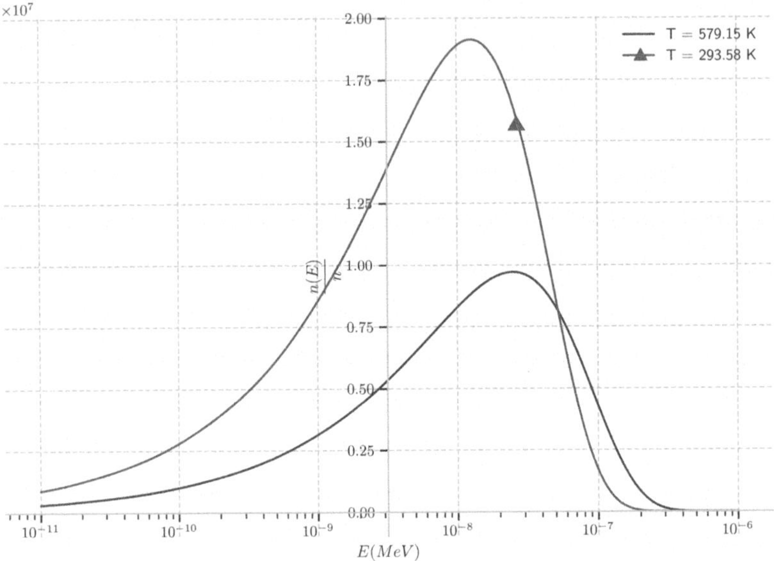

Fig. 3.2 Maxwellian distribution of energies for two temperatures, 20.43 and 306 °C

3.2.2 *Thermal Flux and Maxwellian Neutron Flux Spectrum*

The maxwellian distribution of the flux is defined by:

$$\phi(E) = v(E)n(E) = \sqrt{\frac{2E}{m_n}}n(E) = \sqrt{\frac{2E}{m_n}} \cdot \frac{2}{\sqrt{\pi}}n \cdot \sqrt{\frac{E}{kT}} \cdot \exp\left(\frac{-E}{kT}\right) \cdot \frac{1}{kT}$$

$$\phi(E) = \sqrt{\frac{2E}{m_n}} \cdot \frac{2}{\sqrt{\pi}}n \cdot \sqrt{\frac{E}{E_0}} \cdot \exp\left(\frac{-E}{E_0}\right) \cdot \frac{1}{E_0} = \sqrt{\frac{8E_0}{m_n\pi}} \cdot n \cdot \frac{E}{E_0^2} \cdot \exp\left(\frac{-E}{E_0}\right)$$

We can define $M(E)$:

$$M(E) = \frac{E}{E_0^2} \cdot \exp\left(\frac{-E}{E_0}\right)$$

The total thermal flux, or thermal flux, is defined by:

$$\phi_{th} = \int_0^\infty \phi(E)dE = \sqrt{\frac{8E_0}{m_n\pi}} \cdot n \cdot \int_0^\infty M(E)dE$$

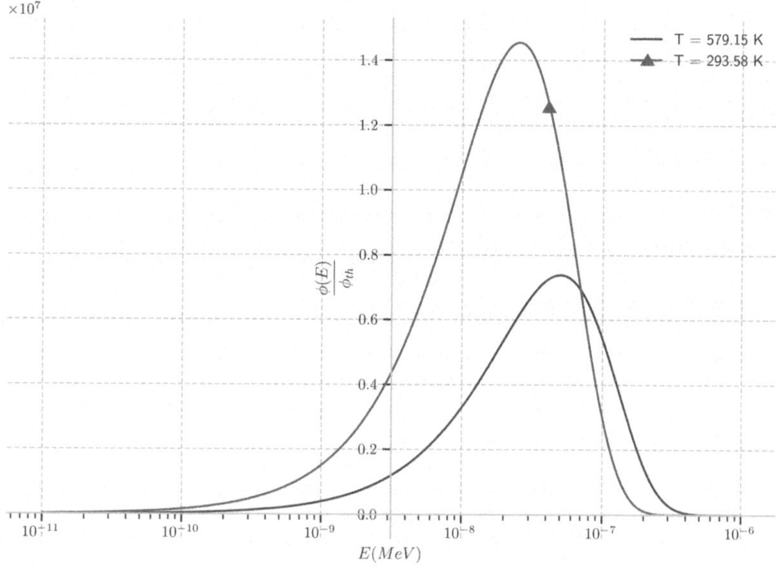

Fig. 3.3 Maxwellian neutron flux spectrum for two temperatures, 20.43 and 306 °C

$$\int_0^\infty M(E)dE = \int_0^\infty \frac{E}{E_0}\exp\left(\frac{-E}{E_0}\right)d\left(\frac{E}{E_0}\right) = \int_0^\infty xe^{-x}dx$$

Since $\int_0^\infty xe^{-x}dx = 1$, we can conclude that $\int_0^\infty M(E)dE = 1$.
And then:

$$\phi(E) = \phi_{th}\cdot M(E)$$

$$\phi_{th} = \sqrt{\frac{8E_0}{m_n\pi}}\cdot n$$

Figure 3.3 represents $\dfrac{\phi(E)}{\int_0^\infty \phi(E)dE} = \dfrac{\phi(E)}{\phi_{th}}$.

We can establish a relationship between the thermal flux and the mean velocity:

$$\frac{\phi_{th}}{n} = \frac{\int_0^\infty \phi(E)dE}{\int_0^\infty n(E)dE} = \frac{\int_0^\infty vn(v)dv}{\int_0^\infty n(v)dv} = \bar{v}$$

$$\phi_{th} = n\bar{v} \tag{3.4}$$

And then conclude that: $\overline{v} = \sqrt{\dfrac{8E_0}{m_n \pi}} = v_0 \dfrac{2}{\sqrt{\pi}}$.

The average velocity is, therefore, greater than the most probable velocity by a factor $\dfrac{2}{\sqrt{\pi}} \approx 1.128$.

3.2.3 *Example of Calculation of Density from Thermal Flux*

When a numerical value of thermal flux is given, it's always a angle and energy integrated flux.

In this example, we calculate the neutron density if the thermal flux at $306\,°C$ is $\phi_{th} = 3.7 \times 10^{13}$ n/cm^2/s, which is a realistic value for a PWR.

At $306\,°C$ ($579.15\,$K), $\overline{v} = 3486.7$ m $\cdot$ s^{-1}.

$$ n = \frac{\phi_{th}}{\overline{v}} = 1.0612 \times 10^8 \text{ n/cm}^3 $$

This is consistent with the rough estimation of the neutron density given in Sect. 2.1.3, 1.0712×10^8 n/cm^3. We can say that, at a givent time, a overwhelming majority of neutrons present in the core are thermal neutrons.

3.2.4 *Remarkable Values*

The most probable neutron speed v_0 for the Maxwellian distribution at temperature T corresponds to the maximum value of $n(v)$:

$$ \frac{1}{2} m_n v_0^2 = kT \tag{3.5} $$

$$ v_0 = \sqrt{\frac{2E_0}{m_n}} \tag{3.6} $$

The mean velocity is defined by:

$$ \overline{v} = \frac{\int v \cdot n(v) dv}{\int n(v) dv} = \frac{2}{\sqrt{\pi}} v_0 $$

The most probable neutron energy for the Maxwellian distribution at temperature T corresponds to the maximum value of $n(E)$:

$$ \frac{1}{2} kT = \frac{1}{2} E_0 $$

The most probable energy, $\frac{1}{2}E_0$ is different from the energy at the most probable velocity, which is E_0.

The average energy is:

$$\overline{E} = \frac{3}{2}E_0 \tag{3.7}$$

3.2.5 Absorption Cross Section

The reaction rate associated to a macroscopic cross section $\Sigma(E)$ in a maxwellian spectrum is:

$$\int_0^\infty \Sigma(E)\phi(E)dE$$

The reaction rate can be written as the product between the total flux ϕ_{th}:

$$\overline{\Sigma}\phi_{th}$$

Then:

$$
\begin{aligned}
\overline{\Sigma} &= \frac{\int_0^\infty \Sigma(E)\phi(E)dE}{\phi_{th}} \\
&= \frac{1}{\phi_{th}} \int_0^\infty \Sigma(E)M(E)\phi_{th}dE \\
&= \int_0^\infty \Sigma(E)\frac{E}{E_0^2} \exp\left(\frac{-E}{E_0}\right)
\end{aligned}
$$

Let's consider the important special case of a $1/v$ absorber:

$$\Sigma_a(E) = \Sigma_a(E_0)\sqrt{\frac{E_0}{E}}$$

$$
\begin{aligned}
\overline{\Sigma_a} &= \Sigma_a(E_0)\int_0^\infty \sqrt{\frac{E_0}{E}}\frac{E}{E_0} \exp\left(\frac{-E}{E_0}\right)\frac{dE}{E_0} \\
&= \Sigma_a(E_0)\int_0^\infty \sqrt{x}e^{-x}dx
\end{aligned}
$$

The integral is the gamma function $\Gamma(\frac{3}{2})$, that is worth $\frac{\sqrt{\pi}}{2}$ (see Sect. B.3). Thus:

$$\overline{\Sigma_a} = \frac{\sqrt{\pi}}{2}\Sigma_a(E_0)$$

Then it's possible to calculate the reaction rate (Beckurts and Wirtz 1964, p. 100):

$$\overline{\Sigma_a \phi_{th}} = \frac{\sqrt{\pi}}{2} \Sigma_a(E_0) \times n\, v_0 \frac{2}{\sqrt{\pi}}$$

$$\overline{\Sigma_a \phi_{th}} = \Sigma_a(E_0) \times n\, v_0$$

3.2.6 Specific Cases

Cross sections are usually given at a temperature of 20.43 °C (293.58 K). At this temperature, the most probable velocity for the neutrons is $v_0 = 2200\,\text{m} \cdot \text{s}^{-1}$. The energy at the most probable velocity is $E_0 = 0.0253\,\text{eV}$. At a temperature of 293.58 K, a non absorbing moderator bring the neutrons in a thermodynamic equilibrium, for which the most probable velocity is $v_0 = 2200\,\text{m} \cdot \text{s}^{-1}$.

Average values for velocity and energy are $\overline{v} = 2482.4\,\text{m} \cdot \text{s}^{-1}$ and $\overline{E} = 0.03795\,\text{eV}$.

3.2.7 Thermal Constants Characterizing Nuclear Cross Sections

In thermal-reactor work, people make very effective use of a few standard thermal constants to characterize nuclear cross sections. These parameters include the cross sections at the standard thermal value of $0.0253\,\text{eV}$ ($2200\,\text{m} \cdot \text{s}^{-1}$), the integrals of the cross sections against a Maxwellian distribution for $0.0253\,\text{eV}$.

The other factors (confer Table 3.1) are not explicited in this document.

The averaged value of a cross section over a Maxwellian spectrum can be calculated by:

$$\overline{\sigma_{th}} = \frac{\displaystyle\int_{10^{-5}eV}^{20E_0} \sigma(E)M(E)dE}{\displaystyle\int_{10^{-5}eV}^{20E_0} M(E)dE} \tag{3.8}$$

Table 3.1 Integral data for nuclide $_{54}{}^{135}$Xe ground state (source: JANIS)

Reaction	Library	Sig(2200)	Sig(E0)	Avg-Sigma	G-fact	Res Integ	Sig(Fiss)	Sig(E14)
MT=102:(z,γ)	JEFF-3.1.1	2.6527×10^6 b	2.6524×10^6 b	3.0679×10^6 b	7663.59 b	1.1574	7.6272×10^{-4} b	2.382×10^{-3} b

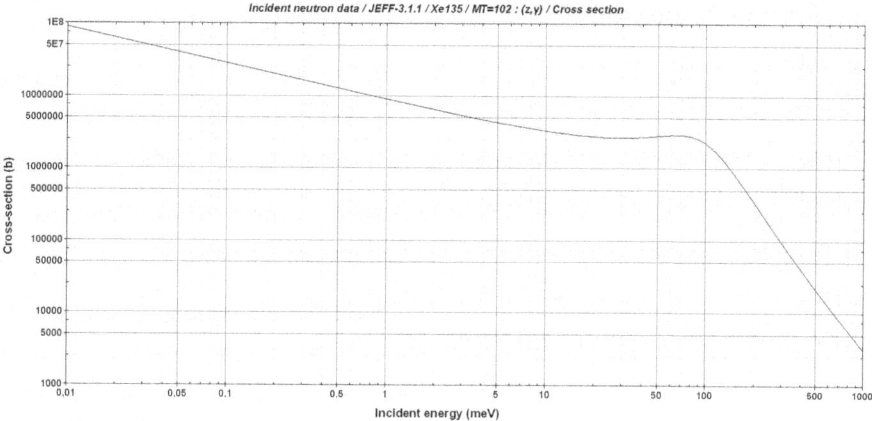

Fig. 3.4   Xenon radiative capture cross section

For example, the absorption of $^{135}_{54}$Xe can be averaged over a Maxwellian spectrum. The radiative capture cross section of $^{135}_{54}$Xe is represented on Fig. 3.4, and a resonance centered on 70 meV can be observed. $^{135}_{54}$Xe is one of a few rare isotopes presenting resonances below 0.5 eV. Xenon is one of the most absorbent fission product in the thermal spectrum, and it weighs considerably in the neutron balance.

Our numerical calculation returns an averaged value for $^{135}_{54}$Xe : $\overline{\sigma_{c,th}} = 2.9968 \times 10^6$ b.

The value displayed by JANIS (Incident neutron data/JEFF-3.1.1/Xe135/General information/Integral data) is slightly different, as show in the Table 3.1.

At 306 °C, our calculation return $\overline{\sigma_{c,th}} = 2.4290 \times 10^6$ b.

Let's explicit two columns from Table 3.1.

- Sig(2200) is the cross section at the standard thermal value of 0.0253 eV $(2200\,\text{m} \cdot \text{s}^{-1})$.

- Res Integ is the epithermal resonance integral weighted over a $1/E$ spectrum, defined as follows:

$$\int_{E=0.5\,\text{eV}}^{E=100\,\text{keV}} \sigma(E)\frac{1}{E}dE$$

Sig(2200) is the cross section at the standard thermal value of 0.0253 eV $(2200\,\text{m} \cdot \text{s}^{-1})$.

3.3 Fission Neutrons

3.3.1 Fission Yield

The independent, or direct, fission yield is the probability of formation in fission of a nuclide of a given mass number after prompt neutron emission and before any radioactive decay.

3.3.2 Neutrons Produced by Fission

After the primary fission fragments have been fully accelerated by their mutual Coulomb repulsion and their shapes have relaxed to their equilibration form, they typically deexcite by (typically sequential) neutron evaporation followed by photon radiation.

Albertsson et al. (2021, p. 7)

The relaxation of the fragment shape from its distorted form at scission to its equilibrium shape converts the change in potential energy, that adds to the excitation received at the time of scission. This total excitation energy, which is a function of the mass number A, is the energy available for neutron evaporation, and the number of neutrons evaporated is indicative of the excitation energy in the emitting fragment (Albertsson et al. 2021, p. 1).

Each fission event of a specific nuclide by a fast or thermal neutron produces various fission fragments, with an associated probability known as the fission yield for the considered nuclide by a fast or thermal neutron. Thus the specific number of neutrons emitted as a result of each fission event can vary from 0 to 7 or more for a given isotope, depending on the fission fragments produced. This distribution, or multiplicity as it is sometimes referred to, has been measured experimentally for many fissionable isotopes, and has been expressed in various empirical correlations. The exact number of neutrons released in any fission event varies from fission to fission, even when the fissions occur under identical circumstances (Nolen 2000, p. 8).

In the case of the fission of ^{235}U by thermal neutrons 4.25% of the fissions give no neutrons, 16.85% give one neutron, 32.46% give two, 29.90% give three and so on.

In almost every book on neutron or reactor physics, the only value mentioned is the mean value designated as $\bar{\nu}$, or for prompt neutrons, $\overline{\nu_p}$. The average number of neutrons released per fission is $\bar{\nu} = 2.4355$ (see Table 3.2) for thermal fission of ^{235}U, and is $\nu = 2.8836$ in the case of ^{239}Pu. The average number of neutrons released per fast fission of ^{238}U is $\bar{\nu} = 2.819$.

If we want to examine the physical reality in more detail, it's important to point out that the average number of prompt neutrons emitted per fission depends not only on the fissile nucleus, but also on the energy of the incident neutron. The report (Division and Gwin 1978) presents the experimental results obtained at the Oak Ridge Linac.[1]

Measurements were taken on the average number of prompt neutrons $\overline{\nu_p}(E)$ emitted in neutron induced fission of ^{235}U and ^{239}Pu relative to $\overline{\nu_p}(E)$ for spontaneous fission of ^{252}Cf over an incident energy range from 0.005 eV to 10 MeV.

The results show that neutron energy dependance of $\overline{\nu_p}(E)$ for ^{235}U and ^{239}U over the neutron energy region 1×10^{-3} to 1.0 eV is very small (about 1%). However, in the case of fission induced by fast neutrons, there are significant variations in the average number of neutrons emitted as a function of incident energy. It increases by around 50% between 1 and 10 MeV.

3.3.3 Delayed Neutrons

The great majority of neutrons are emitted almost along with the fission event, that is, within about 1×10^{-14} s, and these are called prompt neutrons. A small part of neutrons are emitted with some delay after the fission. These are called delayed neutrons.

$$\overline{\nu} = \overline{\nu_p} + \overline{\nu_d}$$

Very neutron-rich nuclei can emit neutrons after β^- decay when their heat reaction Q_{β^-} value is larger than the (one/two/three) neutron separation energy: $Q_{\beta^-} > S_{xn}$. This decay mode is called β^- delayed (one/two/three)-neutron emission and was discovered in 1939 by Roberts et al. (1939), shortly after the discoveries of fission by Meitner, Hahn, and Strassmann in 1938, and the neutron by Chadwick in 1932.   Dillmann et al. (2011, p. 8)

Although the delayed neutrons comprise only a very small part of the total number of neutrons generated from fission, they play a key role during for nuclear reactor control and operation. Operating a reactor would be impossible without delayed neutrons. The values from Table 3.2 enables to calculate the total fraction of delayed neutrons, defined as $\beta = \frac{\overline{\nu_d}}{\overline{\nu_p} + \overline{\nu_d}}$.

For example:

$$\beta_{^{235}\text{U}} = \frac{0.0162}{2.4355} = 0.006651$$

The number of delayed neutrons per fission neutron is designated as β or delayed neutron fraction. Although β is a small number, the delayed neutrons are, however, essential from the point of view of reactor kinetics and safety, as shown in the Chap. 14.

---

[1] The ORNL (Oak Ridge National Laboratory) electron LINAC, ORELA, began operation in 1969 and has been instrumental in providing improved neutron cross section data for many isotopes. The ORELA utilizes a pulsed gridded electron gun, a linear particle accelerator (LINAC), and a water-cooled and moderated tantalum target to generate short neutron pulses (Bigelow et al. 2006).

Table 3.2 Average number of neutrons emitted per fission (IAEA: https://www-nds.iaea.org/sgnucdat/a6.htm)

Nuclide	Type	Total-neutron yield $\bar{\nu}$	Delayed-neutron yield $\overline{\nu_d}$
^{235}U	Thermal spectrum	2.4355 ± 0.0023	0.0162 ± 0.0005
^{238}U	Fast spectrum	2.819 ± 0.020	0.0465 ±0.0024
^{238}Pu	Fast spectrum	3.00 ± 0.14	0.0047 ± 0.0005
^{239}Pu	Thermal spectrum	2.8836 ± 0.0047	0.0065 ± 0.0003
^{240}Pu	Fast spectrum	3.086 ±0.025	0.0090 ± 0.0004
^{241}Pu	Thermal spectrum	2.9479 ± 0.0055	0.0160 ± 0.0008
^{242}Pu	Fast spectrum	3.189 ± 0.035	0.0183 ± 0.0010
^{241}Am	Thermal spectrum	3.239 ± 0.024	0.0043 ± 0.0006

3.3.4 Energy Spectrum of the Fission Prompt Neutrons

The energy distribution of neutrons produced in the fission process is coined as the fission spectrum.

Some data, such as the fission cross section or the average number of neutrons emitted $\bar{\nu}$, are known with very good precision. On the other hand, other key observables, such as the energy spectrum of the fission prompt neutrons (PFNS, Prompt Fission Neutron Spectrum) are much less known.

During the fission of a nucleus induced by low energy neutrons, the prompt neutrons are emitted mainly in flight by the accelerated fission fragments. These neutrons can be thought of as evaporating from the surface of the highly excited compound nucleus during the fission process. They are not monoenergetic but follow a statistically reproducible distribution.

Such an energy distribution is difficult to determine, both experimentally and theoretically. Although no satisfactory theory has been fully developed to describe this energy spectrum theoretically, a number of empirical functions have been adjusted which provide a satisfactory fit to the experimental results.

The energy distribution of these fission spectra is often represented by the so-called Watt Distribution. In the 1950's, B. E. Watt proposed an empirical formulation valid between 0.1 MeV and 18 MeV, and some newer more precise coefficients have been provided by Lamarsh for ^{235}U (Lamarsh 1966, Eq. 3.14, p. 97):

$$\chi(E) = 0.453 e^{-1.036E} \sinh \sqrt{2.29E}$$

In which E is the neutron energy in MeV.

The probability of a neutron from fission having an energy between E and $E + dE$ is $\chi(E)dE$.

Figure 3.5 shows the shape of this function. The possibility that neutrons are emitted with an energy superior to 14 MeV is negligible.

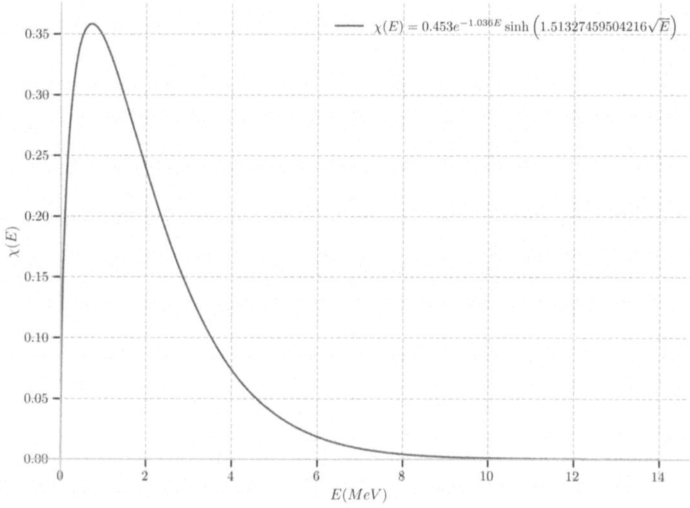

Fig. 3.5 Watt distribution of fission energy, for U-235

A numerical integration enables to check that $\displaystyle\int_0^{14} \chi(E)dE \approx 1$.

The most probable energy of a fission neutron is obtained by the condition:

$$\frac{d\chi(E)}{dE} = 0$$

Where $\frac{d}{dE}$ defines differentiation with respect to energy. In this manner the most probable energy of a neutron appearing as a result of fission can be found to have the value of 0.7238 MeV.

Of course, as soon as a neutron scatters from a nucleus its energy will decrease. The average fission neutron energy is defined by:

$$\overline{E} = \frac{\displaystyle\int_0^{\infty} E \cdot \chi(E)dE}{\displaystyle\int_0^{\infty} \chi(E)dE}$$

A numerical calculation enables to estimate $\overline{E} = 1.98$ MeV.

In a nutshell, the fast neutron spectrum can be described by the following points:

- Almost all fission neutrons have energies between 0.1 and 10 MeV
- The mean neutron energy is about 2 MeV
- The most probable neutron energy is about 0.7 MeV.

References

M. Albertsson, B.G. Carlsson, T. Døssing, P. Möller, J. Randrup, S. Åberg, Correlation studies of fission-fragment neutron multiplicities. Phys. Rev. C **103**(1), 014609 (2021)

K.-H. Beckurts, K. Wirtz, *Neutron Physics* (Springer, Berlin, Heidelberg GmbH, 1964). https://link.springer.com/book/10.1007/978-3-642-87614-1#bibliographic-information

T.S. Bigelow, C. Ausmus, D.R. Brashear, K.H. Guber, J.A. Harvey, P.E. Koehler, R.B. Overton, J.A. White, V.M. Cauley, Recent operation of the orela electron linac at ornl for neutron crosssection research, in *Proceedings of LINAC*, number 21 (2006), pp. 79–81

I. Dillmann, B. Singh, D. Abriola, *Summary Report of Consultants' Meeting on Beta-delayed Neutron Emission Evaluation*. Technical report, International Atomic Energy Agency, Vienna, Austria (2011)

Oak Ridge National Laboratory, Neutron Physics Division, R. Gwin, Measurements of the average number of prompt neutrons emitted per fission of 239 pu and 235u. Technical report (1978)

IAEA (1970) Neutron fluence measurements. Technical Report (1970). https://books.google.fr/books?id=8OAgAQAAIAAJ

J.R. Lamarsh, *Introduction to Nuclear Reactor Theory*. Addison-Wesley Series in Nuclear Engineering (Addison-Wesley Publishing Company, 1966). ISBN 9780201041262. https://books.google.fr/books?id=by5RAAAAMAAJ

S.D. Nolen, The chain-length distribution in subcritical systems (2000)

R.B. Roberts, L.R. Hafstad, R.C. Meyer, P. Wang, The delayed neutron emission which accompanies fission of uranium and thorium. Phys. Rev. **55**(7), 664 (1939)

Chapter 4
Neutron Slowing Down and Thermalization

Abstract Neutrons are born by fission at million electron-volts (MeV) energies, and they slow down (are "moderated") by elastic and inelastic collisions with the materials in the reactor until they reach the thermal range below a few electron-volts (eV). In this range, in addition to losing energy in collisions, they can also gain energy by collisions with atoms and molecules in thermal motion. After some time, the distribution of the neutrons will come into equilibrium with the thermal motion of the atoms or molecules of the material and show a Maxwellian-like shape. In transport theory, a neutron is described completely by its position and velocity vector. The slowing down theory focuses mainly on energetical aspect and leaves aside the spatial aspect. This chapter covers essential concepts: the study of elastic collision, the neutron lethargy, the slowing down current.

4.1 Theory of Classical Collision

4.1.1 The Elastic Collision

A collision or scattering event is said to be elastic if it results in no change in the internal state of any of the particles involved. Thus, no internal energy is liberated or captured in an elastic process, and consequently, both momentum and kinetic energy are conserved. We assume that neutron elastic scattering can be treated by classical "billiard ball" kinematics. Neutrons and target nuclei are considered as homogeneous and rigid spheres.

4.1.2 Calculation of the Neutron Energy After Collision

The mass of the free neutron is set to 1 as a reference, and A is the atomic mass of the target nucleus measured in neutron masses, and it's commonly referred to as the atomic weight ratio. It is important to clearly distinguish the laboratory frame, which is of capital interest to the observer, in which reactor physics calculations are carried

© The Author(s), under exclusive license to Springer Nature Switzerland AG 2026

H. Grard, *Diffusion of Neutrons in Nuclear Reactors*,

https://doi.org/10.1007/978-3-032-05088-5_4

out, from the center of mass frame, which enables certain physical calculations to be simplified. In the laboratory frame, the target is supposed to be a rest. Elastic shock is shown in the laboratory frame on Fig. 4.1, and in the center mass frame on Fig. 4.2.

The motion vectors with a lower index '1' (for example v_1) are expressed in the laboratory frame, and those with a lower index '2' (for example v_2) are expressed in the CM frame. The upper index quote (for example v_1') refers to vectors after the collision.

Small letters refer to the neutron motion vectors (for example v_1), and capital letters refer to the nuclei motion vectors (for example V_1).

The motion vector of the barycenter in the laboratory frame can be written as:

$$v_{G1} = \frac{1}{A+1} v_1$$

The CM frame is the laboratory frame translated with the motion v_{G1}. The motion vectors of neutron and nucleus, in the CM frame, can be expressed as:

$$v_1 - \frac{1}{A+1} v_1 = \frac{A}{A+1} v_1 = v_2$$

$$0 - \frac{1}{A+1} v_1 = \frac{-1}{A+1} v_1 = V_2$$

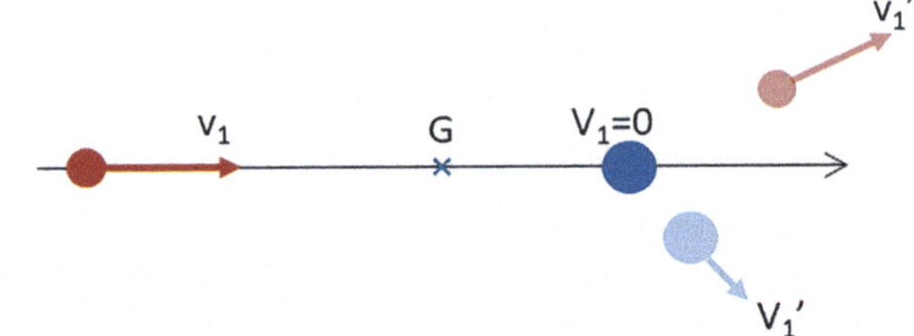

Fig. 4.1 Elastic shock represented in the laboratory frame

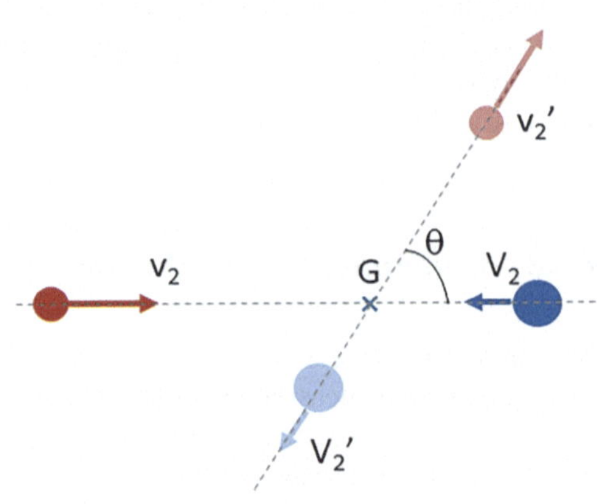

Fig. 4.2 Elastic shock represented in the center mass frame

In the CM frame, the total momentum is null before the collision:

$$p_2 = 1 \times \frac{A}{A+1} v_1 - A \times \frac{1}{A+1} v_1 = 0$$

Total kinetic energy is conserved in the collision process. In the CM frame, this conservation law can be written as:

$$\frac{1}{2} v_2^2 + \frac{1}{2} A V_2^2 = \frac{1}{2} v_2'^2 + \frac{1}{2} A V_2'^2$$

Total momentum is conserved. In the CM frame, this conservation law can be written as: $p_2' = 0$.

Thus:

$$\frac{\|v_2'\|}{\|V_2'\|} = A$$

In the CM frame, after collision $v_2' = A V_2'$ and before collision $v_2 = A V_2$.

As a result $v_2' = v_2$ and $V_2' = V_2$. In the CM frame, the motion of the neutron and the motion of the nucleus remain the same after the collision.

The collision results in a direction change of the straight line connecting neutron-barycenter-nucleus in the CM frame. The initial alignment neutron-nucleus rotates with an angle θ. The value of θ depends on the impact parameter, which is illustrated on Fig. 4.3.

We assume that the elastic collision is isotropic in the CM frame: the impact parameter is equiprobable in the interval $[0, r + R]$ and as a consequence, θ is equiprobable in the interval $[0, \pi]$.

In the laboratory frame, $v_1' = v_2' + v_{G1}$, as shown on Fig. 4.4.

θ is also the angle between v_2' and v_{G1}.

The Generalized Pythagoras Theorem enables us to write:

$$v_1'^2 = \frac{v_1^2}{(A+1)^2} + \frac{A^2 v_1^2}{(A+1)^2} + 2 \frac{A v_1^2}{(A+1)^2} \cos \theta$$

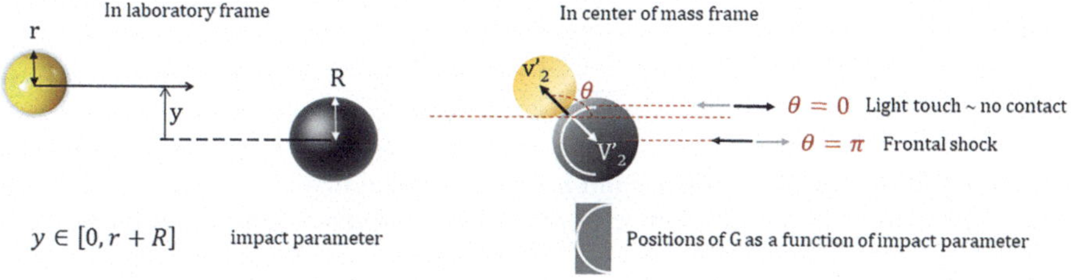

Fig. 4.3 Impact parameter and θ angle in the CM frame

Fig. 4.4 Angle deviation in
the two frames

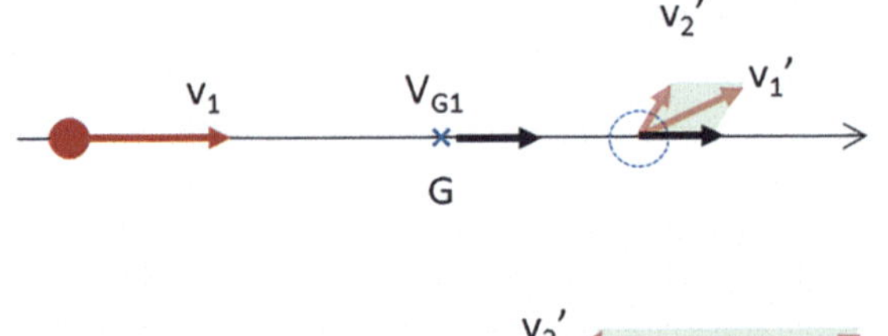

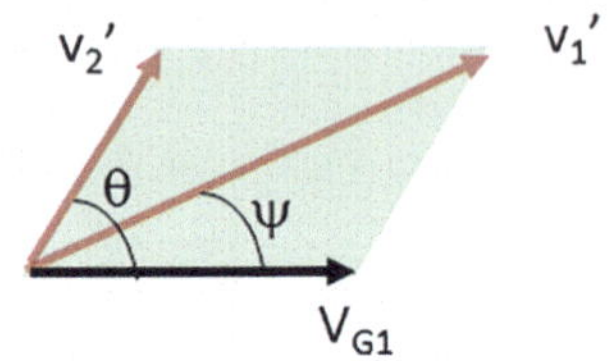

$$v_1'^2 = v_1^2 \frac{A^2 + 2A\cos\theta + 1}{(A+1)^2}$$

ψ is the angle between v_1' and initial motion (direction of the vectors v_1 or v_{G1}), in the laboratory frame. The laboratory scattering angle ψ satisfies the system of equations:

$$\begin{cases} v_1'\cos\psi = v_2'\cos\theta + v_{G1} \\[2mm] v_1'\sin\psi = v_2'\sin\theta \end{cases} \tag{4.1}$$

$$v_1'\cos\psi = \frac{A}{A+1}v_1\cos\theta + \frac{1}{A+1}v_1$$

The previous calculation of $v_1'^2$ enables to write:

$$v_1' = v_1 \frac{\sqrt{A^2 + 2A\cos\theta + 1}}{(A+1)}$$

$$\cos\psi \frac{v_1}{A+1}\sqrt{A^2 + 2A\cos\theta + 1} = \frac{v_1}{A+1}(A\cos\theta + 1)$$

$$\cos\psi = \frac{A\cos\theta + 1}{\sqrt{A^2 + 2A\cos\theta + 1}} \tag{4.2}$$

We remind that Eq. 4.2 is only valid if the target is at rest.

It is interesting to write another relationship between ψ and θ, different than Eq. 4.2.

$$\tan\psi = \frac{v_2'\sin\theta}{v_2'\cos\theta + v_{G1}}$$

Since $\mathbf{v}'_2 = \mathbf{v}_2$, we can write:

$$\tan \psi = \frac{\mathbf{v}_2 \sin \theta}{\mathbf{v}_2 \cos \theta + \mathbf{v}_{G1}}$$

$$\tan \psi = \frac{\sin \theta}{\cos \theta + \dfrac{\mathbf{v}_{G1}}{\mathbf{v}_2}}$$

And then:

$$\tan \psi = \frac{\sin \theta}{\cos \theta + \dfrac{1}{A}} \tag{4.3}$$

The case of hydrogen, for which A is almost equal to 1, is particularly interesting. For energies of neutrons greater than a few eV (it is the case for the slowing down process of fast neutrons), the scattering of neutrons by hydrogen is isotropic in the center of mass system. Equation 4.3 gives:

$$\tan \psi = \frac{\sin \theta}{\cos \theta + 1} = \tan \frac{\theta}{2}$$

And therefore $\psi = \dfrac{\theta}{2}$.

Thus, as θ varies from 0 to π, ψ ranges from 0 to $\frac{\pi}{2}$. In other words, in the laboratory, there is no back scattering of neutrons from hydrogen.

If $A = 1$, Eq. 4.2 can be written as:

$$\cos \psi = \frac{\cos \theta + 1}{\sqrt{2 + 2 \cos \theta}}$$

Which has a limit 0 when θ tends to π, thus the maximal value for ψ is $\frac{\pi}{2}$.

Figure 4.5 shows that the plane defined in the laboratory frame by the two vectors $\mathbf{v}_{G1}$ and $\mathbf{v}'_1$ can be in any rotating position around the collision axis (neutron and target at rest), and all the possible vectors $\mathbf{v}'_1$ with the same deviation ψ define a cone.

The ratio of the kinetic energy after collision over the kinetic energy before collision is:

$$\frac{E'_1}{E_1} = \frac{A^2 + 2A \cos \theta + 1}{(A + 1)^2}$$

Two borderline cases are interesting to consider:

- $\cos \theta = 1 \Rightarrow \dfrac{E'_1}{E_1} = 1$ and thus $E'_1 = E_1$. In this case, the neutron and nucleus just graze each other, this corresponds to the maximal value of the impact parameter

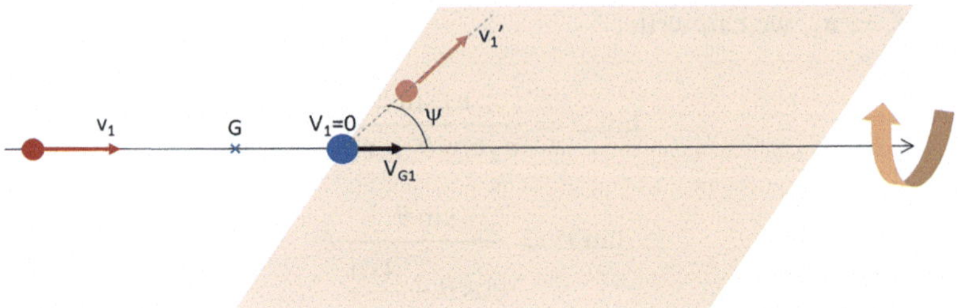

Fig. 4.5 All the possible vectors $\boldsymbol{v}'_1$ with the same deviation ψ define a cone

for a collision, equal to the sum of the radii of the neutron and nucleus. The neutron goes ahead with the same motion.

- $\cos\theta = -1 \Rightarrow \dfrac{E'_1}{E_1} = \left(\dfrac{A-1}{A+1}\right)^2$. In this case, the center of the neutron is directed towards the center of the nucleus, this corresponds to the minimal value of the impact parameter $b = 0$. If $A = 1$ (if the target is a proton, A is almost equal to 1 with A = 0.9991673), then $E'_1 = 0$: the kinetic energy of the neutron is fully transferred to the target.

We define the maximum fractional energy change in elastic scattering as:

$$\alpha \overset{\text{def}}{=} \left(\frac{A-1}{A+1}\right)^2$$

Thus:

$$\frac{E'_1}{E_1} = \frac{1}{2}\left[(1+\alpha) + (1-\alpha)\cos\theta\right]$$

After collision, the energy of the neutron E'_1 is within the range $[\alpha E_1, E_1]$. We can derivate E'_1 as a function of θ:

$$\frac{dE'_1}{d\theta} = \frac{E_1}{2}(-\sin\theta(1-\alpha))$$

$$\frac{dE'_1}{d\theta} = E_1(1-\alpha)\left(\frac{-\sin\theta}{2}\right)$$

$$\frac{dE'_1}{E_1(1-\alpha)} = \frac{-\sin\theta\, d\theta}{2}$$

The higher is θ, the more backscattered is the neutron, the lower is the final energy and the more negative is dE'. The two representations on Fig. 4.6 are equivalent.

After collision, the neutrons are scattered in a equiprobable manner in the 4π steradians around the impact, in the CM frame.

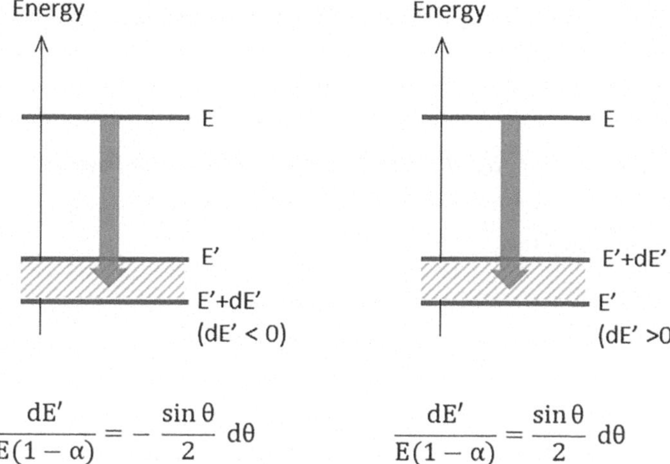

$$\frac{dE'}{E(1-\alpha)} = -\frac{\sin\theta}{2}\, d\theta \qquad\qquad \frac{dE'}{E(1-\alpha)} = \frac{\sin\theta}{2}\, d\theta$$

Fig. 4.6 Final energy E' after a shock, with initial energy E

The solid angle between the cones of angles θ and $\theta + d\theta$, as a proportion of the total solid angle of 4π steradians is:

$$\frac{r\, d\theta\, 2\pi r\, \sin\theta}{4\pi r^2} = \frac{\sin\theta\, d\theta}{2}$$

All the solid angles between the cones of angles θ and $\theta + d\theta$ are equiprobable. As a consequence, the following fractions are equiprobale:

$$\frac{dE_1'}{E_1(1-\alpha)} = \frac{-\sin\theta\, d\theta}{2}$$

$\dfrac{dE_1'}{E_1(1-\alpha)}$ can be interpreted as the probability that after the collision of a neutron with an initial energy E_1, the final kinetic energy of the neutron is in the interval $[E_1', E_1' + dE_1']$.

Which is consistent with the fact that $\displaystyle\int_0^\pi \frac{\sin\theta}{2}\, d\theta = 1$.

The density of probability that an incident neutron of energy E_1 shifts to an energy E_1' after a shock is $p(E_1 \to E_1')$.

$$p(E_1 \to E_1') = \frac{1}{E_1(1-\alpha)}$$

$$\int_{\alpha E_1}^{E_1} p(E_1 \to E_1')\, dE_1' = 1$$

We can conclude that after a collision, the kinetic of the neutron is equiprobable in the interval $[\alpha E_1, E_1]$.

4.1.3 Anisotropy of Scattering in the Laboratory Frame

We already established that:

$$\cos \psi = \frac{A \cos \theta + 1}{\sqrt{A^2 + 2A \cos \theta + 1}}$$

$$\overline{\cos \psi} = \int_0^\pi \frac{A \cos \theta + 1}{\sqrt{A^2 + 2A \cos \theta + 1}} \cdot \frac{\sin \theta \, d\theta}{2}$$

$$= \frac{1}{2} \int_{-1}^1 \frac{1 + Ax}{\sqrt{A^2 + 2Ax + 1}} \cdot dx$$

All calculations done,

$$\overline{\mu} \overset{\text{def}}{=} \overline{\cos \psi} = \frac{2}{3A}$$

Here, we use the symbol μ for the scattering cosine in the laboratory system. Thus, the nominal $\overline{\mu}$ is $2/3$ for ^1_1H and $1/3$ for ^1_2H.

This result tells us that for light target nuclei, scattering occurs forward. The center of mass causes the angular distributions in the laboratory system to be forward-peaked. The fact that neutron scattering with hydrogen is highly forward-peaked strongly impacts the mean square displacement and migration area, and facilitates the penetration in the medium.

4.1.4 Logarithmic Characterization of the Neutrons Slowing Down

The loss of energy during collision can be expressed by the logarithm of the ratio between the two energies, before and after collision: $x = \ln \dfrac{E_1}{E'_1}$.

$dP(E'_1)$ is the probability that the energy after collision is in the interval $[E'_1, E'_1 + dE'_1]$.

$$dP(E'_1) = \frac{dE'_1}{E_1(1 - \alpha)} \tag{4.4}$$

$$\frac{dP(E_1')}{dE_1'} = \frac{1}{E_1(1-\alpha)}$$

Let's define $x \overset{\text{def}}{=} \ln \frac{E_1}{E_1'}$ as the logarithmic energy decrement, or logarithmic loss of kinetic energy, resulting from a collision. $dP(x)$ is the probability that, after collision, the logarithmic energy decrement is in the interval $[x, x + dx]$.

$$\frac{dx}{dE_1'} = \frac{-E_1}{E_1'^2} \frac{E_1'}{E_1} = \frac{-1}{E_1'}$$

$$dx = \frac{-dE_1'}{E_1'}$$

From Eq. 4.4, we can write:

$$dP(x) = \frac{-E_1' \, dx}{E_1(1-\alpha)}$$

$$\frac{dP(x)}{dx} = \frac{1}{1-\alpha} e^{-x} \tag{4.5}$$

Equations 4.4 and 4.5 enable us to draw the density of probability as a function of energy, or as a function of the logarithmic energy decrement, as shown on Fig. 4.7.

$$\overline{x} = \int_0^{\ln \frac{1}{\alpha}} x \, dP(x) dx$$

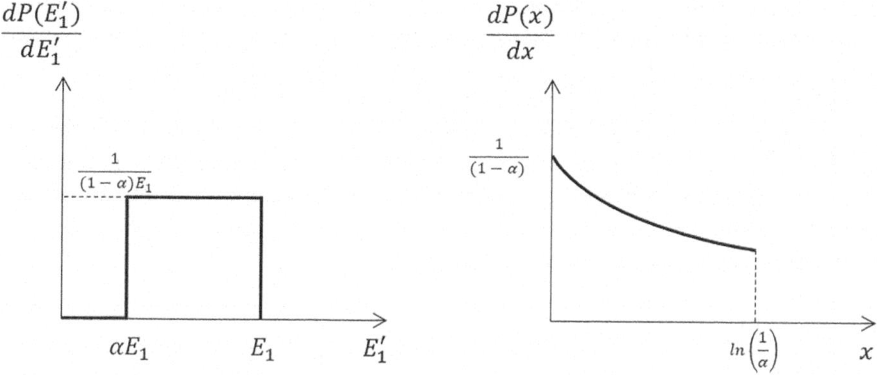

Fig. 4.7 Density of probability as a function of energy (left), and as a function of the logarithmic energy decrement (right)

$$= \frac{1}{1-\alpha} \int_0^{\ln \frac{1}{\alpha}} x e^{-x} dx = \frac{1}{1-\alpha}(\alpha \ln \alpha + (1 - \alpha))$$

$$= 1 + \frac{\alpha}{1-\alpha} \ln \alpha$$

The mean value of the logarithmic loss of kinetic energy, $\overline{x}$ is represented by $\xi \overset{\text{def}}{=} \overline{x}$.

$\xi \overset{\text{def}}{=} \overline{x}$, which is called the average logarithmic energy decrement per collision is a key material constant describing energy transfers during a neutron slowing down.

It is important to note that ξ is identical for each collision, and does not depend on the initial energy before a collision. Since $\xi = 1 + \dfrac{\alpha}{1-\alpha} \ln \alpha$, it depends on the mass of the target nucleus.

Application: calculation of the average number of collisions on protons to slow down a neutron from 2 MeV to 0.025 eV.

$$n\xi = \ln \frac{2 \cdot 10^6}{0.025} = 18.2$$

For protons, $\xi = 0.999993$. Thus, the average number of collisions is almost 18.

4.1.5 0 K Scattering Model

Nota bene: in this subsection, the incident neutron is denoted by E' and the scattered neutron by E.

The 0 K scattering model, which assumes that the target is at rest, enabled us to establish Eq. 4.4, in which neutrons always downscatter into a uniform distribution.

Figure 4.8 shows the energy transfer function $\frac{dP(E)}{dE} = p(E' \to E)$, normalized to 1, in the case of a slowing down on hydrogen nucleir at rest. Since $\alpha = 0$ in the case of protons, the lowest energy of scattered neutrons is zero.

This model is valid if the kinetic energy of the incident neutron is significantly higher than kT, where T is the absolute temperature of the scattering medium. This condition is not met in two cases: for high medium temperature, and for slowed down neutrons.

The differential cross section is defined by:

$$\sigma_s(E' \to E) \overset{\text{def}}{=} \sigma_s(E) p(E' \to E)$$

And then we can write:

$$\frac{\sigma_s(E' \to E)}{\sigma_s(E)} = \frac{dP(E)}{dE}$$

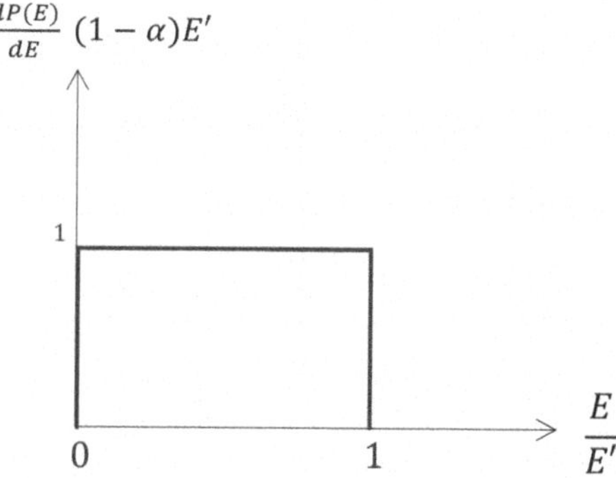

Fig. 4.8 Energy transfer function for 0 K proton gas

The elastic scattering cross section is supposed to be independant of energy, which is true in the case of ^{1}H as shown on Fig. 4.18, thus $\sigma_s(E)$ is denoted σ_{sf}, as the energy independant scattering cross section. And in this case, we can write:

$$\frac{\sigma_s(E' \to E)}{\sigma_{sf}} = \frac{dP(E)}{dE}$$

4.1.6 Scattering on Target Nuclei with a Maxwell-Boltzmann Velocity Distribution

A calculation of the differential cross section $\sigma_s(E' \to E)$, i.e. the outgoing energy distribution of a neutron elastically scattered by a target nuclei with a Maxwell-Boltzmann velocity distribution for an ideal, monatomic gas has been carried out by Beckurts and Wirtz (1964, p. 182) and later by Bell and Glasstone (1970, p. 335).

With $\eta = \frac{A+1}{2\sqrt{A}}$ and $\rho = \frac{A-1}{2\sqrt{A}}$, it can be established (Beckurts and Wirtz 1964, p. 182) that:

$$\frac{dP(E)}{dE} = \frac{\eta^2}{2E'}\left\{ e^{\frac{E'-E}{kT}} \operatorname{erf}\left(\eta\sqrt{\frac{E'}{kT}} - \rho\sqrt{\frac{E}{kT}}\right) + \operatorname{erf}\left(\eta\sqrt{\frac{E}{kT}} - \rho\sqrt{\frac{E'}{kT}}\right) \right.$$
$$\left. - \left| e^{\frac{E'-E}{kT}} \operatorname{erf}\left(\eta\sqrt{\frac{E'}{kT}} + \rho\sqrt{\frac{E}{kT}}\right) - \operatorname{erf}\left(\eta\sqrt{\frac{E}{kT}} + \rho\sqrt{\frac{E'}{kT}}\right) \right| \right\} \quad (4.6)$$

Where erf is the error function, see Eq. B.10. Equation 4.6 enables to plot the energy transfer function for various kinetic energies of the incident neutron, characterized by $\frac{E'}{kT}$, as shown on Fig. 4.9a.

The two following situations: any energy neutrons scattering on a proton at rest, and fast neutrons with $E' \gg kT$ scattering on thermal protons, are equivalent and the energy transfer function is a constant function. When $\frac{E'}{kT}$ tends to infinite, the thermal

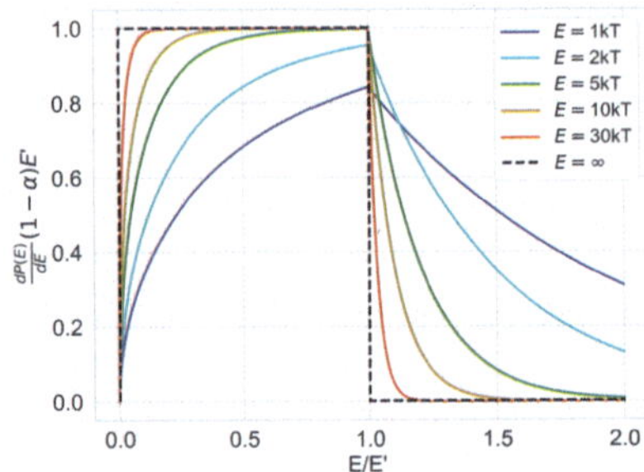

(a) Energy transfer for proton gas at temperature T (K)

(b) Energy transfer for a gas of oxygen nuclei at temperature T (K)

Fig. 4.9 a Energy transfer for proton gas at temperature T (K). **b** Energy transfer for a gas of oxygen nuclei at temperature T (K)

velocities of the scattering nuclei can be neglected compared to the velocity of the neutron, and energy transfer function tends to a constant function. This is equivalent to say that, for a given low neutron energy, the energy transfer function tends to a constant function when $kT \to 0$.

An important obervation is the possibility, and increasing likelihood of upscattering (Friant et al. 2023, p. 1994) with increasing target nuclei temperature, or decreasing neutron velocity (decreasing $\frac{E'}{kT}$).

Figure 4.9a and b allow us to estimate that upscattering occurs more or less if $\frac{E'}{kT} < 100$. In the case of a PWR, if the average coolant temperature at rated power is for example 306.4 °C, this correspond to a most probable energy $kT = 0.05$ eV; and thus upscattering occurs for incident neutron energy below 5 eV.

4.1.7 Lethargy and Average Lethargy Increase

The lethargy is defined as: $u \overset{\text{def}}{=} \ln \dfrac{E_0}{E}$ where $E_0 = 10$ MeV.

From the definition, we write $du = -\dfrac{dE}{E}$.

$\phi(E)dE$ is the total lengths of neutron paths cm^3/s in the energy band dE. This total lengths of neutron paths can be written in lethargy units, in the same band of energy:

$$\phi(E)dE = -\phi(u)du$$

The negative sign is due to the fact that if dE is a positive variation of energy, the corresponding lethargy variation du is negative. And thus, we can write:

$$\phi(u) = E \cdot \phi(E)$$

Lethargy increases as energy decreases (during successive collisions) and is zero for a chosen reference energy E_0. To obtain only positive values of lethargy, the reference energy may be taken as the maximum energy of neutrons emitted by fission (10 MeV).

Since the lethargy gain resulting from a collision is the logarithmic energy decrement, we can write the average lethargy increase per collision:

$$\xi \overset{\text{def}}{=} \overline{x} = 1 + \frac{\alpha}{1-\alpha}\ln\alpha$$

Another interesting parameter γ can be calculated from the mean square value of the lethargy gain $\overline{x^2}$, which is also written $\overline{\Delta u^2}$.

$$\gamma \overset{\text{def}}{=} \frac{1}{2\xi}\cdot\overline{x^2} = \frac{1}{2\xi}\cdot\int_{\ln\frac{1}{\alpha}}^{0} x^2 dP(x)dx$$

$$\int_{\ln\frac{1}{\alpha}}^{0} x^2 dP(x)dx = \frac{1}{1-\alpha}\int_{\ln\frac{1}{\alpha}}^{0} -x^2 e^{-x}dx$$

$$= \frac{1}{1-\alpha}\left(\left[x^2 e^{-x}\right]_{\ln\frac{1}{\alpha}}^{0} + 2\int_{0}^{\ln\frac{1}{\alpha}} x\cdot e^{-x}dx\right)$$

$$= \frac{1}{1-\alpha}\left[x^2 e^{-x}\right]_{\ln\frac{1}{\alpha}}^{0} + 2\xi$$

$$= \frac{1}{1-\alpha}\left(-\ln\frac{2}{\alpha}\cdot e^{\ln\alpha}\right) + 2\xi$$

$$= \frac{-\alpha}{1-\alpha}\left(\ln\frac{1}{\alpha}\right)^2 + 2\xi$$

Thus:

$$\gamma = 1 - \frac{\alpha}{1-\alpha}\cdot\frac{\left(\ln\frac{1}{\alpha}\right)^2}{2\xi}$$

Figure 4.10 shows that for heavy nuclei, the average lethargy increase tends to be null.

If the slowing down medium is a compound (e.g. light water H_2O, heavy water D_2O), a mean value of $n\xi$ can be calculated. Example for light water:

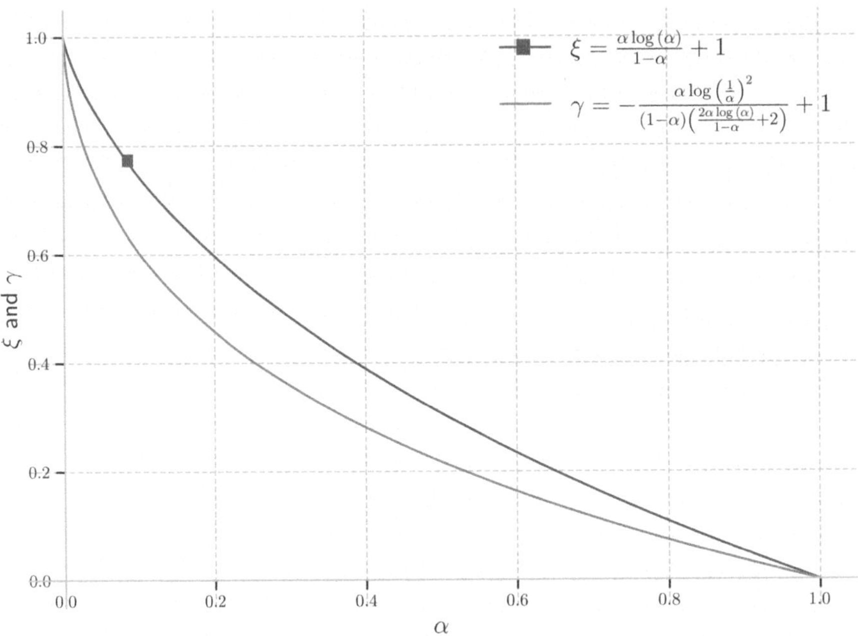

Fig. 4.10 Average lethargy gain by collision as a function of α

$$\bar{\xi} = \frac{\xi_H \Sigma_s^H + \xi_O \Sigma_s^O}{\Sigma_s^H + \Sigma_s^O} = \frac{\xi_H \sigma_s^H \times 2 + \xi_O \sigma_s^O}{2 \times \sigma_s^H + \sigma_s^O}$$

In the case of water, scattering cross sections in the range $1\,\text{eV}{-}10\,\text{keV}$ are almost constant, and it is also the region of resonance. With cross sections values at $10\,\text{keV}$, we calculate:

$$\bar{\xi}(\text{H}_2\text{O}) = \frac{1 \times 2 \times 19.22 + 0.120 \times 3.83}{42.27} = 0.92$$

The same calculation with heavy water gives:

$$\bar{\xi}(\text{D}_2\text{O}) = \frac{0.725 \times 2 \times 3.367 + 0.120 \times 3.83}{10.564} = 0.506$$

4.1.8 Transfer Probability and Arrival Density

It is interesting to write the Eq. 4.5 under a new form, introducing lethargy. The transfer probability is:

$$p(u \rightarrow u') = \frac{e^{-(u'-u)}}{1-\alpha} \tag{4.7}$$

Equation 4.7 is valid in the case of a monoatomic (all target nuclei have the same mass) slowing down, and isotropic in the center of mass system.

The number of neutrons scattering within the interval of lethargy du, per unit volume and per unit time is:

$$\Sigma_s(u)\phi(u)du$$

And the density of arrival at energy u', from neutrons scattering within the interval of lethargy du, per unit volume and per unit time:

$$\Sigma_s(u)\phi(u)du \times p(u \rightarrow u')$$

An integration gives the density of arrival at lethargy u', from neutrons scattering at any lower lethargy:

$$\rho(u') = \int_{-\infty}^{u'} \Sigma_s(u)\phi(u)p(u \rightarrow u')du \tag{4.8}$$

$\rho(u')du'$ is the number of neutrons arriving in the lethargy interval $[u', u' + du']$ from scattering events occuring at any lethargy u below u', per unit time and per unit volume. The lethargy differential cross section is defined by:

$$\Sigma_s(u \rightarrow u') = \Sigma_s(u)p(u \rightarrow u')$$

4.1.9 Examples: Slowing Down in Water and Heavy Water

4.1.9.1 Slowing Down in Water: Point Fission Source in a Water Sphere

To begin with, we consider the slowing down of fast fission neutrons emitted in an infinite hydrogen medium with the energy spectrum presented in Sect. 3.3.4, and we take the average energy 2 MeV as an example of fission neutron kinetic energy.

The neutron undergoes its first collision after a straight-line travel that is, on the average, one mean free path, i.e., the distance $^1/_{\Sigma_s^H\,(2\,\text{MeV})}$. At this point, since $\xi_H = 1$, its lethargy increases by one unit, on the average, which is equivalent to a division of its energy by a factor $\exp(1)$.

Figure 4.18 shows that from 1 eV to about 10 keV, the cross section σ_s^H is almost constant, and thereafter decreases rapidly with energy. The cross section for a second collision is thus much larger, and the second mean free path is considerably shorter than the first. This will be repeated during the first two or three collisions, until the energy of the neutron reaches the cross section constant area. As a consequence,

Table 4.1 Slowing down of fission neutrons in hydrogen

Collision no	Energy (eV)	$\lambda_s = \frac{1}{\Sigma_s^H}$ (cm)	$\overline{\mu}_{H_2O}$
	2.0000×10^6	5.1375	0.66667
1	7.3576×10^5	2.9862	0.66667
2	2.7067×10^5	1.7858	0.66667
3	9.9574×10^4	1.1711	0.66667
4	3.6631×10^4	0.90264	0.66667
5	1.3476×10^4	0.79845	0.66667
6	4957.5	0.75503	0.66667
7	1823.8	0.74301	0.66667
8	670.93	0.73516	0.66667
9	246.82	0.73326	0.66667
10	90.800	0.73151	0.66667
11	33.403	0.73127	0.66667
12	12.288	0.73073	0.66667
13	4.5207	0.72937	0.57737
14	1.6631	0.72581	0.53081
15	0.61180	0.71642	0.48118
16	0.22507	0.69226	0.37879
17	0.082799	0.63431	0.28248
18	0.030460	0.52235	0.17138

after the first two or three first collisions, each collision occurs near the site of the previous one.

It is also interesting to observe the evolution of the average cosine of the neutrons scattering angle in water during the slowing down process: Fig. 4.17 shows that scattering tends to be isotropic when the neutron slows down.

The two facts, that the scattering is virtually no more peaked forward when the neutron is thermalized, and that the mean free path are significantly reduced after a few collisions as demonstrated with hydrogen, explain why, after the first collisions, the neutron's movements are confined to a small volume, until it is captured by hydrogen. Table 4.1 shows the evolution of the mean free path in hydrogen and the mean cosine of the scattering angle in water during successive shocks, each time increasing the lethargy by 1.

4.1.9.2 Slowing Down in Water: Point Fission Source in a Water Sphere

A point fission source of neutrons emits neutrons within water at 20 °C and 1 bar. The trajectories of neutrons are represented on Fig. 4.11. The trajectories illustrate that the neutrons are thermalized away from the source, which is characterized by

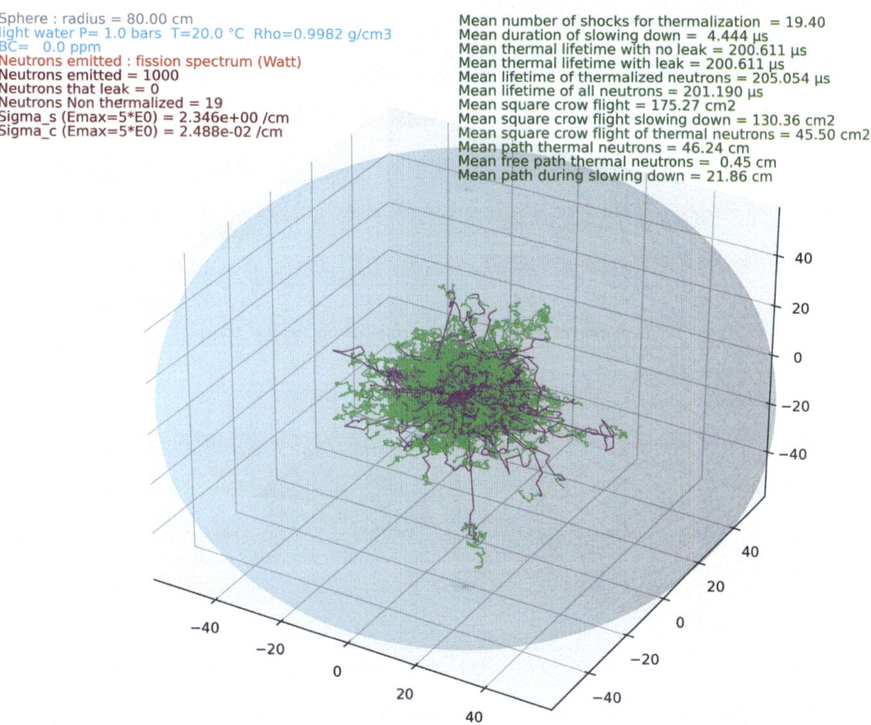

Fig. 4.11 Trajectories of a batch of neutrons emitted by a fission source. Thermalized neutrons are represented by green paths

the mean squared crow flight distance between the source and the point at which a neutron is thermalized. Once thermalized in light water, neutrons remain confined in a small volume, until absorbption.

Section 9.4 will show that, although hydrogen is primarily responsible for the kinetic energy decrease of neutrons in water, oxygen nuclei play an important role in determining the average squared crow flight distance between the point where the fission neutron is emitted and the point where it is thermalized.

Figure 4.12 enables to vizualize the slowing down process of a small batch of 1000 neutrons. In this calculation, the mean number of shocks for thermalization is estimated to be 19, greater than the average number of collisions on protons previously calculated (18.2) due to the smaller average logarithmic energy decrement per collision of oxygen nuclei.

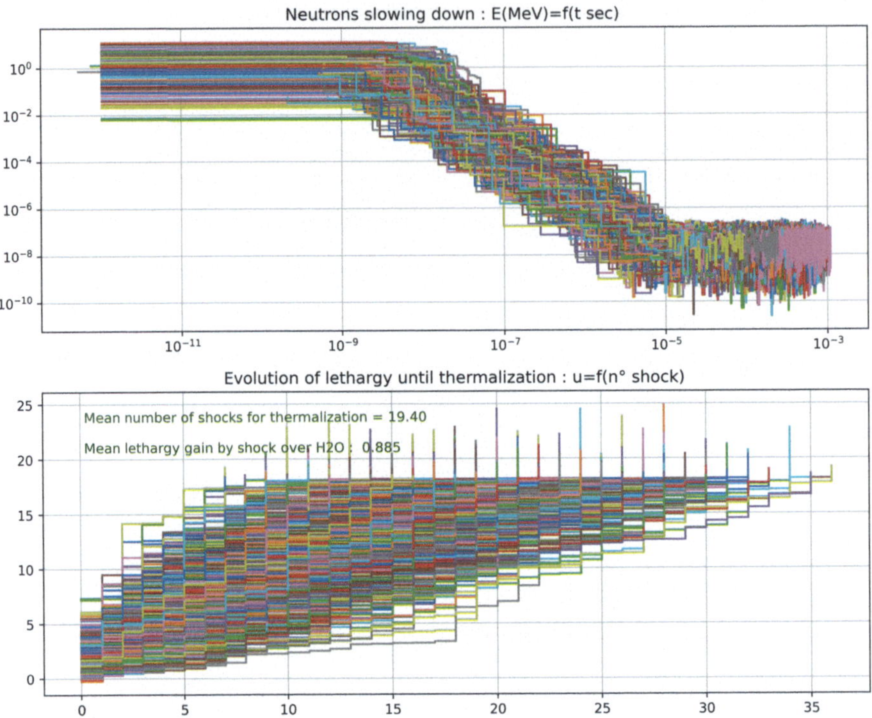

Fig. 4.12 Energy loss as a function of time until absorption; lethargy gain as a function of the number of shocks (lethargy curves are drawn until thermalization is achieved)

4.1.9.3 Slowing Down in Heavy Water: Point Fission Source in a Heavy Water Sphere

4.1.10 Other Key Constants to Compare Moderators

A good moderator must be:

- Light. A small value of the nucleus mass A implies a small α, hence a high ξ.
- Diffusive. That means a high σ_s. The more scattering collisions we have, the faster the neutron will be thermalized.
- Non absorbant. This means a low σ_a. We want to keep the neutrons for fission, so we don't want them to be absorbed during slowing down.
- Dense. Higher density means higher $\Sigma_s = N\sigma_s$, this increases the occurence of collision (cm^{-1}).

We have defined the probability of elastic scattering reaction and the average energy loss during the reaction. The product of these variables (the logarithmic energy decrement and the macroscopic cross section for scattering in the material) is the slowing down power (SDP).

Table 4.2 Different industrial moderators

Moderator	State	$\xi\Sigma_s$ (cm^{-1})	Σ_a (cm^{-1})	$\dfrac{\xi\Sigma_s}{\Sigma_a}$
Water (298 K)	Liquid	1.5	0.02	75
Heavy water	Liquid	0.18	8×10^{-6}	2.2500×10^4
Graphite	Solid	0.063	4×10^{-4}	157

$$SDP = \xi\Sigma_s$$

The SDP describes the ability of a given material to slow down neutrons and indicates how rapidly a neutron will slow down in the material. Still, it does not fully reflect the effectiveness of the material as a moderator. In fact, a material with high SDP can slow down neutrons with high efficiency, but it can be a poor moderator because of its high probability of absorbing neutrons (Borio et al. 2004, p. 177). It is typical, for example, for boron, which has a high slowing down power but is absolutely inappropriate as a moderator. The complete measure of the effectiveness of a moderator is the Moderating Ratio (MR), also called specific slowing-down power, where:

$$MR = \xi\frac{\Sigma_s}{\Sigma_a}$$

MR characterizes a moderator by taking into account both the scattering and absorption phenomena. This ratio is around 70 for light water but is greater than 20000 for heavy water.

Table 4.2 summarizes the properties of the various moderators used in the nuclear industry. A light-water reactor cannot become critical with natural uranium (0.711% of $^{235}_{92}$U).

4.2 Shape of the Equilibrium Spectrum

4.2.1 Calculation of the Flux as a Function of Energy

Our objective is to calculate how the neutrons are distributed, according to their energy, in an infinite and scattering medium at equilibrium (with no time dependence).

The slowing down current $q(E)$ (/cm^3/s), is the number of neutrons crossing the energy E by slowing down, per unit volume and per unit time. $q(u)$ or $q(E)$ is sometimes called slowing down density. This is quite inaccurate, since $q(u)$ is not a density with respect to u, $q(u)$ has a physical sense and not $q(u)du$.

$F(E)$ is the scattering reaction rate density, or collision density (Glasstone and Edlund 1952, p. 148): $F(E) = \Sigma_s(E) \cdot \phi(E)$, which is the number of scattering reactions, at energy E, per unit time and per unit energy.

The scattering reaction rate in the energy band dE is:

$$F(E) \cdot dE = \Sigma_s(E) \cdot \phi(E) \cdot dE \quad (/\text{cm}^3/\text{s})$$

The meaning is the number of neutrons scattered out of the dE band, per unit volume and per unit time.

The neutrons from the interval $\left[E, \dfrac{E}{\alpha} \right]$ have the possibility to cross the level of energy E after one single collision.

The probability that a neutron of energy $E' \in \left[E, \dfrac{E}{\alpha} \right]$ crosses the level of energy E is:

$$\frac{E - \alpha \cdot E'}{E'(1 - \alpha)}$$

The neutrons scattering out of the band $dE' : F(E' \cdot dE')$.
Thus the crossing of E by neutrons coming from dE':

$$\frac{E - \alpha \cdot E'}{E'(1 - \alpha)} \cdot F(E')dE'$$

This may be integrated along the interval $\left[E, \dfrac{E}{\alpha} \right]$:

$$q(E) = \int_E^{\frac{E}{\alpha}} F(E') \cdot \frac{E - \alpha \cdot E'}{E' \cdot (1 - \alpha)} dE'$$

We now assume that the neutrons are not directly coming from the fission source, and have already undergone a first collision. This means that the variable E is significantly below the inferior limit of the fission spectrum, and we can solve the problem and write the asymptotic solution. The neutrons scattered towards the band dE from higher energies are:

$$\int_E^{\frac{E}{\alpha}} F(E') \cdot \frac{dE}{E'(1 - \alpha)} dE'$$

Which is also $F(E)dE$.
Thus the asymptotic scattering reaction rate density:

$$F(E) = \int_E^{\frac{E}{\alpha}} \frac{F(E')}{E'(1 - \alpha)} dE'$$

One can easily check that $F(E) = \frac{C}{E}$ is a solution of this previous equation:

$$\int_E^{\frac{E}{\alpha}} \frac{C}{E'^2(1-\alpha)} \cdot dE' = \frac{C}{\alpha-1}\left[\frac{1}{E'}\right]_E^{\frac{E}{\alpha}} = \frac{C}{\alpha-1}\left[\frac{\alpha}{E} - \frac{1}{E}\right] = \frac{C}{E}$$

We can conclude that $F(E) = \dfrac{C}{E}$.

Then, the neutrons slowing down current can be written as:

$$q(E) = \int_E^{\frac{E}{\alpha}} \frac{C}{E'} \cdot \frac{E - \alpha E'}{E'(1-\alpha)} dE'$$

$$q(E) = \int_E^{\frac{E}{\alpha}} \frac{C}{1-\alpha} \cdot \frac{E - \alpha E'}{E'^2} dE'$$

$$E \int_E^{\frac{E}{\alpha}} \frac{1}{E'^2} dE' = E\left[\frac{-1}{E'}\right]_E^{\frac{E}{\alpha}} = 1 - \alpha$$

$$-\alpha \int_E^{\frac{E}{\alpha}} \frac{E'}{E'^2} dE' = -\alpha\left[\ln E'\right]_E^{\frac{E}{\alpha}} = \alpha \ln \alpha$$

We thus have:

$$q(E) = \frac{C}{1-\alpha}(1 - \alpha + \alpha \ln \alpha) = C(1 + \frac{\alpha \ln \alpha}{1-\alpha}) = C\xi$$

$$\frac{q(E)}{\xi} = E \cdot F(E) = \Sigma_s(E)\,\phi(E)\,E$$

And then:

$$\phi(E) = \frac{q(E)}{\Sigma_s(E)E\xi} \tag{4.9}$$

With lethargy unit, the flux can be written as a function of the asymptotic slowing down current:

$$\phi(u) = \frac{q(u)}{\Sigma_s(u)\xi} \tag{4.10}$$

4.2.2 Conditions for $^1/_E$ Shape

We consider an infinite and non absorbing medium supplied by an homogeneous neutron source. This last assumption is very important since its consequence it that net transfers of neutrons in space can be disregarded, which is not the case in Fermi's theory of age presented in the Sect. 9.1.5.

Given the assumptions (no boundary, non absorbing medium and homogeneous source of neutron), we can consequently write that:

$$\frac{\partial q(r, E)}{\partial E} = 0$$

If there are no losses, by leakage or absorption, during the slowing down process, the same number of neutrons will slow down per unit time past all energies in a steady state, and this number is equal to the number of source neutrons.

If Q_0 neutrons/cm^3/s are produced in the system:

$$\phi(E) = \frac{Q_0}{\Sigma_s(E)\xi} \frac{1}{E} \tag{4.11}$$

The previous Eq. 4.11 for $\phi(E)$ can also be written in the case of an homogeneous absorption small as compared to scattering. In a strong absorber, the flux will depart from the pure $1/E$ law because of the decrease of q with decreasing E.

If the medium is hydrogen, then $\xi = 1$ and Eq. 4.11, written in lethargy units, becomes:

$$F(u) = \sigma_s(u)\phi(u)$$

The collision density per unit lethargy in hydrogen is constant at all values of the lethargy (or energy) and is equal to the source strength (Glasstone and Edlund 1952, p. 150).

The epithermal or intermediate energy region (between the fast and thermal regions), with the above assumptions, gives a shape close to $1/E$ at equilibrium. This $1/E$ dependence is caused by the nature of the slowing down process. In this region, $\Sigma_s(E)$ varies only a little and the elastic scattering removes a constant fraction of the neutron energy per collision, independent of energy. Since the neutrons lose a constant fraction of kinetic energy per collision, the energy-dependent neutron flux does "pile up" at lower energies.

Below fission energies, the fission spectrum turns into a $1/E$ curve which makes junction with Maxwellian flux shape. At thermal energies, we see typical Maxwellian flux shapes from thermalization. This is illustrated by the drawing on Fig. 4.13.

4.2.3 The $1/E$ Tail Added to the Maxwellian

If we consider a real reactor, for example a PWR, the neutrons undergo absorption during the slowing down due to the resonance region of ^{238}U (cf. Sect. 5.1) and hence the epithermal region does not show a flux with a $1/E$ shape.

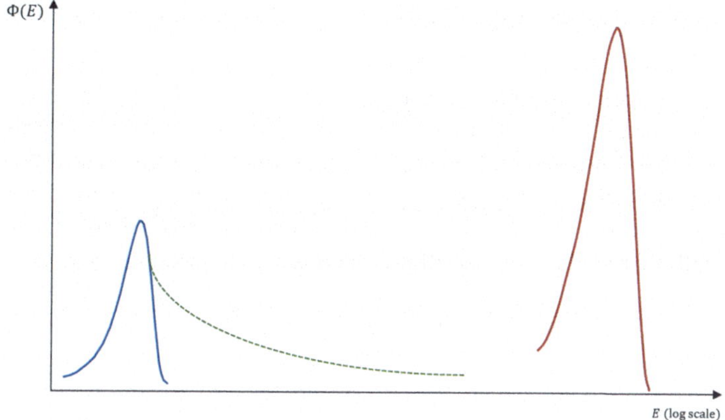

Fig. 4.13 Drawing of the Maxwellian shape connected to $1/E$ slowing down and fission spectrum, in the case of no absorption during slowing down

Fig. 4.14 Comparison of a real spectrum with a Maxwell distribution (Marguet 2018, Fig. 7.5)

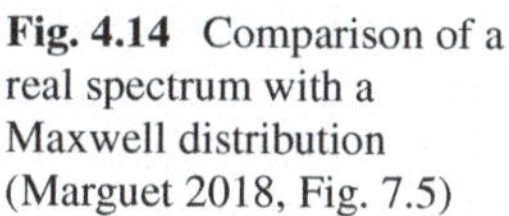

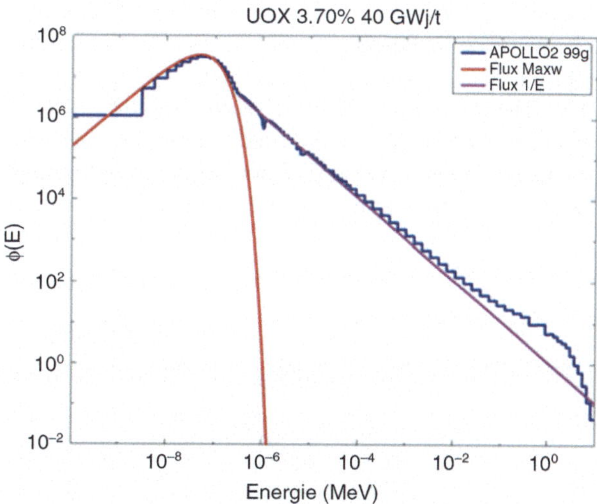

However, below the resonance region, the absorbing cross section of ^{238}U is very small, not above 1 barn. This is an energy range in which neutrons keep losing energy in successive collisions, without being significantly absorbed. Therefore they have a tendency to populate lower energies, following a $1/E$ like shape, until connection with the thermal spectrum as illustrated by Fig. 4.14. This figure from Marguet (2018, p. 409)compares the spectrum of a UOX 3.70% assembly (as calculated by the French spectral code APOLLO2) with a Maxwell distribution and a $1/E$ epithermal flux.

4.2.4 Case of a Ponctual Source

The case of a ponctual source emitting fission neutrons, within a scattering and non absorbing medium, does not meet the assumption of a medium supplied by an homogeneous neutron source of neutrons. This case will be studied in the Chap. 9, which does address Fermi's theory of age. The theory of age enables to establish that, in a specific point space away from the source, the slowing down current is an increasing function of the age, and thus a decreasing function of energy.

$$\text{At a point } \boldsymbol{r} : q(\boldsymbol{r}, E) \nearrow \text{ if } E \searrow$$

Even if the denominator term $\Sigma_s(E)\xi$ of the Eq. 4.11 is almost constant, $\phi(E)$ does increase more sharply than $\frac{1}{\Sigma_s(E)\xi}\frac{1}{E}$ along the slowing down due to the increase of the slowing down current $q(E)$. From the theory of age, as shown by Fig. 9.3, we can estimate the increase of the slowing down current in heavy water à 20 °C at 30 cm from the source: from 0.1 MeV down to 1 $times$ 10^{-6} MeV, $q(E)$ is multiplied by a factor of 60. However, this estimation is not accurate since applying theory of age to heavy water is flawed.

Figure 4.15 shows the flux spectrum at 30 cm from a source in heavy water at 20 °C. The piling up of neutrons according to Eq. 4.11 with a constant slowing down current is represented by the blue continuous line which has an $^1/_E$ shape, and the green dashed line is the flux shape. This figure highlights that $\phi(E)$ increases more sharply due to the increase of the slowing down current $q(E)$.

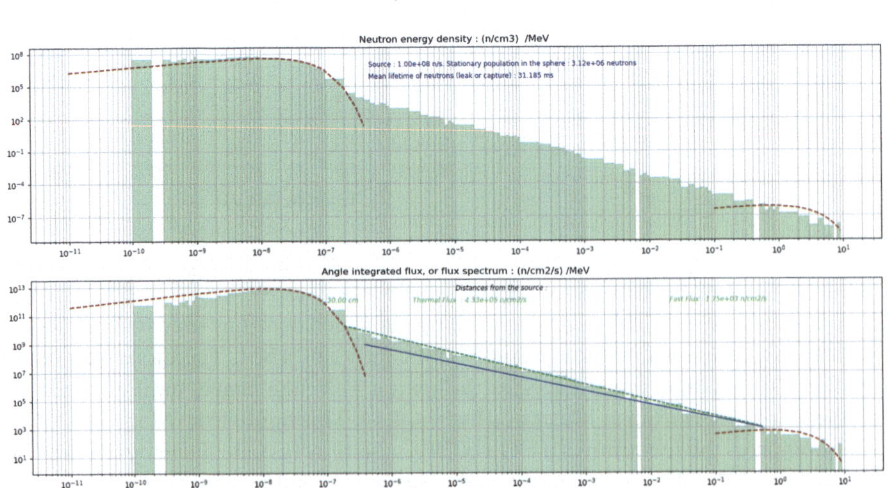

Fig. 4.15 Flux spectrum calculated at 30 cm of a fission source in D$_2$O

4.3 Some Interesting Properties of ^{1}H and ^{16}O Concerning Slowing Down

The important materials for neutron slowing down in PWRs are: ^{1}H, ^{16}O.

4.3.1 Elastic and Inelastic Scattering

In potential scattering, the incident neutron leaves, while in resonant scattering, any neutron of the compound nucleus is ejected.

- The potential scattering is always elastic. It corresponds to a single diffusion of the wave associated with the neutron by the potential field of the nucleus. Its cross section is of the order of a few barns.
- The resonant scattering corresponds to the absorption of the incident neutron, the formation of a compound nucleus and then the re-emission of a neutron.

  - If, after ejection of the neutron, the target nucleus is at the fundamental level (back to initial state), the resonant scattering is elastic. The hypothesis of elastic scattering is based on the notion that the energy of a neutron does not in theory exceed the excitation energy of the first excited state energy level of the target atom.
  - Else, it is inelastic. Scattering is termed inelastic if at least one of the objects involved undergoes excitation, with removal of energy from the kinetic energy balance: some of the energy of the incident particle has the effect of modifying the internal state of the target. The sufficient energy necessary to produce inelastic scattering may have very different thresholds depending on target. For the lighter isotopes, the first inelastic threshold is in the MeV range, but for heavy isotopes, it can come down to tens of keV or lower. NB: inside a reactor, no neutrons produced from fission have an energy greater than 14 MeV and the experimental mean of the fission spectrum is 2 MeV. So, in reactors, inelastic scattering will mainly be observed in the fuel materials (^{238}U particularly).

4.3.2 Scattering at High Energy

- If the energy of an incident neutron is large compared with the chemical binding energy of the atoms in a molecule, the molecule is generally disrupted when the neutron is scattered by one of the atoms. In this case, the fact that the atoms are bound within a molecule can be ignored and it is possible to use free gas cross sections. Because the neutron wavelength is very small at the higher energies, neutron elastic scattering can be treated by classical "billiard ball" (classical collisions) kinematics.

- At high incident energies, inelastic neutron scattering becomes possible. The excited compound system resulting from the initial collision has had time to come to equilibrium; thus, the emitted neutron has "forgotten" the direction of the incident neutron.

4.3.3 Thermal Scattering

At thermal energies, the wavelengths of neutrons approach the size of molecules and the spacing of crystalline lattices (UO_2). Scattering becomes a quantum mechanical problem.

In Monte Carlo codes, the thermal cross section of chemically bound atoms (for example hydrogen bound in light water H_2O, and deuterium bound in heavy water D_2O) is different from the free-gas cross section. Figure 4.16 illustrates that the cross section of free atoms is constant while the cross section of the water molecule increases at low energies.

At thermal neutron energies, the binding of the scattering nucleus in a solid or liquid affects not only the cross section, but also the angular and energy distributions of the scattered neutrons.

For ^1H bound in H_2O, the hydrogen atom is not as free to recoil as the free atom. At low energies ($E < 5\,eV$), the interaction of the scattered neutron with the scattering nucleus is influenced by dynamics of water molecules, translation, rotation, also bending and stretching vibrations. The effect of the binding of hydrogen in H_2O is to make the scattering almost isotropic below 1 meV. Figure 4.17 shows the transition

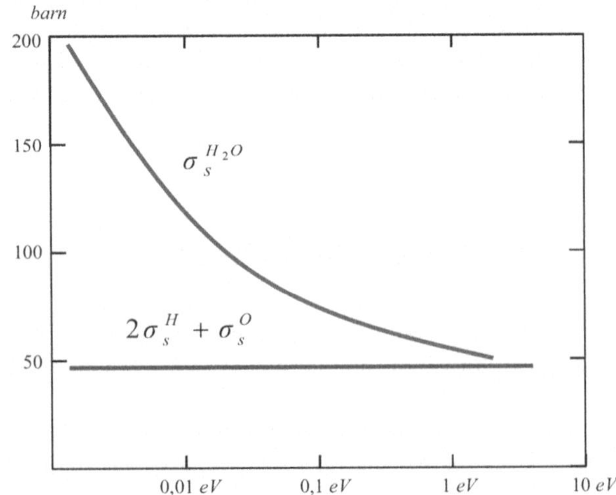

Fig. 4.16 Effect of binding of water molecule on the scattering cross section (Marguet 2018, Fig. 2.21, p. 124), see also Lamarsh (1966, Fig. 2.18, p. 48)

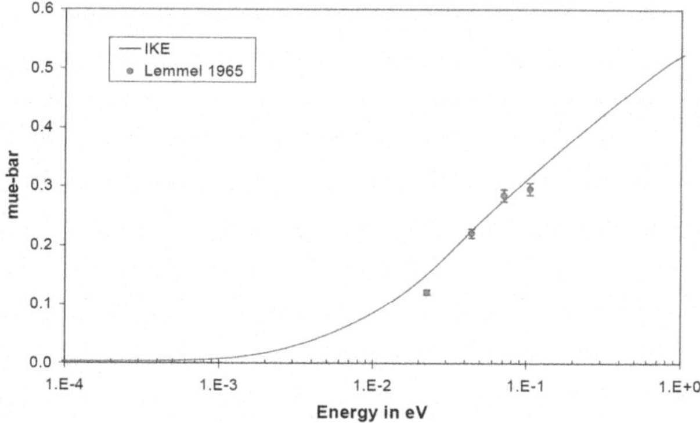

Fig. 4.17   Average cosine of the neutron scattering angle in light water at 473 K (Mattes and Keinert 2005, Fig. 4.19, p. 29)

of the average cosine of the neutron scattering angle in water, from a forward-peaked scattering tending to $\frac{2}{3}$ at high energies, towards a quasi-isotropic scattering for low energy neutrons.

It is interesting to note that at 306 °C, a typical average coolant temperature in a PWR, the average energy of neutrons is $^3\!/_2\, E_0 = 0.075\,\text{eV}$ (Eq. 3.7). Thus, we can roughly say that scattering is close to be isotropic for thermalized neutrons.

4.3.4   A Few Interesting Remarks About Scattering in Water

Elastic scattering plays the most important role in neutrons slowing down, in thermal neutron reactors containing a moderator.

- ^{1}H only supports elastic scattering and (n, γ) capture. The elastic cross sections for ^{1}H is basically constant overmuch of the neutrons energy range (around 20 barn up to 10 keV, and it plummets above 10 keV), and the capture cross section is small with a $^1\!/_v$ energy dependence.
  The scattering and capture cross sections of ^{1}H and D are shown on Figs. 4.18 and 4.19.
- It is assumed that scattering from the oxygen atom can be treated as a free gas of mass 16. Below 1 MeV, the elastic cross section is almost constant, around 4 barn. ^{16}O has the additional complication of resonant behavior and inelastic scattering at the higher energies, as shown on Fig. 4.20.
  Because of the higher atomic mass, the angular distribution in the laboratory system for elastic scattering is fairly isotropic up to the 10 keV region, with $\overline{\mu} = 0.042$.
  Fast neutrons, of a few MeV, undergo scattering reactions with oxygen (which removes a small fraction of energy) and hydrogen, both with modest cross sections. This enable fast neutrons to move away from the point of emission. From the

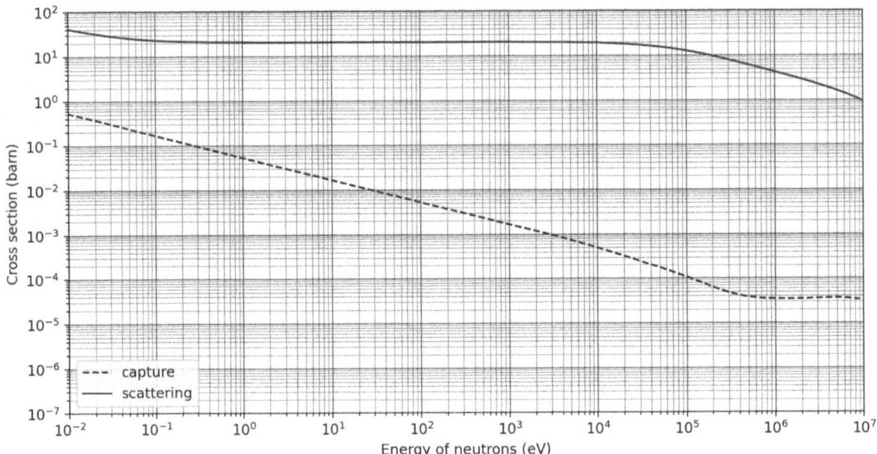

Fig. 4.18 Elastic and capture (dashed) cross sections for H-1 from JEFF3.1.1

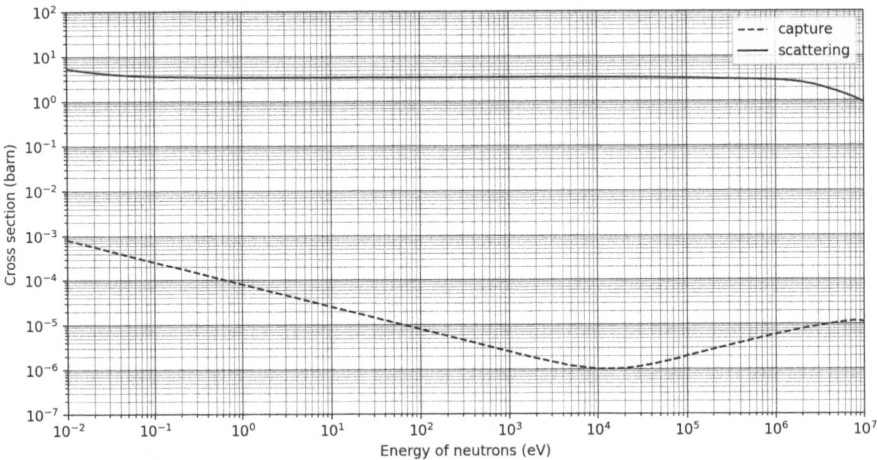

Fig. 4.19 Elastic and capture (dashed) cross sections for D from JEFF3.1.1

moment when a first shock on hydrogen occurs, the mean energy loss is half the initial energy, and therefore hydrogen scattering cross section becomes greater and the neutron slows down in a small volume.

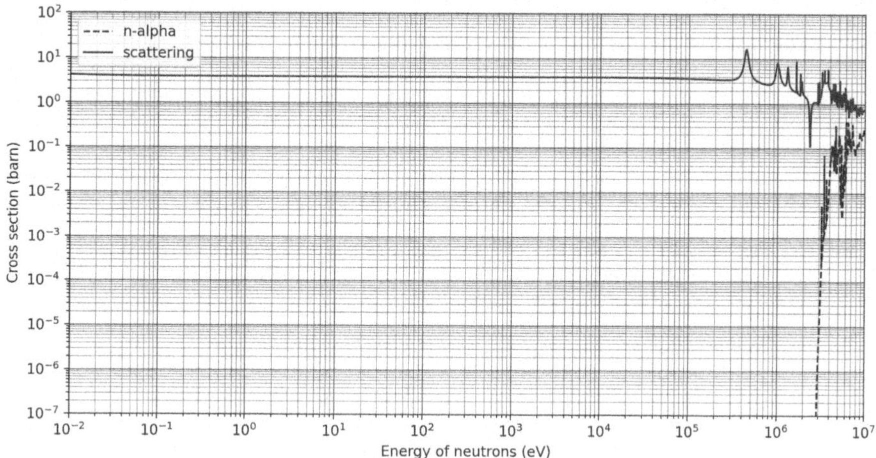

Fig. 4.20 Elastic (solid) and (n, α) (dashed) cross sections for O-16 from JEFF3.1.1

4.3.5 (n, α) Captures by Oxygen in UO$_2$

The (n, γ) capture cross section for ^{16}O is extremely small. However, at higher energies, an (n, α) channel opens as shown on Fig. 4.20, and this absorption cross section, although small, does have some effect on criticality in water systems. The (n, α) captures by oxygen from UO$_2$ can be seen on the fast fission spectrum of a PWR, as shown on Fig. 5.1. Fast neutrons emitted within the fuel can initiate (n, α) captures by oxygen from UO$_2$. This cannot occur with oxygen from H$_2$O, since neutrons are slowed down in water and do not have sufficient energy for (n, α) captures by oxygen.

The fission spectrum may be taken as a first approximation to the neutron spectrum, close to the neutron source, for energies above about 1.5 MeV.

References

K.-H. Beckurts, K. Wirtz, *Neutron Physics* (Springer, Berlin, Heidelberg GmbH, 1964). https://link.springer.com/book/10.1007/978-3-642-87614-1#bibliographic-information

G.I. Bell, S. Glasstone, *Nuclear Reactor Theory* (Van Nostrand Reinhold Company, 1970). https://books.google.fr/books?id=RNQmAQAAMAAJ

A. Borio, M. Cagnazzo, F. Marchetti, P. Pappalardo, A. Salvini, Moderating ratio parameter evaluation for different materials by means of monte carlo calculations and reactivity direct measurements, in *2nd World TRIGA Users Conference. Conference volume*, number INIS-AT–0076 (2004), pp. 177–182

D. Friant, D. Bernard, P. Blaise, State of the art on doppler broadening: modern developments on the fuel temperature coefficient. Nuclear Sci. Eng. **197**(8), 1991–2006 (2023)

S. Glasstone, M.C. Edlund, *The Elements of Nuclear Reactor Theory* (D. Van Nostrand Company, 1952). ISBN 9780598464224. https://books.google.fr/books?id=MSRRAAAAMAAJ

J.R. Lamarsh, *Introduction to Nuclear Reactor Theory*. Addison-Wesley Series in Nuclear Engineering (Addison-Wesley Publishing Company, 1966). ISBN 9780201041262. https://books.google.fr/books?id=by5RAAAAMAAJ

S. Marguet, *The Physics of Nuclear Reactors* (Springer International Publishing, 2018). ISBN 9783319595603. https://books.google.fr/books?id=9DlODwAAQBAJ

M. Mattes, J. Keinert, *Thermal neutron scattering data for the moderator materials h 2 o, d 2 o and zrh x in endf-6 format and as ace library for mcnp (x) codes*. Technical report (International Atomic Energy Agency, 2005)

Chapter 5
Resonant Absorption. Doppler Feedback Effect

Abstract The resonance phenomenon is a fundamental subject in nuclear physics and in reactor physics. This chapter explains how resonances affect the probability that a neutron will be slowed down to thermal energy, known as resonance escape probability. The cases of an homogeneous and an heterogeneous medium are presented. An heterogeneous geometry also produces an effect called spatial self shileding, which improves the resonance escape probability. The last part of this chapter introduces the phenomenon of Doppler broadening, which constitutes a temperature effect and is of major importance for the Safety of reactors.

5.1 Capture Cross Section of ^{238}U

The resonance structure of the neutron cross section of most heavy nuclei of mass A arises from the properties of the compound nucleus $A + 1$, formed when the incident neutron the incident neutron enters the target nucleus A. The compound-nucleus formation favors specific neutron energies so that the excitation energy of the compound nucleus matches one of the quantized energy levels and thus the interaction probability, i.e. the cross section, increases sharply. For example, in the case of ^{238}U, a peak in the capture cross section occurs when the kinetic energy of the incident neutron, added to the separation energy of the neutron in the compound nucleus ^{239}U, i.e. the excitation energy, is close to a quantified energy level of ^{239}U. The capture cross section for ^{238}U is shown in Fig. 5.1.

Starting from low energies, $E \lesssim 1\,\mathrm{eV}$, there is the $1/v$ region wherein the cross section follows such a trend. This can be explained by the fact that the probability of the absorption is proportional to the time the neutron spends in the vicinity of the nucleus (Friant et al. 2023, p. 1993). Thus, the absorption cross section is generally inversely proportional to the velocity.

For $1\,\mathrm{eV} < E < 1\,\mathrm{MeV}$, one finds the resonance region. This region is generally dominated by large resonances wherein the cross section can reach many thousands of barns. It is also subdivided into two regions: the resolved and unresolved regions. The energy range in which resonance data for a given nuclide are experimentally resolved, is generally referred to as the resolved energy region. In the resolved

H. Grard, *Diffusion of Neutrons in Nuclear Reactors*,
https://doi.org/10.1007/978-3-032-05088-5_5

region, the resonances are spaced sufficiently far apart that they may be individually characterized. At higher energies, resonance data are not resolved due to difficulties resulting from instruments resolution, since the resonance are too close and overlap. This region is referred to as the unresolved energy region and requires a statistical treatment (Hwang et al. 2016, p. 10).

^{238}U has 2126 resonance peaks in its resonance region, ranging from 4.40 eV to 20.6 keV, and the low-energy resonances have relatively high peak cross section (Ju and Kim 2017).

The resolved resonance region is the most important with regard to the Doppler coefficient.

5.2   Introduction to Resonance Escape Probability

5.2.1   Infinite Medium with Homogeneous Absorption

We consider an infinite medium with homogeneous absorption, that is supplied by an homogeneous neutron source of neutrons.

The slowing down, if absorption is taken into account shows a reduced pile up of neutrons, as compared to $1/E$.

We combine the two following equations, below the fission neutron spectrum:

$$\begin{cases} \phi(E) = \dfrac{q(E)}{\Sigma_s(E)E\xi} \\ \Sigma_a(E)\phi(E) = \dfrac{dq}{dE}(E) \end{cases} \qquad (5.1)$$

We eliminate ϕ between the two equations and thus:

$$\frac{dq}{dE}(E) = \frac{\Sigma_a(E)}{E\xi\Sigma_s(E)} \cdot q(E)$$

After integration:

$$q(E) = Q_0 \exp\left[-\int_E^{E_0} \frac{\Sigma_a(E')}{\xi\Sigma_s(E')} \cdot \frac{dE'}{E'}\right]$$

And the flux:

$$\phi(E) = \frac{Q_0}{\xi\Sigma_s(E)E} \exp\left[-\int_E^{E_0} \frac{\Sigma_a(E')}{\xi\Sigma_s(E')} \cdot \frac{dE'}{E'}\right] \qquad (5.2)$$

Where Q_0 is the homogeneous neutron source emissivity (ncm^3/s) at the energy E_0 (origin of the lethargy).

Taking into account that $dE = -E\,du$ and $\phi(u) = E\,dE$, Eq. 5.2 can be written as a function of lethargy u:

$$\phi(u) = \frac{Q_0}{\xi \Sigma_s(u)} \exp\left[-\int_{-\infty}^{u} \frac{\Sigma_a(u')}{\xi \Sigma_s(u')} \cdot du' \right] \tag{5.3}$$

The exponential describes the gradual decrease in flux $\phi(u)$ due to absorption, as neutrons progress through the slowdown, corresponding to an increase in lethargy u.

We can introduce $p(E)$ which is the number of neutrons that survive until energy E, or $p(u)$ which is the number of neutrons that survive until lethargy u, for one emitted neutron.

$$p(E) = \exp\left[-\int_{E}^{E_0} \frac{\Sigma_a(E')}{\xi \Sigma_s(E')} \cdot \frac{dE'}{E'} \right]$$

With $\Sigma_a \ll \Sigma_s$, we can replace Σ_s by Σ_t. The equations becomes:

$$p(E) = \exp\left[-\int_{E}^{E_0} \frac{\Sigma_a(E')}{\xi \Sigma_t(E')} \cdot \frac{dE'}{E'} \right]$$

It is important to note that this equation constitutes an approximation, even if cross sections have resonances with very high peak values, provided that these resonances are widely spaced and narrow (Glasstone and Edlund 1952, p. 168). For this reason, $p(E)$ is commonly known as antitrap factor, or resonance escape probability.

Because of the complicated structure of the absorption cross section in the epithermal range, it is very difficult to calculate p precisely.

If the absorption cross section does not vary rapidly with the neutron energy in the resonance region, it can be demonstrated (Greuling–Goertzel model) (Glasstone and Edlund 1952, p. 168) that:

$$p(u) = \exp\left[-\int_{0}^{u} \frac{\Sigma_a(u')}{\xi \Sigma_s(u') + \gamma \Sigma_a(u')} du' \right]$$

and

$$\phi(u) = \frac{q(u)}{\xi \Sigma_s(u) + \gamma \Sigma_a(u)}$$

In lethargy units, the slowing down current is given by the equation:

$$q(u) = p(u) Q_0$$

5.2.2 *Crossing of a Narrow Isolated Black Trap*

Black trap means that if the lethargy of the neutron is within the trap, the neutron
is absorbed. The width of the trap is Γ, and is narrow with respect to the average
energy loss per collision. The lethargy interval of the trap is set to $[0, \Gamma]$.

If the width is narrow ($\Gamma \ll \xi$) then the probability that a neutron crossing the
lethargy 0 falls into the trap is $\frac{\Gamma}{\xi}$. Then:

$$p = 1 - \frac{\Gamma}{\xi}$$

The resonance at $6.67\,\mathrm{eV}$ of $^{238}\mathrm{U}$ (see Figs. 5.1 and 5.2) can be considered as
narrow in the case neutrons are slowed down by water ($\xi = 0.92$, calculated in
Sect. 4.1.7): an estimation of the width Γ at $T = 1800\,\mathrm{K}$, calculated at middle height
is $\ln \frac{6.9}{6.5} = 0.06$.

Thus $1 - p = \frac{0.06}{0.92} = 0.0646$ and $p = 93.5\%$.

5.2.3 *Crossing of a Black Isolated Trap*

We consider an isolated black resonance, that is the absorption cross section at any
energy except within the black resonance is zero, and all the neutrons scattered into

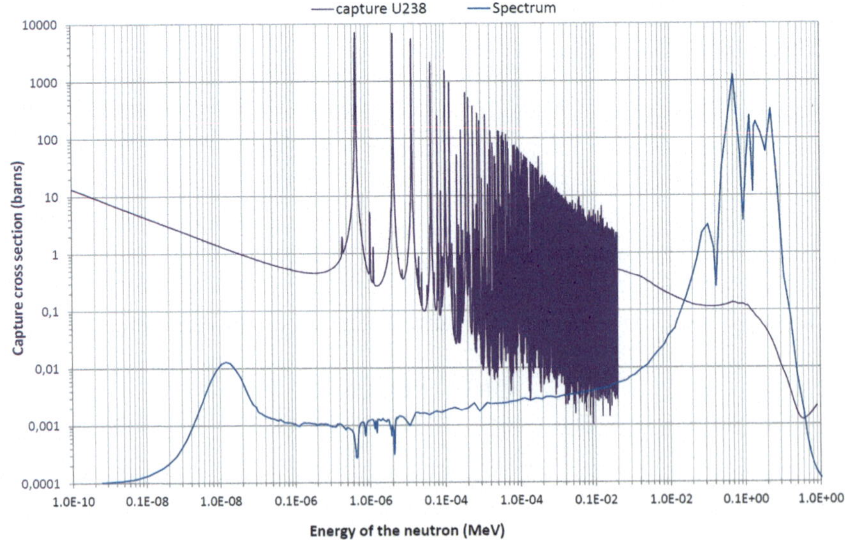

Fig. 5.1 Radiative capture cross section for U-238 and flux spectrum

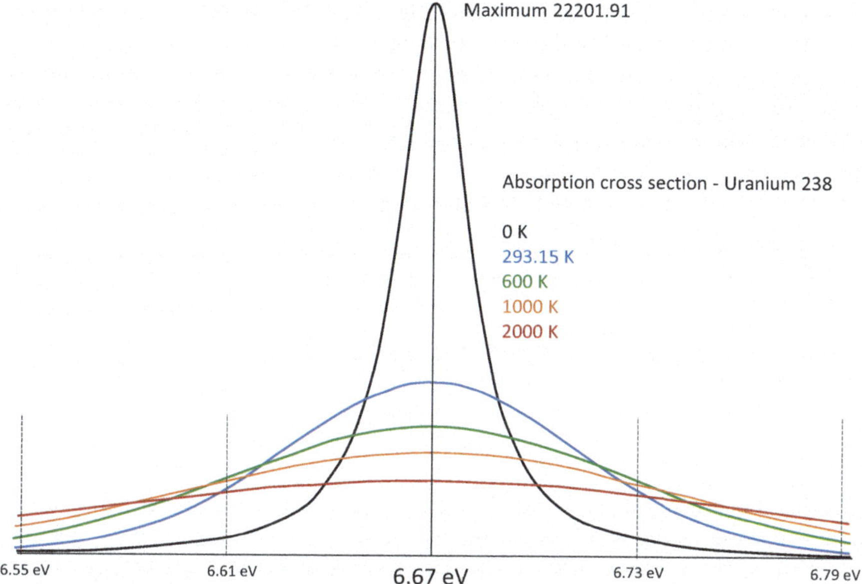

Fig. 5.2 The large resonance at 6.67 eV of U-238 absorption cross section, at different temperatures

the resonance, an interval of lethargy width Γ, are absorbed. In the energy range above the resonance, the flux has its asymptotic value:

$$\phi(u) = \frac{q(u)}{\xi\, \Sigma_s(u)}$$

The number of absorptions in the black resonance, per unit volume and per unit time, is:

$$\int_{\Gamma} \rho(u')\,du'$$

$\rho(u')$ is the density of arrivals at lethargy u', defined in Sect. 4.1.8.

$$\rho(u') = \int_{u'+\ln\alpha}^{u'-\epsilon} \frac{q(u)}{\xi}\, \frac{e^{-(u'-u)}}{1-\alpha}\,du$$

Where ϵ is the lethargy difference between u' and the lethargy at the entrance in the resonance.

$$\int_{\Gamma} \rho(u')\,du' = \int_{\Gamma} \int_{u'+\ln\alpha}^{u'-\epsilon} \frac{q(u)}{\xi}\, \frac{e^{-(u'-u)}}{1-\alpha}\,du\,du'$$

Since at energies above the resonance, the absorption is zero, $q(u)$ is constant below the lethargy interval of the resonance, and is the number of neutrons crossing the lower lethargy border of the black resonance per unit volume per unit time. We denote $q(u) = q$ below the resonance.

Thus the number of absorptions in the black resonance divided by the number of neutrons crossing the lower lethargy border of the resonance is is the probability of absorption $1 - p$, where p is the antitrap factor.

$$1 - p = \frac{1}{\xi(1 - \alpha)} \int_\Gamma du' \tag{5.4}$$

$$= \frac{1}{\xi(1 - \alpha)} \int_\Gamma (e^{-\epsilon} - \alpha) du' \tag{5.5}$$

$$= \frac{1 - e^{-\Gamma} - \alpha\Gamma}{\xi(1 - \alpha)} \tag{5.6}$$

During the slowing down in the resonance region, a key parameter is the ratio $\frac{\Gamma}{\xi}$ where Γ represents the width of a resonance and ξ is the average lethargy increase per collision. If a resonance is narrow with respect to the average energy loss per collision in the moderator, then very few neutrons are captured.

5.2.4   Crossing of a Narrow Isolated Grey Trap

Grey trap means that if the lethargy of the neutron is within the trap, it may not be absorbed. The trap is in the lethargy interval $[0, \Gamma]$. The probability that a neutron crossing the lethargy 0 falls into the lethargy range du within the trap is $\frac{du}{\xi}$.

The probability that a nuclear reaction leads to an absorption within the grey trap is:

$$\frac{\Sigma_a(u)}{\Sigma_s(u) + \Sigma_a(u)} \quad \text{where } u \in [0, \Gamma]$$

The non-absorbed neutrons are scattered out of the trap, because it is assumed to be narrow.

Thus the probability of absorption by the trap is obtained by the integration:

$$\int_0^\Gamma \frac{1}{\xi} \cdot \frac{\Sigma_a(u)}{\Sigma_s(u) + \Sigma_a(u)} du$$

And:

$$p = 1 - \int_0^\Gamma \frac{1}{\xi} \cdot \frac{\Sigma_a(u)}{\Sigma_s(u) + \Sigma_a(u)} du$$

If grey trap tends to be black for any lethargy inside the trap: $\lim\limits_{\Sigma_a \to \infty} p = 1 - \frac{\Gamma}{\xi}$.

5.2.5 Slowing Down Through a Succession of Narrow Isolated Grey Trap

The traps are supposed to be "isolated", which means that they do not interact one with another and between two traps, the flux is asymptotic. The assumption narrow traps enables to write that p_i is close to 1 and then the probability to cross the trap $[u_i, u_i + \Gamma_i]$ is

$$p_i \simeq \exp\left[-\int_{u_i}^{u_i+\Gamma_i} \frac{1}{\xi} \cdot \frac{\Sigma_a(u)}{\Sigma_s(u) + \Sigma_a(u)} du \right]$$

The probability to cross all the traps is

$$p \simeq \Pi p_i \simeq \exp\left[-\sum_i \int_{u_i}^{u_i+\Gamma_i} \frac{1}{\xi} \cdot \frac{\Sigma_a(u)}{\Sigma_s(u) + \Sigma_a(u)} du \right]$$

All the traps are within the lethargy interval $[0, U]$ then

$$p \simeq \exp\left[-\int_{0}^{U} \frac{1}{\xi} \cdot \frac{\Sigma_a(u)}{\Sigma_s(u) + \Sigma_a(u)} du \right]$$

5.3 Qualitative Discussion About the Effects of Heterogeneity on 4 Factors

5.3.1 A Higher Resonance Escape Probability in Heterogeneous Systems

The reader should refer to the Chap. 12, Sect. 13.1.3 for a presentation of the four factors and the non leakage probabilities.

The resonance escape probability, sometimes called antitrap factor and symbolized by p, is the probability that a neutron will be slowed down to thermal energy and escape resonance capture. This probability is defined as the ratio of the number of neutrons that reach thermal energies to the number of fast neutrons that slow down. A few resonances are substantially responsible for the absorption rates over the resonance energy range which, in turn, determines the fraction of neutrons that can reach the thermal energy via the elastic scattering.

We have previously considered an infinite homogeneous medium. Thermal reactors are heterogeneous, which means that a reactor lattice is composed of a handful of light nuclide with relatively constant cross sections serving as coolant and/or moderator and cladding along with fuel element consisting of a mixture of fissile and nonfissile isotopes dominated by ^{238}U. A large variety of fuel elements have been used in heterogeneous reactors, these often consist of plates or rods of natural uranium or enriched uranium which are wrapped in a cladding material. Thermal reactors are heterogeneous, for obvious technological reasons, the necessity to extract heat with a circulating fluid and to moderate neutrons with a medium that may be solid or liquid leads to separate fissile material, coolant and moderator in separate areas.

There are also important reasons related to neutronics. In thermal reactors, the neutron balance requires to limitate as much as possible the resonant captures by ^{238}U. If fuel and moderator are separated, neutrons can go through the energy positions associated to traps while scattering out of the fuel (Bussac and Reuss 1985, p. 279). This was conceptualized by the French scientist Frédéric Joliot in 1939.[1]

Therefore, the resonance escape probability calculated in an heterogeneous reactor core is higher than in an homogeneous core. The fuel lumping which was implemented by Enrico Fermi in his pile CP1 (Chicago, first divergence in the world, in December 2, 1942), using natural uranium and graphite, enabled to achieve criticality.

The fuel material in a PWR is in pellet form encapsulated within tubes. This arrangement increases the probability that the fast neutron escapes the traps and the neutron slows down in the moderator where there are no nuclei of ^{238}U and ^{240}Pu present.

The fast fission factor ϵ is increased by an heterogeneous design given that there is a higher probability that a fast neutron undergoes a collision with a fuel heavy nucleus while its energy is still above the fast fission. If fuel heavy nuclei were mixed with light nuclei, a higher proportion of neutrons would loose some kinetic energy with with scattering reactions on light nuclei before collisions with fuel heavy nuclei. However, this geometrical effect is of second importance to self shielding (Duderstadt and Hamilton 1976, pp. 401–404).

5.3.2   *Moderator-to-Fuel Ratio*

We can define the moderator-to-fuel ratio as the ratio of the number of moderator nuclei within the reactor core volume to the number of fuel nuclei. Another possible ratio is the moderator volume to the fuel volume. Wathever is the point of view for the definition of the moderator-to-fuel ratio (nuclei ratio or volumes ratio), this ratio has a major impact on the antitrap factor p and also the thermal utilization factor f. An increase in the moderator-to-fuel ratio causes an increase in resonance escape probability p. As more moderator molecules are added relative to the number of fuel nuclei, then it becomes easy for neutrons to slow down to thermal energies without encountering a resonance absorption at the resonance energies. An increase in the moderator-to-fuel ratio causes a decrease in the thermal utilization factor f. The value of the thermal utilization factor is given by the ratio of the number of thermal neutrons absorbed in the fuel (all nuclides) to the number of thermal neutrons absorbed in all the material that makes up the core.

---

[1] Joliot draw up five patents, the first one on May 1, 1939 (Von Halban et al. 1939, 1940) and the fifth patent on May 1, 1940.

By design, PWR are under-moderated. This means that the moderator-to-fuel ratio is less than the optimum value that gives the highest product $p \times f$. Thanks to under-moderation, an increase in moderator temperature decreases the k_{eff} of the reactor. This provides a negative feedback, required for a stability of the coupling between the Net Steam Supply System and the conventionnal island extracting thermal power in the steam generators.

5.3.3 Heterogeneity and Self Shielding

Heterogeneity provides a spatial self shielding effect, that combines with energy self shielding. Self shielding is presented in the next Sect. 5.4.

5.4 Self Shielding

In an infinite homogeneous medium, the resonances produce flux depressions in energy, referred as energy self shielding effect. In the case of an infinite lattice consisting of repeated cells with fuel lumps surrounded by cladding and moderator, self shielding effects are both in energy and in space.

5.4.1 Energy Self Shielding

Energy self shielding of the resonance occurs in homogeneous or heterogeneous systems. This phenomenon is characterized by a correspondence between the peaks of a cross-section and dips in the flux. Flux and the resonance absorption cross section are anti-correlated, the energy flux decreases at the locations of the resonances and thus fewer neutrons are lost than might be expected if the flux were not depressed. This is particularly visible on Fig. 5.1 below the large resonance at 6.67 eV. Large, absorbing resonances are typically highly self shielding, meaning that they depopulate the neutron spectrum in their vicinity.

In non resonant regions, or between separated resonances, we can write that:
$$\phi(E) = \frac{q(E)}{\Sigma_s E \xi}.$$

Within a resonance, an approximation for the flux is: $\phi(E) = \dfrac{q(E)}{(\Sigma_s + \Sigma_a) E \xi}.$

In a PWR, for neutron energies near resonances, in particular the large resonance peak of ^{238}U at 6.67 eV, the depression of the energy-dependent flux is very strong inside the fuel rod (Hwang et al. 2016, Fig. 2).

5.4.2 *Spatial Self Shielding*

> Owing to the so-called selfshielding effects with regard to both energy and space : the greater
> the cross section, the smaller the neutron flux and consequently the reaction rate (product of
> cross section by flux), particularly the capture rate, remains small even if the cross section
> goes to infinity. So despite the many large uranium 238 resonances and the generally high
> content of this material in nuclear fuels, the so-called antitrap factor p (resonance escape
> probability during neutron slowing down) is reasonably high, e.g. about 0.75 with PWRs.
>
> Reuss and Coste-Delclaux (2003)

Spatial self shielding is a phenomenon primarily connected with this heterogeneity
of the reactor core. The fission neutrons are born in the fuel, but they are primarily
moderated (slowed down) in the moderator. Neutrons thermalized in the moderator
have to diffuse back into the fuel to produce a fission. However, the highly absorbing
nuclei encountered near the surface of the fuel tend to absorb the thermal neutrons
and hence shield the fuel nuclei in the interior of the fuel rod. The surface layers of
the fuel geometrically shield the inner layers from neutron flux, leading to a relatively
lower neutron flux inside the fuel rod. This is why this effect is called spatial self
shielding (Lamarsh 2001, p. 319).

Since the outer layers of the fluel tend to shield its interior from resonance
energy neutrons, decreasing the resonance abosrption, this is another reason why
heterogeneity increases the resonance escape probability.

It is interesting to note that since spatial self shielding leads to a thermal flux
depression in the fuel, the thermal utilization factor is reduced. The thermal utilization
of an heterogeneous system is less than that of an homogeneous mixture of the same
material.

A thermal disadvantage factor is defined as the ratio of the average flux in the
moderator to that in the fuel (Lamarsh 1966, p. 373):

$$\zeta = \frac{\overline{\phi}_M}{\overline{\phi}_D} > 1$$

5.5 Resonant Absorption and Slowing Down in a PWR

A PWR reactor is a mixture of light scattering nuclei, and heavy nuclei from the fuel,
very absorbing. Two domains may be considered:

- From fission energy down to a few keV area, the absorption in the core can be
  considered as weak and continuous.
- In the resonance region, between eV and a few keV, the probability for a neutron
  to escape the numerous resonances, specially of ^{238}U, during its slowing down
  and to reach the lethargy u is $p(u)$. The order of magnitude is 0.75 for a PWR.
  It can be demonstrated that the probability to escape each resonance, considered
  separately, is:

$$p_i = 1 - \frac{N_{238_U}}{\xi \Sigma_{s,m}} \cdot I_{\text{eff},i}$$

Where $\Sigma_{s,m}$ is the moderator macroscopic scattering cross section, $I_{\text{eff},i}$ is the effective integral of the resonance i, N_{238_U} is the concentration of the resonant isotope. The effective resonance integral $I_{\text{eff},i}$ is defined for a given resonance as the integral of the effective cross section over lethargy, which is such that the absorption rate is verified for the asymptotic flux $\psi(u)$ (flux outside the resonance).

$$I_{\text{eff},i} = \int_{u_i}^{u_i+\Gamma_i} \sigma_a^{\text{eff}}(u)du \quad \text{where} \quad N_{238_U}\sigma_a^{\text{eff}}(u)\psi(u) = N_{238_U}\sigma_a(u)\phi(u)$$

The assumption narrow traps enables to write that p_i is close to 1 and then the probability to cross the traps is:

$$p_i = \exp\left[-\frac{N_{238_U}}{\xi \Sigma_{s,m}} \cdot I_{\text{eff},i}\right] \tag{5.7}$$

And, if the traps are considered to be "isolated", which means that they do not interact one with another and between two traps the flux is asymptotic, then we can write for the whole of the traps:

$$p = \exp\left[-\frac{N_{238_U}}{\xi \Sigma_{s,m}} \cdot I_{\text{eff}}\right] \tag{5.8}$$

Where $I_{\text{eff}} = \sum_i I_{\text{eff},i}$.

In the heterogeneous case, it is assumed that the cell consists of the fuel rod surrounded by moderator. Slowing-down is effective only in the moderator and outside the resonances. A new Eq. 5.9 can be established from Eq. 5.8:

$$p = \exp\left[-\frac{V_U \, N_{238_U} \, I_{\text{eff}}^{\text{het}}}{V_m \, N_m \, \xi \Sigma_{s,m}}\right] \tag{5.9}$$

In Eq. 5.9, the numerator and the denominator voice the competition between the two process governing the slowing down, the capture by the resonances of the fuel material (mainly ^{238}U) and the scattering mainly by the moderator. A new form of the resonance integral $I_{\text{eff}}^{\text{het}}$ has to be established, taking into account the heterogeneity of the core.

The resonance structure of the fuel affects slowing down, because of the losses due to absorption in the fuel, and also because some effect of the strong absorption resonances in the fuel shows up as dips in the flux. The absorption dips in the fuel are very large and lead to the self shielding effects.

The $\frac{1}{E}$ shape is modified because of the losses to absorption in the fuel.

At thermal energies, we see typical Maxwellian flux shapes from thermalization, but the thermal flux inside the fuel is depressed because of the absorption by capture and fission.

5.6 Doppler Broadening

The Doppler effect consitutes a passive and instantaneous negative reactivity feedback that limits peak power excursion during reactivity-initiated accidents. It also accounts for a significant proportion of the reactivity loss when the reactor goes from hot standby to rated power, due to fuel temperature increase.

The cross section or probability of a neutron interacting with any given nucleus depends on the relative speed between the neutron and the nucleus. Free-atom cross sections are not temperature dependent. That is, the cross sections for the same relative speed between a projectile (neutron) and a target atom are independent of temperature (Cacuci 2010, p. 319).

In evaluated nuclear data library, the neutron energy is measured relative to stationary target nuclei. When the temperature of the medium is 0 K, the target nucleus is stationary, so that the speed of the neutron in the laboratory system is equal to the relative speed in the relative frame of reference.

The temperature of the nucleus has a significant impact on the relative speed between a neutron and the target, and any material in the reactor, is at much higher and varying temperature. This for any given neutron energy in the laboratory system, there is an entire spectrum of relative energies.

In order to solve the transport equation in the laboratory frame of reference, cross sections have to be defined in the same frame of reference: this means that the shape of resonances in the laboratory frame of reference takes into account the effect of thermal agitation on the relative speed. This step is called Doppler broadening (Cacuci 2010, p. 320). The broadened cross section is calculated by averaging the 0 K cross section over the target velocity distribution. Commonly it is assumed that the target nuclei behave like atoms of a free gas and that their velocity distribution is then given by a Maxwell Boltzmann distribution, since fortunately, the differing solid-state models generally tend toward a Maxwell-Boltzmann gas model with increasing temperature (Friant et al. 2023, p. 1991).

Figure 5.2 illustrates the broadening of the large absorption resonance at 6.67 eV of ^{238}U, as a consequence of temperature increase. As temperature increases, the peak of the resonance lowers and the resonance widens, however the resonance integral (area under the curve) remains constant: this behavior is known as Doppler broadening. Doppler broadening extends the resonances into populated regions without meaningfully reducing their absolute ability to absorb neutrons in their original bounds, hence reducing the self shielding effect (Friant et al. 2023, p. 1993).

The Doppler temperature coefficient α_D quantifies the reactivity change as a function of fuel temperature change:

$$\alpha_D = \frac{1}{k_{\text{eff}}} \frac{\delta k_{\text{eff}}}{\delta T}$$

A fuel temperature increase results in a lower factor p and also a lower factor f which also contributes, to a lesser extent, to the reactivity decrease (Friant et al. 2023, p. 1998).

The approximate formula Eq. 14.31 enables to estimate a value of the Doppler temperature coefficient.

The effective cross section is defined as the cross section which, applied to the macroscopic flux, gives the same absorption rate as the cross section applied to the fine structure flux (the fine structure flux is depressed at the resonances, which is called self shielding). The effective integral I_{eff} is defined as the integral over the entire resonance domain of the effective cross sections of each resonance.

References

J. Bussac, P. Reuss, *Traité de neutronique: physique et calcul des réacteurs nucléaires avec application aux réacteurs à eau pressurisée et aux réacteurs à neutrons rapides*, vol. 25. (Hermann, 1985)

D.G. Cacuci, *Handbook of Nuclear Engineering: Vol. 1: Nuclear Engineering Fundamentals; Vol. 2: Reactor Design; Vol. 3: Reactor Analysis; Vol. 4: Reactors of Generations III and IV; Vol. 5: Fuel Cycles, Decommissioning, Waste Disposal and Safeguards*. Number vol. 1 in Handbook of Nuclear Engineering (Springer, 2010). ISBN 9780387981307. https://books.google.fr/books?id=pu9BWuf2gdkC

J.J. Duderstadt, L.J. Hamilton, *Nuclear Reactor Analysis* (Wiley, 1976)

David Friant, David Bernard, Patrick Blaise, State of the art on doppler broadening: Modern developments on the fuel temperature coefficient. Nuclear Sci. Eng. **197**(8), 1991–2006 (2023)

S. Glasstone, M.C. Edlund, *The Elements of Nuclear Reactor Theory* (D. Van Nostrand Company, 1952). ISBN 9780598464224. https://books.google.fr/books?id=MSRRAAAAMAAJ

R.N. Hwang, R.N. Blomquist, L.C. Leal, W.S. Yang, Neutron resonance theory for nuclear reactor applications: modern theory and practices. Technical report, Argonne National Lab.(ANL), Argonne, IL (United States) (2016)

K. Ju, Y. Kim, A comparative investigation of doppler broadening of neutron absorption and photon-induced nuclear resonance fluorescence reactions, in *The Korean Nuclear Society Autumn Meeting* (2017)

J.R. Lamarsh, *Introduction to Nuclear Reactor Theory*. Addison-Wesley Series in Nuclear Engineering (Addison-Wesley Publishing Company, 1966). ISBN 9780201041262. https://books.google.fr/books?id=by5RAAAAMAAJ

J.R. Lamarsh, *Introduction to Nuclear Engineering* (Prentice-Hall, 2001)

Paul Reuss, Mireille Coste-Delclaux, Development of computational models used in France for neutron resonance absorption in light water lattices. Prog. Nuclear Energy **42**(3), 237–282 (2003)

H.H. Von Halban, J.-F. Joliot, L. Kowarski, Dispositif de production d'énergie (1939), 1950. https://worldwide.espacenet.com/publicationDetails/originalDocument?CC=FR&NR=976541&KC=&FT=E

H.H. Von Halban, J.-F. Joliot, L. Kowarski, Perfectionnements apportés aux dispositifs de production d'énergie (1940), 1951. https://worldwide.espacenet.com/publicationDetails/originalDocument?CC=FR&NR=971386&KC=&FT=E

Chapter 6
Diffusion Current and Flux Gradient.
Fick's Law Applied to Neutronics

Abstract In diffusion phenomena, such as the diffusion of gas molecules or of heat, it is found that the diffusing substance tends to move from regions of high density toward those of low density. Neutrons behave similarly, since there are more collisions per unit volume in regions of high density, and after colliding the neutrons move away from the collision centers. Thus neutrons tend to diffuse from regions of high neutron density to low neutron density. Fick's first law of diffusion describes diffusion from high to low concentration as proportional to the particle's concentration gradient, and were first posited by Adolf Fick in 1855 on the basis of his experimental results. In this chapter, we will introduce a Fick's Law for neutrons, applied to the total flux, and discuss its validity.

6.1 The Law of Fick

The general problem of neutron transport is a very complicated one. However, it will now be shown that the neutron flux and current are related in a simple way if certain conditions are met. When this is the case it is possible to obtain elementary solutions to transport problems. This relationship between flux and current is identical in form with Fick's law, which has been used for many years to describe diffusion phenomena in liquids and gases. For this reason, the use of Fick's law in reactor theory leads to what is known as the diffusion approximation.

Lamarsh (1966, p. 125)

6.1.1 Introduction and Assumptions

We consider an absorbing and scattering medium, Σ_a and Σ_s, and monokinetic neutrons. The neutron flux is stationary: $\Phi(r)$.

H. Grard, *Diffusion of Neutrons in Nuclear Reactors*,
https://doi.org/10.1007/978-3-032-05088-5_6

6.1.2 Calculation of the Current of Neutrons Coming from Half-Space and Crossing dS

Our objective is to calculate the neutron current crossing dS in the xOy-plane and coming from the half-space defined by $z > 0$, as shown on Fig. 6.1.

Our calculation will be based on the preliminary equations:

- The rate of neutron emission from the volume dV is: $\Sigma_s \Phi dV$, where Φ is the flux in the element volume dV about r.
- The probability that a neutron emitted by dV is emitted in the direction of dS is $\dfrac{\cos\theta dS}{4\pi r^2}$. Implicitly, this means that the flux is isotropic, the neutrons have no privileged direction.
- The probability that a neutron emitted towards dS reaches dS without absorption nor scattering is: $\exp\left(-r(\Sigma_a + \Sigma_s)\right)$.

We can deduce that the rate of neutrons emitted by dV and crossing dS is:

$$\Phi\Sigma_s \cdot \exp\left(-r(\Sigma_a + \Sigma_s)\right) \cdot \frac{\cos\theta}{4\pi r^2} dS dV \tag{6.1}$$

Equation 6.1 can be integrated over the half-space $dz > 0$, so as to calculate the rate of neutrons crossing dS and coming from the half-space. Per unit surface, it can be written as:

Fig. 6.1 Element volume dV scattering neutrons

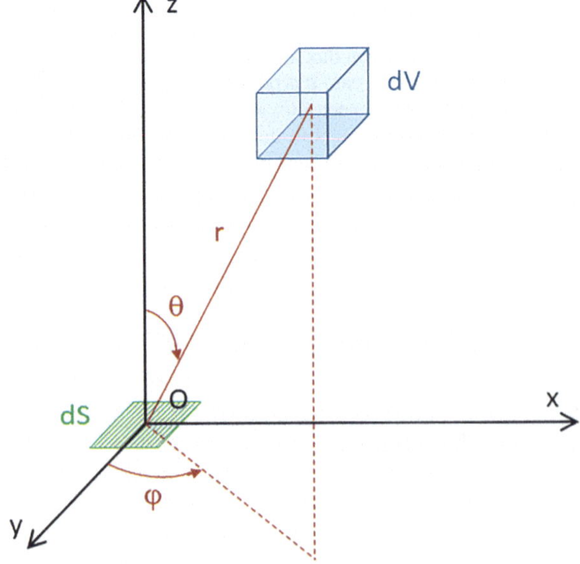

$$i = \iiint \Phi(\mathbf{r})\Sigma_s \cdot \exp\left(-r(\Sigma_a + \Sigma_s)\right) \cdot \frac{\cos\theta}{4\pi r^2} dV \ \ (\text{n/cm}^2/\text{s}) \tag{6.2}$$

In spherical coordinates, $dV = r^2 \sin\theta\, d\theta\, dr\, d\varphi$. Then Eq. 6.2 can be written as:

$$i = \frac{1}{4\pi} \int_{\varphi=0}^{2\pi} \int_{\theta=0}^{\frac{\pi}{2}} \int_{r=0}^{\infty} \Phi(r,\theta,\varphi)\Sigma_s e^{-r(\Sigma_a+\Sigma_s)} \cos\theta \sin\theta\, d\theta\, dr\, d\varphi \tag{6.3}$$

In cartesian coordinates, $\Phi(x, y, z)$ can be developped around the origin of the axes with the Mac Laurin formula:

$$\Phi(x, y, z) = \underbrace{\Phi_0}_{\Phi(0,0,0)} + x\left(\frac{\partial\Phi}{\partial x}\right)_{\text{origin}} + y\left(\frac{\partial\Phi}{\partial y}\right)_{\text{origin}} + z\left(\frac{\partial\Phi}{\partial z}\right)_{\text{origin}} \tag{6.4}$$

We have to inject each term of Eq. 6.4 into Eq. 6.3 for integration.

$$i\{\Phi_0\} = \frac{\Phi_0\Sigma_s}{4\pi} \int_{\varphi=0}^{2\pi} d\varphi \int_{\theta=0}^{\frac{\pi}{2}} \cos\theta \sin\theta\, d\theta \int_{r=0}^{\infty} e^{-r(\Sigma_a+\Sigma_s)}\, dr$$

- It comes immediately that $\displaystyle\int_{\varphi=0}^{2\pi} d\varphi = 2\pi$
- The integral over θ can be easily calculated:

$$\begin{aligned}
\int_{\theta=0}^{\frac{\pi}{2}} \cos\theta \sin\theta\, d\theta &= \int_{\theta=0}^{\frac{\pi}{2}} \frac{1}{2}\sin(2\theta)\, \frac{d2\theta}{2} \\
&= \frac{1}{4} \int_{2\theta=0}^{2\theta=\pi} \sin(2\theta)d(2\theta) \\
&= \frac{1}{4}[-\cos]_0^\pi = \frac{1}{2}
\end{aligned}$$

- The integral over r:

$$\int_{r=0}^{\infty} e^{-r(\Sigma_a+\Sigma_s)}\, dr = \left[\frac{-1}{\Sigma_a+\Sigma_s} e^{-r(\Sigma_a+\Sigma_s)}\right]_{r=0}^{r=\infty} = \frac{1}{\Sigma_a+\Sigma_s}e^0 = \frac{1}{\Sigma_a+\Sigma_s}$$

We can conclude that:

$$i\{\Phi_0\} = \frac{\Phi_0\Sigma_s}{4\pi} \cdot 2\pi \cdot \frac{1}{2} \cdot \frac{1}{\Sigma_a+\Sigma_s} = \frac{\Phi_0}{4}\frac{\Sigma_s}{\Sigma_a+\Sigma_s}$$

If $\Sigma_a \ll \Sigma_s$ and if the flux is uniform, then $i = \dfrac{\Phi_0}{4}$.

Fig. 6.2 Net currents at the interface

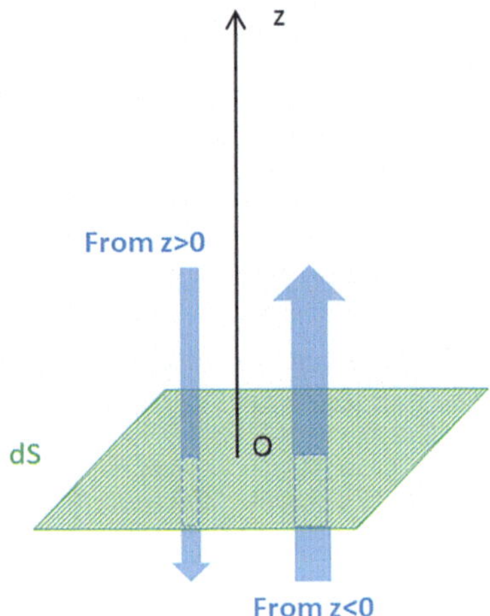

From Eq. 6.2, with $\dfrac{z}{r} = \cos\theta$, we can write:

$$i\left\{z\left(\frac{\partial\Phi}{\partial z}\right)_{(0,0,0)}\right\} = \left(\frac{\partial\Phi}{\partial z}\right)_{(0,0,0)} \cdot \frac{\Sigma_s}{4\pi} \underbrace{\int_{\varphi=0}^{2\pi} d\varphi}_{2\pi} \underbrace{\int_{r=0}^{\infty} r e^{-r(\Sigma_a+\Sigma_s)}\, dr}_{\frac{1}{(\Sigma_a+\Sigma_s)^2}} \underbrace{\int_{\theta=0}^{\frac{\pi}{2}} \cos^2\theta \sin\theta d\theta}_{[\frac{-\cos^3\theta}{3}]_0^{\pi/2}=\frac{1}{3}}$$

$$= \frac{1}{6}\left(\frac{\partial\Phi}{\partial z}\right)_{(0,0,0)} \frac{\Sigma_s}{(\Sigma_s+\Sigma_a)^2}$$

It can be easily verified that $i\left\{x\left(\frac{\partial\Phi}{\partial x}\right)_{(0,0,0)}\right\} = i\left\{y\left(\frac{\partial\Phi}{\partial y}\right)_{(0,0,0)}\right\} = 0$.

6.1.3   Current Assessment at the Interface

We consider a surface dS on the xOy-plane and centered on the point O, as shown on Fig. 6.2. The neutron current at the point O is $J(O) = J_x(O) + J_y(O) + J_z(O)$.

The vector $J_z(O)$ results from the addition of two neutron current vectors:

the one relating to the neutrons scattered by the half plane $z > 0$ ($J_z^-(O)$) and the one relating to the neutrons scattered by the half plane $z < 0$ ($J_z^+(O)$).

- Neutrons scattered by the half plane $z > 0$:

$$J_z^-(O) = -\frac{\Phi_0}{4}\frac{\Sigma_s}{\Sigma_a+\Sigma_s} - \frac{1}{6}\left(\frac{\partial\Phi}{\partial z}\right)_0 \cdot \frac{\Sigma_s}{(\Sigma_s+\Sigma_a)^2} \tag{6.5}$$

- Neutrons scattered by the half plane $z < 0$:

$$J_z^+(O) = \frac{\Phi_0}{4}\frac{\Sigma_s}{\Sigma_a+\Sigma_s} - \frac{1}{6}\left(\frac{\partial\Phi}{\partial z}\right)_0 \cdot \frac{\Sigma_s}{(\Sigma_s+\Sigma_a)^2} \tag{6.6}$$

If there is no gradient of flux along z axis, then:

$$\left(\frac{\partial \Phi}{\partial z}\right)_0 = 0 \text{ and } \boldsymbol{J}_z(\boldsymbol{O}) = \boldsymbol{J}_z^-(O) + \boldsymbol{J}_z^+(O) = 0$$

The net current $\boldsymbol{J}_z(O)$ is null.

It is important to understand that the net current is null, although neutrons are crossing dS on two opposite direction, compensating for one another. And in the same time, the flux $\Phi_0 \neq 0$, the flux is a notion dealing with volumes and not surfaces.

The net current at the interface, in O, is:

$$\boldsymbol{J}_z(O) = - \underbrace{\frac{1}{3} \cdot \frac{\Sigma_s}{(\Sigma_s + \Sigma_a)^2}}_{\text{Diffusion coefficient } D} \cdot \left(\frac{\partial \Phi}{\partial z}\right)_0$$

Taking into account all the directions, we can write Fick's law applied to reactors:

$$\boldsymbol{J} = -D \cdot \overrightarrow{\text{grad}}\Phi \tag{6.7}$$

With:

$$D = \frac{1}{3} \frac{\Sigma_s}{(\Sigma_s + \Sigma_a)^2} \tag{6.8}$$

Diffusion coefficient in a weakly-absorbing medium is $D \approx \dfrac{1}{3} \dfrac{1}{(\Sigma_s + \Sigma_a)}$.

6.1.4 Physical Analysis of Fick's Law

In diffusion phenomena, such as the diffusion of gas molecules or of heat, it is found that the diffusing substance tends to move from regions of high density toward those of low density. Neutrons behave similarly, since there are more collisions per unit volume in regions of high density, and after colliding the neutrons move away from the collision centers.

Glasstone and Edlund (1952, p. 91)

In a similar way to diffusive phenomena encountered in other areas of physics, neutrons diffuse from higher to lower densities (or flux) as shown by the gradient operator. Absorption and scattering reactions oppose this neutron diffusion. This is obvious for absorption. Scattering disperses neutrons by orienting them in different directions and hinders the scattering movement towards decreasing density gradients. Isotropic scattering in the laboratory reference frame is the most unfavorable to diffusion in the direction of decreasing gradients.

If scattering in the laboratory frame is anisotropic and peaked forward, the diffusion is improved. In this case, the diffusion coefficient is calculated according to the Eq. 9.19:

$$D_{tr} = \frac{1}{3\left[\Sigma_a + \Sigma_s - \overline{\mu}\Sigma_s\right]} = \frac{1}{3\,\Sigma_{tr}} \qquad (6.9)$$

In this equation, the term $\overline{\mu}\Sigma_s$ increases the value of the diffusion coefficient D which means a better diffusion.

6.2 Implementation

6.2.1 Taking into Account Anisotropy

In the previous calculations, we supposed that the scattering collisions exhibited spherical symmetry (isotropic) in the laboratory. This is true only if scattering is from heavy nuclei at lower energies, and the reality is that scattering over light nuclei is anisotropic. An improvement toward scattering anisotropy is accounted by the average cosine of the scattering angle: the anisotropy of a collision is quantified by the average cosine of the scattering angle $\overline{\mu} = \overline{\cos \psi}$.

6.2.2 Validity of Fick's Law

In reactor theory, Fick's law is used to derive the relation between neutron current and flux under certain conditions. Neutrons diffuse from a region of high concentration to a region of lower concentration like diffusion of gas atoms in a medium. Fick's law is derived based on various assumptions. Therefore, it has certain limitations, and requires that the angular flux varies sufficiently slowly with angle.

Fick's law is valid:

1. up to the point which is away from the edges or boundaries of the medium by few mean free paths.
2. at a few mean free paths away from strong sources or sinks. Fick's law has been established assuming that all neutrons contributing to the net current $\mathbf{J}$ originated in scattering collisions. If the sources are located at a few mean free paths, very few of the source neutrons survive as such to contribute to the value of $\mathbf{J}$ thus Fick's law is valid.
3. if the medium is not too absorbing in nature.
4. if the diffusion coefficient is modified with transport correction. Indeed, the assumption of isotropic scattering in the laboratory, in general, is not true.
5. if the neutron flux is slowly varying with respect to position.
6. if the neutron flux is not a function of time.

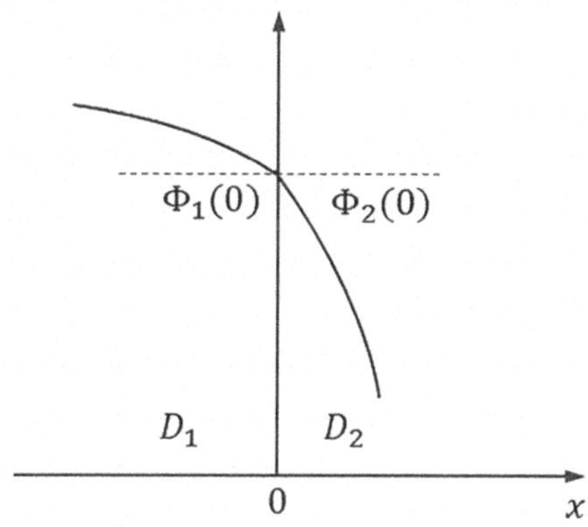

Fig. 6.3 Interface between two media

6.3 The Principle of Continuity of Flux and Current

The important point to bear in mind is that the number of neutrons in a packet [1] is not changed merely by crossing a physical interface. This means that the neutron angular density must be continuous in r as the interface is crossed or, more formally, $\nu(r + s\Omega, \Omega, E, t + \frac{s}{v})$ must be a continuous function of s, where s is a distance along Ω (..) It is equally applicable to the angular flux.   Bell and Glasstone (1970, p. 15)

Diffusion theory implies continuity of flux and net current normal to any surface.

The continuity of flux is: $\Phi_1(r) = \Phi_2(r)$ at each point r located on the interface.

The continuity of diffusion current is: $-D_1(r) \cdot \overrightarrow{\mathrm{grad}}\Phi_1(r) = -D_2(r) \cdot \overrightarrow{\mathrm{grad}}\Phi_2(r)$

In a 1D geometry, as show on Fig. 6.3, it can be written as:

$$D_1 \frac{d\Phi_1}{dx}(x_0) = D_2 \frac{d\Phi_2}{dx}(x_0).$$

References

G.I. Bell, S. Glasstone, *Nuclear Reactor Theory* (Van Nostrand Reinhold Company, 1970). https://books.google.fr/books?id=RNQmAQAAMAAJ

S. Glasstone, M.C. Edlund, *The Elements of Nuclear Reactor Theory* (D. Van Nostrand Company, 1952). ISBN 9780598464224. https://books.google.fr/books?id=MSRRAAAAMAAJ

J.R. Lamarsh, *Introduction to Nuclear Reactor Theory*. Addison-Wesley Series in Nuclear Engineering (Addison-Wesley Publishing Company, 1966). ISBN 9780201041262. https://books.google.fr/books?id=by5RAAAAMAAJ

---

[1] $\nu(r, \Omega, E, t)dVd\Omega dE$ is the probable number of neutrons at time t in a volume element dV having energies in dE about E and directions within a narrow beam $d\Omega$ about Ω.

Chapter 7
Equations Expressing the Conservation of Neutrons

Abstract In this short chapter, associated with the Appendix A, we first establish the monokinetic diffusion theory mathematical description. Such a relatively simple description has the great advantage of illustrating the balance between production, absorption, leakage and rate of change of neutron density without the complexity that is introduced by the treatment of important effects associated with the neutron energy spectrum and direction of the velocity vector. The neutron transport equation is a far more fundamental and exact description of the neutron population in a reactor than the neutron diffusion equation, and maybe easier to understand. The understanding of the transport equation provides a deeper and more thorough understanding of the diffusion theory and of the so-called "transport corrections". This is sufficient reason for including an introduction to the transport equation in this elementary nuclear reactor book.

7.1 Equation of Diffusion for Monokinetic Neutrons

The underlying principle for any theory of nuclear chain reactors is what may be called the conservation or balance of neutrons; thus, in a given volume, the time rate of change of neutron density is equal to the rate of production minus the rate of leakage and the rate of absorption.

Glasstone and Edlund (1952, p. 90)

  83
H. Grard, *Diffusion of Neutrons in Nuclear Reactors*,
https://doi.org/10.1007/978-3-032-05088-5_7

Table 7.1 Analogy between neutronics and fluid mechanics

Fluid mechanics	Neutronics
ρ: g/cm^3	n: neutrons/cm^3
ρV: g/cm^2/s	J: neutrons/cm^2/s
$\iint_S \rho V.dS$	$\iint_S J.dS$
$\frac{\partial \rho}{\partial t} + \mathrm{div}\,\rho V = 0$	$\frac{\partial n}{\partial t} + \mathrm{div}\,J = 0$

7.1.1 *Analogy Between Neutronics and Fluid Mechanics*

The neutron current J is analog to ρV as illustrated by Table 7.1.

We remind that:

$$\mathrm{div}\,\rho V = \lim_{d\tau \to 0}\left(\frac{1}{\tau}\underbrace{\iint_S \rho V.dS}\right)$$

$$\text{Mass balance per unit volume}$$
$$\text{and time at a point (g/cm}^3\text{/s)}$$

The divergence of a neutron current density is similar:

$$\mathrm{div}\,J = \lim_{d\tau \to 0}\left(\frac{1}{\tau}\iint_S J.dS\right)$$

7.1.2 *General Case: Non Homogeneous Reactor*

We admit the validity of the law of Fick (cf Chap. 6), represented by Eq. 7.1.

$$J(r) = -D(r) \cdot \overrightarrow{\mathrm{grad}}\,\Phi(r) \tag{7.1}$$

The law of Fick (Eq. 7.1) combined with Eq. A.7 gives the equation of diffusion for monokinetic neutrons (Eq. 7.2).

$$\frac{\partial n}{\partial t}(r) = \mathrm{Source}(r) - \Sigma_a(r)\Phi(r) + \mathrm{div}(D(r) \cdot \overrightarrow{\mathrm{grad}}\,\Phi(r))$$

Instead of Source(r), we can write $\nu\Sigma_f(r)\Phi(r)$.

Then

$$\frac{\partial n}{\partial t}(r) = \nu\Sigma_f(r)\Phi(r) - \Sigma_a(r)\Phi(r) + \mathrm{div}(D(r) \cdot \overrightarrow{\mathrm{grad}}\,\Phi(r)) \tag{7.2}$$

$-\text{div}\,\boldsymbol{J}(\boldsymbol{r}) = \text{div}(D(\boldsymbol{r}) \cdot \overrightarrow{\text{grad}}\Phi(\boldsymbol{r}))$ is the neutron leakage per unit volume per unit time (Glasstone and Edlund 1952, p. 101).

Due to the use of Fick's law, this Eq. 7.2 is valid at a sufficient distance (a few λ_s) from intense neutron sources, strong absorbers, and from interfaces between two mediums having very different properties.

7.1.3 Equation for an Homogeneous Reactor

If the reactor is homogeneous, $D(\boldsymbol{r})$ is constant and does not anymore depend on the position within the medium. Σ_a is also constant and the source is homogeneously distributed in the medium. This enables to write Eq. 7.3:

$$\frac{\partial n}{\partial t} = \nu\Sigma_f\Phi - \Sigma_a\Phi + D\Delta\Phi \tag{7.3}$$

This can be expressed in the following way:

Production $-$ Absorption $-$ Leakage $=$ rate of change of neutron density n with time

7.2 The Integro-Differential Form of Neutron Transport Equation

7.2.1 A Balance in Element Space

The integro-differential form of neutron transport equation, and the previously established diffusion equation for monokinetic neutrons Eq. 7.2 are conceptually similar and both express the neutron balance:

Production $-$ Absorption $-$ Leakage $=$ rate of change of neutron density n with time

In the diffusion equation for monokinetic neutrons, neutrons are differentiated only by their position in space, and the balance is established in an element volume. In the integro-differential form of neutron transport equation, neutrons are characterized by their positions, and their vectorial velocities. As a result, the balance is established in an element space which comprises the neutrons in an element volume about $\boldsymbol{r}$, having directions within $d\Omega$ about $\boldsymbol{\Omega}$ and energies in dE about E. Since it contains both derivatives in space and times as well as integrals over angle and energy, it is known as "integrodifferential" equation.

This transport equation uses the physical quantities described in Chap. 2. Simplifications and numerical resolution methods of the integro-differential form of neutron transport equation are not in the scope of the present book, nevertheless it is interesting to study this equation to sharpen the physical understanding of neutronics and to better understand the inherent limitations of diffusion theory.

7.2.2   *The Equation and Its Terms*

We express the time rate of change of neutron angular density, $\nu(r, \Omega, E, t)$:

$$\frac{\partial \nu(r, \Omega, E, t)}{\partial t} = \tag{7.4}$$

$$- \Omega.\overrightarrow{\mathrm{grad}}\varphi(r, \Omega, E, t) \qquad\qquad \text{(Streaming)}$$

$$- \Sigma_t(r, E)\varphi(r, \Omega, E, t) \qquad\qquad \text{(Collision)}$$

$$+ \int_0^\infty \int_{4\pi} \Sigma_s(r, E' \to E, \Omega'.\Omega)\varphi(r, \Omega', E', t)d\Omega dE' \qquad\qquad \text{(Transfer)}$$

$$+ \frac{1}{4\pi} \sum_j \int_0^\infty \nu_{p,j}(E')\Sigma_{f,j}(r), E')\psi(r, E', t)\chi_{p,j}(E' \to E)dE'$$

$$\text{(Prompt neutrons)}$$

$$+ \frac{1}{4\pi} \sum_{i=1}^8 \chi_{d,i}(E)\lambda_i\, C_i(r, t) \qquad\qquad \text{(Delayed neutrons)}$$

$$+ S_{\mathrm{indep}} \qquad\qquad \text{(Independant source)}$$

Transport equation is a linear equation in the unknown dependant variable:

$$\nu(r, \Omega, E, t)$$

With seven independant variables: three variables for the position in space, two variables for the direction of velocity vector, one variable energy and one variable time. It requires initial condition: the initial value of angular density for all positions, energies and directions, and also boundary conditions.

Streaming is the rate of loss per unit time, by movements, of neutrons having the position r, the energy E and moving in the direction Ω per unit energy, per unit solid angle, per unit volume (see Eq. A.6).

Collision is the rate at which neutrons collide. It is a total cross section, including absorption and scattering events removing neutrons from element space.

Transfer represents the arrivals of neutrons in element space by scattering.

$\varphi(r, \Omega, E, t)$ is the angle dependent flux, or angular neutron flux defined by the Eq. 2.1, and $\Sigma_s(r, E' \to E, \Omega'\Omega)$ is a differential cross section which, multiplied by φ, gives the rate at which neutrons in the element volume about r scatter into the element space, i.e. after scattering they have directions within $d\Omega$ about Ω and energies in dE about E.

All directions forming a cone of revolution around the incident direction Ω' are equiprobable. This is why differential scattering cross sections can be written with a scalar product: $\Sigma_s(r, E' \to E, \Omega'.\Omega)$ instead of $\Sigma_s(r, E' \to E, \Omega' \to \Omega)$ (Coste-Delclaux 2013, p. 46).

Prompt neutrons represents the production of new neutrons in the element space, immediately after fission reactions. The new neutrons produced by fission are emitted equally in all directions (isotropic fission), which is why it is possible to multiply the macroscopic fission cross section $\Sigma_{f,j}$ by the angle integrated flux $\psi(r, E', t)$ (see Eq. 2.2). The fission cross section depends on the incident neutron energy, so it is written as a function of E'. The index j enables to take into account the various fissile nuclei involved. The fission spectrum of prompt neutrons depends on the nucleus involved j, and on the energy of the incident neutron E'; it is written $\chi_{p,j}(E' \to E)$. The average number of prompt neutrons emitted per fission depends not only on the fissile nucleus, but also on the energy of the incident neutron (see Sect. 3.3.2), consequently it is written $\nu_{p,j}(E')$. The division by 4π enables to express the result per unit solid angle, and implicitly assumes that the emission of prompt neutrons is isotropic.

$C_i(r, t)$ is the concentration of the precursors group i, λ_i is the probability of decay by unit time and $\chi_{d,i}$ is the energy spectrum of delayed neutrons emitted by the precursors group i. The concept of precursors group is developed in the Chap. 14, in the Sects. 14.1.2–14.1.5.

The independent source of neutrons is made up of all neutron productions that are not neutron induced fissions (see Sect. 14.8.1).

We now need to add the equations of precursors groups balance:

$$\frac{dC_i(r, t)}{dt} = \tag{7.5}$$

$$- \lambda_i\, C_i(r, t) + \sum_j \beta_i^j \int_0^\infty \nu^j(E')\Sigma_{f,j}(r), E')\psi(r, E', t)dE' \tag{7.6}$$

Where $C_i(r, t)$ is the concentration at the position r of the precursor group i. $\nu^j(E')$ is the total number of neutrons (prompts + delayed) emitted by the fission of the fissile nucleus j induced by a neutron of energy E' and β_i^j is the fraction of delayed neutrons resulting from the fission of nucleus j.

References

M. Coste-Delclaux, C. Diop, A. Nicolas, B. Bonin, *La neutronique* (CEA Saclay, Groupe Moniteur, 2013)

S. Glasstone, M.C. Edlund, *The Elements of Nuclear Reactor Theory* (D. Van Nostrand Company, 1952). ISBN 9780598464224. https://books.google.fr/books?id=MSRRAAAAMAAJ

Chapter 8
Source Neutrons in a Scattering and Absorbing Medium

Abstract The solutions of the Diffusion Equation for monokinetic neutrons in simple cases, the plane source and the point source in infinite medium, highlight the significance of the diffusion length and and how the diffusion length is related to the migration area of monokinetic neutrons.

8.1 The Diffusion Equation in an Homogeneous Scattering and Absorbing Medium

We consider the diffusion equation for the special case in which the system is an homogeneous scattering and absorbing medium in a steady state and the time rate of change of the neutron density is zero. The diffusion equation then becomes:

$$- D\Delta\Phi + \Sigma_a \Phi = S_0 \tag{8.1}$$

When the neutron source is either a point, a line, or a plane, the production term S_0 or source term, as it is frequently called, is zero except at the particular source position. To solve the problem of the flux distribution in these cases, the differential equation is first solved with $S_0 = 0$ outside the source region (Glasstone and Edlund 1952, p. 106). Everywhere except at the source, the Eq. 8.1 becomes:

$$- D\Delta\Phi + \Sigma_a \Phi = 0$$

$$\Delta\Phi - \chi^2 \Phi = 0 \tag{8.2}$$

With the reciprocal length squared $\chi^2 = \dfrac{D}{\Sigma_a}$, expressions of the form of (8.2) are similar to the Laplacien eigen value problem, that enables to calculate the shape of standing waves confined to a finite region of space. Readers are invited to refer to the Appendix E.

H. Grard, *Diffusion of Neutrons in Nuclear Reactors*,
https://doi.org/10.1007/978-3-032-05088-5_8

8.2　Study of a Plane Source

We consider an infinite plane source (slab of null thickness), emitting mono-kinetic thermal neutrons at $x = 0$, at the center of an homogeneous scattering and absorbing medium. This plane source consists in a volumic density source of neutrons $S_0(x)$ $(\text{n/cm}^3/\text{s})$ which is distributed along the x-axis according to a Dirac delta function (see Appendix B in Sect. B.1). Then:

$$\int_{-\epsilon}^{\epsilon} S_0 \delta(x) dx = S_0 \ (\text{n/cm}^3/\text{s})$$

Thus this plane source emits S_0 neutrons by unit surface of the plane source and unit time $(\text{n/cm}^2/\text{s})$.

In the diffusion equation, the source is introduced as a source density function $S_0(x)$, which is a number of neutrons emitted per unit volume per unit time. The thickness of the planar source is null, thus $S_0(x \neq 0) = 0$.

8.2.1　Resolution of the Diffusion Equation in an Infinite Slab

The source and the planar medium are both infinite in the (x, y) plane, this is a 1D configuration, and along the x-axis. For symmetry reasons, Eq. 8.1 becomes:

$$\frac{d^2}{dx^2} \Phi(x) - \frac{\Sigma_a}{D} \Phi(x) = -\frac{S_0(x)}{D}$$

And:

$$\frac{d^2}{dx^2} \Phi(x) - \chi^2 \Phi(x) = 0 \quad \text{for } x \neq 0 \tag{8.3}$$

The absorbing and scattering medium is supposed to be infinite along the x-axis. The general solution of Eq. 8.3 is $B\, e^{\chi x} + C\, e^{-\chi x}$.

We can immediately see that $B = 0$, if not the solution has no physical meaning.

We consider two planes, P_1 and P_2, one on each side of the plane source, located at the abscissa x_1 and $x_2 = -x_1$.

x_1 has an order of magnitude of a few λ_s.

The neutrons emitted by the plane source cross P_1, P_2 with no absorption, if we make the assumption that $\lambda_a \ll \lambda_s$

The diffusion current on the surface of the planes P_1 is $\boldsymbol{J} = -D \cdot \overrightarrow{\text{grad}\Phi}$

Combined with the general solution of the diffusion equation, the current can be written as:

$$J(x_1) = -D\,(-\chi)\, C\, e^{-\chi x_1}$$

Half of the neutrons from the plane source are emitted towards P_1.

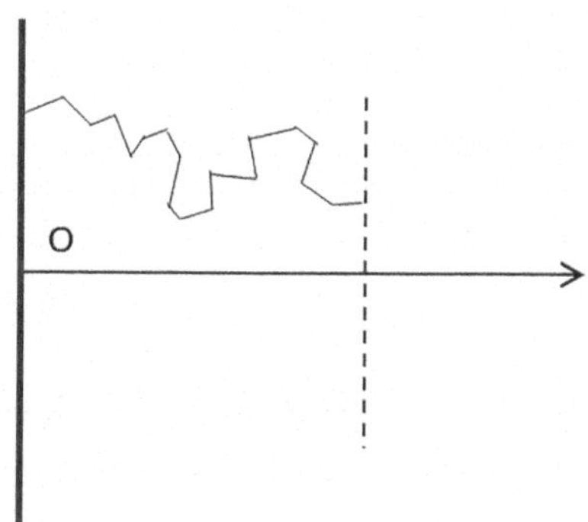

Fig. 8.1 Absorption of a neutron emitted by a plane source

$$\frac{S_0}{2} = D \chi C e^{-\chi x_1}$$

Since $x_1 \chi \ll 1$, we can approximate $e^{-\chi x_1} \approx 1$. Thus $C = \dfrac{S_0}{2D\chi}$.
And finally

$$\Phi(x) = \frac{S_0}{2D\chi} e^{-\chi x} \tag{8.4}$$

We define an attenuation coefficient L_D:

$$L_D = \frac{1}{\chi} = \sqrt{\frac{D}{\Sigma_a}}$$

Equation 8.4 can also be written as:

$$\Phi(x) = \frac{S_0 L_D}{2D} \exp -\left(\frac{x}{L_D}\right) \tag{8.5}$$

Equation 8.4 is an exponential attenuation function: exponential decrease with distance from the plane source. We know that $\dfrac{1}{\lambda}$ is the mean lifetime of a radioactive nucleus, thus $\dfrac{1}{\chi} = L_D$ is the mean projected distance on x-axis at which the neutrons emitted by a plane source are absorbed, as illustrated on Fig. 8.1

$$L_D^2 = \frac{D}{\Sigma_a} \approx \frac{1}{3\,\Sigma_a \Sigma_t}$$
$$\gg \lambda_s^2$$

The length of diffusion is far greater than the diffusion mean free path.

If the thickness of a medium is worth a few λ_s, then it can be considered as transparent to neutrons. If the order of magnitude of the thickness is L_D, it can be considered as opaque.

Length of diffusion is usually defined for fully thermalized neutrons at an energy $0.0253\,\text{eV}$.

8.2.2 Calculation of the Mean Squared Distance from the Plane Source at Which the Neutron Is Absorbed

The probability that a neutron moves away from the source at a distance x on X axis, and the be absorbed in the interval $[x, x + dx]$ is $\chi\, e^{-\chi x} dx$.

The mean squared distance from the plane source at which the neutron is absorbed:

$$\overline{x^2} = \int_0^\infty x^2 \chi\, e^{-\chi x} dx = \chi \int_0^\infty x^2 e^{-\chi x} dx$$

We can also write that:

$$\int_0^\infty x^2 e^{-\chi x} dx + \int_0^\infty 2x \frac{-1}{\chi} e^{-\chi x} dx = \int_0^\infty \frac{d}{dx}\left(x^2 \frac{-1}{\chi} e^{-\chi x}\right) dx$$

$$= \underbrace{\left[x^2 \frac{-1}{\chi} e^{-\chi x}\right]_0^\infty}_{=0}$$

We remind that:

$$\int_0^\infty t \cdot e^{-\lambda t} \cdot dt = \frac{1}{\lambda^2}$$

Thus:

$$\int_0^\infty x^2 e^{-\chi x} dx = -2 \cdot \int_0^\infty x \frac{-1}{\chi} e^{-\chi x} dx = \frac{2}{\chi} \cdot \frac{1}{\chi^2} = \frac{2}{\chi^3}$$

Then

$$\overline{x^2} = \chi \frac{2}{\chi^3} = \frac{2}{\chi^2} = 2L_D^2.$$

8.2.3 *Example of a Plane Source in Water*

We consider an infinite plane source, emitting mono-kinetic thermal neutrons at $x = 0$, within a wall of water at 20 °C, 1 bar. The width of the water wall is infinite.

In a plane geometry, the flux is devided by e when the distance from the source increases by the diffusion length. With $\Sigma_a = 0.021249\,\text{cm}^{-1}$ and $D = 0.158\,\text{cm}$, the diffusion length is $L_D = \sqrt{\frac{D}{\Sigma_a}} = 2.73\,\text{cm}$.

8.2.4 *Example of a Plane Source in a Medium of Finite Thickness*

Unlike in Sect. 8.2.1, the thickness along x-axis of the scattering and absorbing medium considered here is not infinite: a plane source is in the center of a infinite (in the (x, y) plane) medium, of width d. The diffusion equation within the water slab is the wave Eq. 8.2:

$$\Delta\Phi - \chi^2\,\Phi = 0$$

And two boundary conditions have to be respected:

- $\Phi(\frac{d}{2}) = 0$ Dirichlet boundary condition; a first insight into boundary conditions, with the case of Dirichlet boundary condition, is given in 10.1.4
- $\lim_{x \to 0} J = \frac{S_0}{2}$

The solutions of the diffusion equation are $\Phi(x) = A\cosh(\chi x) + B\sinh(\chi x)$.

Lamarsh (1966, p. 143) points out that hyperbolic functions are solutions of problems involving finite media, such as the present problem of a plane source in a finite water slab, and exponential functions are solutions for infinite media. This is because $\sinh x$ has a zero, which is a useful property when dealing with finite media, whereas the exponentials do not. Similarly, one of the exponential solutions goes to zero at infinity, a useful property for problems involving infinite media.

The term $A\cosh(\chi x)$ has to be eliminated in order to respect the boundary condition $\Phi(\frac{d}{2}) = 0$. Hence, the solution of the equation with Dirichlet boundary condition is:

$$\Phi(x) = B\sinh\left(\chi\left(\frac{d}{2} - x\right)\right)$$

Respect of the boundary condition at $x = 0$ sets the value of B:

$$\lim_{x \to 0} = -D \cdot \overrightarrow{\text{grad}}\Phi(x) = \frac{S_0}{2}$$

$$-D\frac{d\Phi}{dx} = -D\,B\cosh\left(\chi\left(\frac{d}{2} - x\right)\right)(-\chi)$$

$$\frac{S_0}{2} = B\,\chi D \cosh\left(\chi\frac{d}{2}\right)$$

And then:

$$B = \frac{S_0}{2\chi D \cosh\left(\chi\frac{d}{2}\right)}$$

Then the solution for the flux, as a function of the distance from the plane source:

$$\Phi(x) = \frac{S_0}{2\chi D \cosh\left(\chi\frac{d}{2}\right)} \sinh\left(\chi\left(\frac{d}{2} - x\right)\right)$$

Figure 8.2 illustrates this problem for various half thickenesses of the water wall (20 °C, 1 bar), ranging from 10 to 30 cm. The ratio of the flux to the source S_0, expressed in (n/cm^2/s), is plotted as a function of the distance x from the plane source in an infinite medium. Up to the vicinity of the boundary with vacuum, the flux vanishes sharply to zero for the smaller thicknesses of the water wall. Figure 8.2 shows that, if the width of the water slab is $> \frac{3}{\chi}$ (which is 8.181 cm in the considered water wall), except near the boundary the flux tends to an pure exponential flux shape (Glasstone and Edlund 1952, p. 113), which is the behavior within an infinite medium If the thickness is at least three times the diffusion length $L_D = \frac{1}{\chi}$, the medium may be treated as an infinite medium at distances greater than one diffusion length from the boundary, where the flux vanishes.

It is easy to check that:

$$\lim_{d\to 0} \Phi(x) = \frac{S_0}{2\chi D} e^{-\chi x}.$$

8.3   Study of a Point Source

We consider a neutron source emitting mono-kinetic thermal neutrons within an homogeneous scattering and absorbing medium. In the diffusion equation, the source can be described as a volumic density source $S_0(r)$ (number of neutrons emitted per unit volume per unit time), distributed according to a Dirac function. The integration over an infinitely small spherical volume gives a point isotropic source, emitting S_0 neutrons per unit time. The size of the point source is null, thus, in the diffusion equation, $S_0(r \neq 0) = 0$.

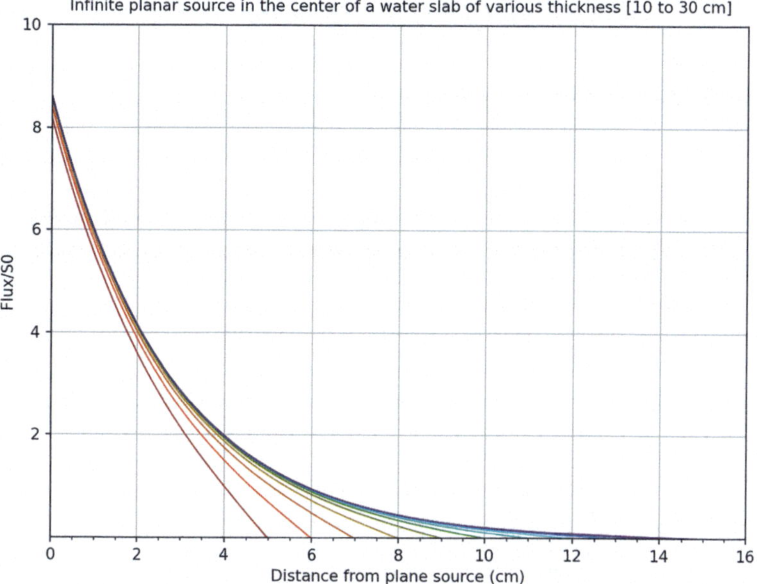

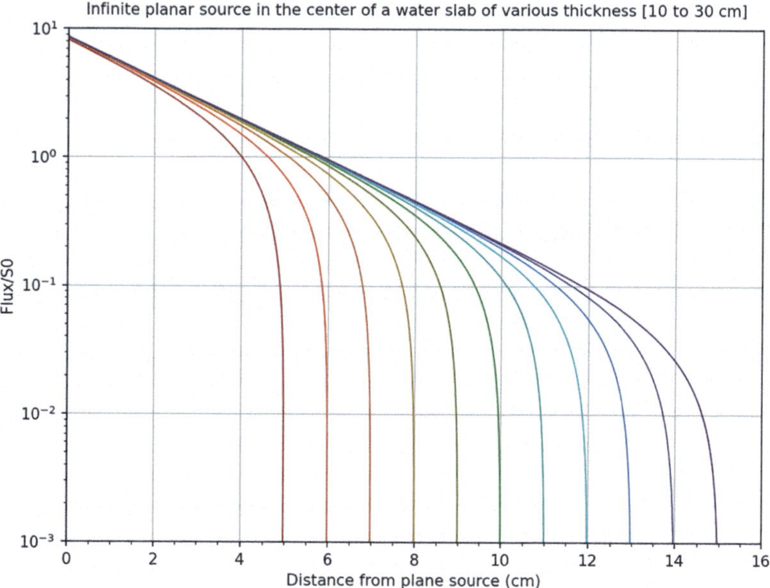

Fig. 8.2 Thermal flux resulting from a plane source of mono-kinetic neutrons in a water wall of various thicknesses (cm)

8.3.1   Resolution of the Diffusion Equation

The equation of diffusion car be written as:

$$- D\Delta\Phi + \Sigma_a\Phi = S_0 \tag{8.6}$$

Due to symmetry of the problem, the flux only depends on the distance from the point source r, and thus the Laplacian in spherical coordinates can be simplified:

$$\Delta\Phi(r) = \frac{\partial^2}{\partial r^2}\Phi(r) + \frac{2}{r}\frac{\partial}{\partial r}\Phi(r)$$

This equation does not apply, of course, at the source where r is zero.

Let's write that $\dfrac{\Sigma_a}{D} = \dfrac{1}{L^2}$.

Equation 8.6 can be written with a dependency on r:

$$\frac{d^2}{dr^2}\Phi(r) + \frac{2}{r}\frac{d}{dr}\Phi(r) - \frac{1}{L^2}\Phi(r) = \frac{S_0(r)}{D} \tag{8.7}$$

We define a new function $u(r)$ by the relation $\Phi(r) = \dfrac{u(r)}{r}$

$$\frac{d}{dr}\left(\frac{u(r)}{r^2}\right) = \frac{1}{r^2}\frac{d}{dr}u(r) + 2u(r)\frac{1}{r}\cdot\frac{-1}{r^2}$$

Then:

$$\frac{d^2}{dr^2}\Phi(r) = \frac{d^2}{dr^2}\left(\frac{u(r)}{r}\right)$$

$$= \frac{-1}{r^2}\frac{d}{dr}u(r) + \frac{1}{r}\frac{d^2}{dr^2}u(r) - \frac{d}{dr}\left(\frac{u(r)}{r^2}\right)$$

$$= \frac{d}{dr}u(r)\left(\frac{-1}{r^2} + \frac{-1}{r^2}\right) + \frac{1}{r}\frac{d^2}{dr^2}u(r) + 2u(r)\frac{1}{r^3}$$

$$\Delta\Phi(r) = \frac{d}{dr}u(r)\left(\frac{-2}{r^2}\right) + \frac{1}{r}\frac{d^2}{dr^2}u(r) + 2u(r)\frac{1}{r^3} + \frac{2}{r}\left(\frac{1}{r}\frac{d}{dr}u(r) - \frac{u(r)}{r^2}\right)$$

$$= \frac{1}{r}\frac{d^2}{dr^2}u(r)$$

Then, if $r \neq 0$, $S_0(r) = 0$ and from Eq. 8.7 we can now write:

$$\frac{1}{r}\frac{d^2}{dr^2}u(r) - \frac{1}{L^2}\frac{1}{r}u(r) = 0$$

Fig. 8.3 Monokinetic
neutrons $2200\,\mathrm{m\cdot s^{-1}}$
emitted in the center of a
heavy water sphere ($20.4\,^\circ$C)
of radius $r_0 = 10\,\mathrm{cm}$

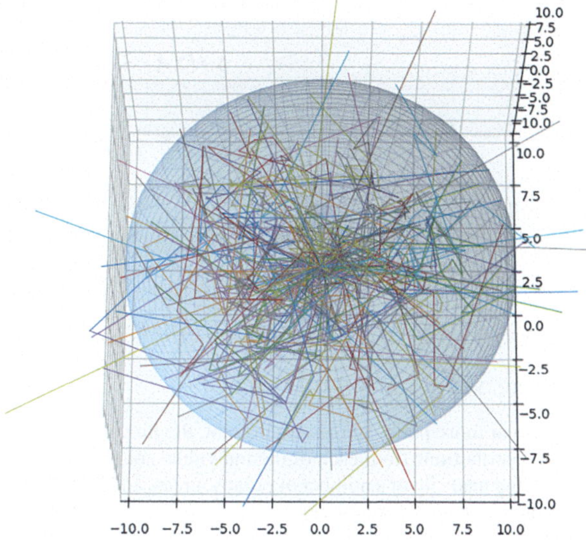

$$\frac{d^2}{dr^2}u(r) - \frac{1}{L^2}u(r) = 0 \qquad\qquad (8.8)$$

The general solution of Eq. 8.8 is $u(r) = A_1 \exp\left(-\frac{r}{L}\right) + A_2 \exp\left(\frac{r}{L}\right)$
Then

$$\Phi(r) = \frac{A_1}{r}\exp\left(\frac{-r}{L}\right) + \frac{A_2}{r}\exp\left(\frac{r}{L}\right)$$

A_2 has to be equal to zero, if not $\lim\limits_{r\to\infty} u(r) = \infty$.

We now consider a sphere with a radius r_0 of the order of a few λ_s and make the assumption that $\lambda_s < r_0 \ll \lambda_a$: the absorption within the sphere is negligible, and all the neutrons produced by the source cross the sphere, as shown on Fig. 8.3. On this example, $\lambda_s < r_0 \ll \lambda_a$ since $\lambda_s = 2.41$ cm and $\lambda_a = 29700$ cm.

Then $4\pi r^2 J(r_0) = S_0$. Since we consider a source in a infinite medium, we have to be aware that neutrons can cross the r_0 sphere in two opposite directions: the corresponding currents compensate one for each other and the previous equality is relevant. In a stationary state, there cannot be accumulation of neutrons.

At the surface of the sphere, supposing that r_0 is a sufficient distance from the source, we can admit the validity of the Fick's law: $\boldsymbol{J}(r_0) = -D \cdot \overrightarrow{\mathrm{grad}}\,\Phi(r_0)$

$$J(r_0) = -D\frac{d}{dr}\left(\frac{A_1}{r_0}\exp\left(\frac{-r_0}{L}\right)\right) = \frac{1}{r_0^2}DA_1\exp\left(\frac{-r_0}{L}\right)\left(1+\frac{r_0}{L}\right)$$

$$4\pi r_0^2\frac{1}{r_0^2}DA_1\exp\left(\frac{-r_0}{L}\right)\left(1+\frac{r_0}{L}\right) = S_0$$

Since $r \ll L$, we can approximate $\exp\left(\frac{-r_0}{L}\right) \approx 1$. Then

$$4\pi D\, A_1 = S_0$$

$$A_1 = \frac{S_0}{4\pi D}$$

Finally, the flux $\Phi(r)$ is:

$$\Phi(r) = \frac{S_0}{4\pi D}\, \frac{\exp\left(\frac{-r}{L_D}\right)}{r} \tag{8.9}$$

Where $S_0 : n/s$ and $\Phi(r) : \text{n/cm}^2/\text{s}$.

If two or more point sources are present, the flux at any position can be obtained by adding the contributions from each of the individual sources. In general, any source can be treated as being made up of a number of point sources, and the solution is given by the superposition of the point source solutions.

Glasstone and Edlund (1952, p. 108)

Equation 8.9 does not have physical sense if $r \to 0$, considering that Fick's law is not valid close to the source: it requires that r is at least a few λ_s. Indeed, $\lim\limits_{r \to 0} \Phi(r) = +\infty$. Note that the flux itself has no physical meaning when radius is lower than the size of a neutron. This is not due to the Diffusion approximation with the Fick's law

If the medium is not absorbing, then Eq. 8.6 becomes $D\Delta\Phi(r) + S_0(r) = 0$, which is similar to the heat equation.

8.3.2 Calculation of the Mean Squared Distance of a Neutron from a Source to the Point Where It Gets Absorbed

The probability that a neutron is absorbed between two spheres of radius r and $r + dr$ is:

$$P(r)dr = \frac{4\pi r^2\, dr\, \Phi(r)\Sigma_a}{S_0}$$

Then

$$\overline{r^2} = \int_0^\infty r^2 4\pi r^2 \Sigma_a \frac{S_0}{4\pi D\, r} \exp\left(\frac{-r}{L_D}\right) \frac{1}{S_0} dr$$

$$= \int_0^\infty r^3 \frac{1}{L_D^2} \exp\left(\frac{-r}{L_D}\right) dr$$

$$= 3\, L_D \frac{1}{L_D^2} \underbrace{\int_0^\infty r^2 \exp\left(\frac{-r}{L_D}\right) dr}_{2\,L_D^3}$$

$$= 6 L_D^2$$

The diffusion length squared is one-sixth of the mean squared distance, as the crow flies, that a (thermal) neutron would travel from its source to where it is absorbed (Glasstone and Edlund 1952, pp. 115, 116). That result is important, since even if thermal neutrons do not all have the same energy, they behave approximately like monoenergetic neutrons in weakly absorbing media, with properly averaged absorption cross section and mean free path. The thermal scattering area is presented in Sect. 9.2.1.

We can also write:

$$\frac{1}{6}\overline{r^2} = \frac{D}{\Sigma_a}$$

It has been established in the Chap. 6, Eq. 6.8, that $D = \dfrac{1}{3}\dfrac{\Sigma_s}{(\Sigma_s + \Sigma_a)^2} = \dfrac{1}{3}\dfrac{\Sigma_s}{\Sigma_t^2}$. Then, from the theory of diffusion, we can write:

$$\frac{1}{6}\overline{r^2} = \frac{1}{3}\frac{\Sigma_s}{\Sigma_a\,\Sigma_t^2} \approx \frac{1}{3\Sigma_a\,\Sigma_t}\ \ \text{if } \Sigma_a \ll \Sigma_s$$

This result can be compared to the rigorous migration area for monokinetic neutrons in an infinite homogeneous medium (Chap. 1.1): $\dfrac{1}{6}\overline{r^2} = \dfrac{1}{3\Sigma_t\,\Sigma_a}$.

References

S. Glasstone, M.C. Edlund, *The Elements of Nuclear Reactor Theory* (D. Van Nostrand Company, 1952), ISBN 9780598464224, https://books.google.fr/books?id=MSRRAAAAMAAJ

J.R. Lamarsh, *Introduction to Nuclear Reactor Theory*. Addison-Wesley Series in Nuclear Engineering (Addison-Wesley Publishing Company, 1966), ISBN 9780201041262, https://books.google.fr/books?id=by5RAAAAMAAJ

Chapter 9
The Theory of Age

Abstract In the previous chapters, we discussed two special cases: diffusion without moderation (equation of diffusion for mono-kinetic neutrons in Chaps. 7 and 8) and moderation without diffusion (Chap. 4). In this chapter we shall calculate the space and energy distribution of neutrons during the slowing down process, as a result of diffusion. The study of this distribution is relevant since it determines the probability of loss by leakage. Our goal will be to obtain the slowing-down current $q(r, E)$ arising from given sources, in various geometries.

9.1 Spatial Distribution During Slowing Down and the Fermi Age

9.1.1 The Equation of Fermi

λ_s is the mean free scattering path (mean distance travelled by a neutron between two successive elastic collisions). λ_s is supposed to be constant during the slowing down.

$$\lambda_s = \frac{1}{\Sigma_s}$$

The mean elapsed time between two collisions is $\dfrac{1}{\Sigma_s v}$.

During the time interval δt, the mean number of elastic collisions is $\delta t \cdot \Sigma_s v$.

We now consider an elastic collision slowing down a neutron: before shock the energy of the neutron is E and its lethargy is u, after shock $(E + \delta E)$ and $(u + \delta u)$ with $\delta E < 0$ and $\delta u > 0$.

By definition, $u = \ln \left(\dfrac{E_0}{E} \right)$ and $u + \delta u = \ln \left(\dfrac{E_0}{E + \delta E} \right)$.

Then:

$$\delta u = \ln \left(\frac{E_0}{E + \delta E} \cdot \frac{E}{E_0} \right)$$

H. Grard, *Diffusion of Neutrons in Nuclear Reactors*,
https://doi.org/10.1007/978-3-032-05088-5_9

$$\delta u = \ln\left(\frac{E + \delta E - \delta E}{E + \delta E}\right)$$

$$\delta u = \ln\left(1 - \frac{\delta E}{E + \delta E}\right)$$

We assume that $\delta E \ll E$. Graphite fits this assumption, but light nuclei (hydrogen, deuterium) do not. This means that the steps of slowing down are so small that it may be considered as a continuous process. This enables to write the variations consecutively to a shock:

$$\delta u \simeq \frac{-\delta E}{E + \delta E}$$

$$\delta u \simeq \frac{-\delta E}{E} \tag{9.1}$$

This equation links energy variations and lethargy variations in the case of a quasi-continuous slowing down. It is consistent with the differential calculus of the function $u(E) = \ln\dfrac{E_0}{E}$:

$$du = -\frac{dE}{E}$$

During the time interval δt, the mean gain of lethargy is $\delta t \cdot \Sigma_s v \xi$.

This lethargy variation during δt can be linked to the corresponding energy variation:

$$\delta t \cdot \Sigma_s v \xi = -\frac{\delta E}{E} \tag{9.2}$$

Note that δE has not the same meaning in the Eq. 9.2 than in the Eq. 9.1: energy variation during δt instead of consecutively to a shock.

Now we combine two equations in the system (9.3). The first one is Eq. 4.9 established in the Chap. 4 for an infinite and scattering medium at equilibrium. The second one is not demonstrated but it may be intuitively grasped as the gradient of slowing down current due to spatial transfer of neutrons.

$$\begin{cases} \Phi(E) = \dfrac{q(E)}{\Sigma_s(E)E\xi} \\ \dfrac{\partial q(r, E)}{\partial E} = -D(E)\Delta\Phi(r, E) \end{cases} \tag{9.3}$$

Where $D(E)$ is the diffusion coefficient, which is introduced in the Chap. 6. And thus:

$$\frac{\partial q(r, E)}{\partial E} = -\frac{D(E)}{\Sigma_s(E)E\xi} \cdot \Delta q(r, E)$$

We define a new variable $\tau(E)$, coined age of Fermi, by:

$$d\tau = -\frac{D(E)}{\Sigma_s(E)E\xi}dE$$

The same reference E_0 is chosen for the age of Fermi than for lethargy.

$$\tau(E) = \int_E^{E_0} \frac{D(E')}{\Sigma_s(E')E'\xi}dE'$$

And in lethargy units:

$$\tau(u) = \int_0^u \frac{D(u')}{\Sigma_s(u')\xi}du' \tag{9.4}$$

With this new variable, we can write the equation in a new form, the equation of age of Fermi:

$$\frac{\partial q}{\partial \tau}(\boldsymbol{r}, \tau) = \Delta q(\boldsymbol{r}, \tau) \tag{9.5}$$

The age of Fermi of a neutron increases when the neutron's energy decreases. The dimension of τ is not a time, but a surface, which is consistent with the Eq. 9.5.

τ is linked to the duration of the slowing down: τ starts at 0 at the fission energy, and then increases along with an energy decrease and an increase of path's length.

The path's length of the neutron slowing down from E to $E - dE$ is:

$$\frac{-dE}{\xi\Sigma_s(E)E} = v \cdot dt$$

$$d\tau = vD(v)dt \text{ then } \tau = \int_0^t vD(v)dt$$

t is the chronological age of neutrons.

An average value of the diffusion coefficient, which is a function of the neutron's energy, can be calculated by:

$$\overline{D} = \frac{\int_0^t vD(v)dt}{\int_0^t vdt}$$

The path of the neutron, during the slowing down interval of time $[0, t]$ is $d = \int_0^t vdt$.

Thus:

$$\tau = \overline{D} \cdot d$$

Consequently, the Fermi age τ is equal to the path of the neutrons multiplied by the average diffusion coefficient.

9.1.2   Solving the Equation of Fermi

Our objective is now to solve the Fermi equation 9.5, which is a second order differential equation as a function of space, and also first order as a function of age, which from a formal point of view, has to be considered as surface. It is interesting to notice that the equation of age is formally analog to the heat diffusion equation with a constant conductivity, and no heat generation:

$$\frac{\partial T}{\partial t} = \alpha \Delta T + q$$

Where T is the temperature in K, α is the thermal diffusivity (thermal conductivity divided by the product of the density and the specific heat capacity in m^2/s, Δ is the Laplace operator and q is the rate of heat input, known function varying in space and time.

Without heat generation, the transient heat equation is:

$$\frac{\partial T}{\partial t} = \alpha \Delta T$$

The knowledge of initial temperature enables to calculate the temperature as a function of time, at each point of the medium.

The Fermi age equation, $\frac{\partial q}{\partial \tau}(\mathbf{r}, \tau) = \Delta q(\mathbf{r}, \tau)$, enables to calculate the slowing down current $q(E)$, at each point of the medium and as a function of age.

9.1.3   Planar Source in an Infinite Slab of Finite Thickness

We consider an infinite plane source (slab of null thickness) defined in the same way as in Sect. 8.2.

The age equation is:

$$\frac{\partial q(x, \tau)}{\partial \tau} = \frac{\partial^2 q(x, \tau)}{\partial x^2} + S_0 \delta(x)$$

We admit Dirichlet boundary condition for the flux and:

$$q(\mathbf{r}, u) = \xi \Sigma_s(u) \underbrace{\varphi(\mathbf{r}, u)}_{=0}$$

Thus the boudary condition for the age equation is:

$$q(\pm \frac{d}{2}, \tau) = 0$$

The initial counditions at $\tau = 0$ (analog to initial conditions at $t = 0$ for the heat equation) is:

$$q(x, 0) = S_0 \delta(x)$$

We will use the eigenfunction method to solve the age equation. The first step is to find the eigenfunctions that are the solutions of the eigenvalue problem with Dirichlet boundary condition:

$$\Delta f = \lambda f$$

This is the eigen value problem (10.4) applied to a parallel wall in Sect. 10.2.1. There are an infinity of eigenvalues λ_i, all negative in the Dirichlet case. The eigenfunctions corresponding to distinct eigenvalues are the following:

$$f_n(x) = \cos B_n x$$

with:

$$\lambda_0 = -B_0^2 \qquad\qquad B_0 = \frac{\pi}{d}$$

$$\lambda_1 = -B_1^2 \qquad\qquad B_1 = \frac{3\pi}{d}$$

$$\lambda_2 = -B_2^2 \qquad\qquad B_2 = \frac{5\pi}{d}$$

$$(..)$$

The functions $f_n(x)$ form an orthogonal basis for the space of functions satisfying the boundary condition.

At each fixed age τ, the function $q(x, \tau)$ regarded as a function of x can be expressed in term of the basis, and thus there are coefficients $\varphi_n(\tau)$ such that:

$$q(x, \tau) = \sum_{n=0}^{\infty} \varphi_n(\tau) \cos B_n x \tag{9.6}$$

The next step is the determination of the coefficients $\varphi_n(\tau)$.

$$\sum_{n=0}^{\infty} \frac{\partial}{\partial \tau} \varphi_n(\tau) \cos B_n x = \sum_{n=0}^{\infty} -B_n^2 \cos B_n x \, \varphi_n(\tau) \quad \text{for } \tau \neq 0$$

$$\sum_{n=0}^{\infty} \left(\frac{\partial \varphi_n(\tau)}{\partial \tau} - \lambda_n \varphi_n(\tau) \right) \cos B_n x = 0$$

If $\cos B_n x$ are a basis, then the coefficient of each basis function must be zero at all ages τ. It follows that for each n:

$$\frac{\partial \varphi_n(\tau)}{\partial \tau} - \lambda_n \varphi_n(\tau) \quad \text{for } \tau \neq 0 \tag{9.7}$$

Equation 9.8 is a first order linear ODE for $\varphi_n(\tau)$ that is easy to solve. The general solution of Eq. 9.8 is:

$$\varphi_n(\tau) = \varphi_n(0) e^{\lambda_n \tau} \tag{9.8}$$

The last missing piece is the initial condition, the values $\varphi_n(\tau = 0)$. In order to find $\varphi_n(\tau = 0)$, we will write the initial condition in terms of the eigen function orthogonal basis:

$$q(x, 0) = S_0 \delta(x) = \sum_{n=0}^{\infty} S_n \cos B_n x \tag{9.9}$$

We also know that:

$$q(x, 0) = \sum_{n=0}^{\infty} \varphi_n(0) \cos B_n x \tag{9.10}$$

Since $\cos B_n x$ constitute a basis, the two sums (9.9) and (9.10) must be equal term by term, so:

$$\varphi_n(0) = S_n$$

From Eqs. 9.6, 9.8, and the equality $\varphi_n(0) = S_n$, we can write:

$$q(x, \tau) = \sum_{n=0}^{\infty} S_n e^{-B_n^2} \cos B_n x$$

From Eq. 9.9, we can calculate the inner product defined in the vector space of real functions as an integral (see Appendix G in Sect. G.1). The orthogonality of the basis enables to remove all the terms $\cos B_{m \neq n}$:

$$S_n \int_{-\frac{d}{2}}^{-\frac{d}{2}} \cos^2 B_n x \, dx = \int_{-\frac{d}{2}}^{-\frac{d}{2}} S_0 \delta(x) \cos B_n x \, dx$$

Since $B_n = \frac{\pi}{d} + n \frac{2\pi}{d}$, we can easily calculate for all n:

$$\int_{-\frac{d}{2}}^{-\frac{d}{2}} \cos^2 \left(\frac{\pi}{d} + 2n \frac{\pi}{d} \right) x \, dx = \frac{d}{2} \quad \forall n$$

Then, from the inner product calculation:

$$S_n = \underbrace{\frac{2}{d} \int_{-\frac{d}{2}}^{-\frac{d}{2}} S_0 \delta(x) \cos B_n x \, dx}_{=S_0 \cos(0) = S_0}$$

Taking into account that the delta function has the property that its convolution with any function equals the value of this function at zero (see Sect. B.3), it comes:

$$S_n = \frac{2S_0}{d}$$

$$q(x, \tau) = \frac{2S_0}{d} \sum_{n=0}^{\infty} e^{-B_n^2} \cos B_n x$$

Figure 9.1 shows the slowing down current as a function of the distance to the plane source, for different values of the age. For the age $\tau = 2 \, \text{cm}^2$, the first harmonics that make up the slowing current are plotted in dotted lines. When the age tends to zero, the slowing down current tends to the delta function $S_0 \delta(x)$, Which means that, by unit time, all the neutrons produced by the source increase their age over $\tau = 0$. When the age increases enough, the spatial distribution of the age become a $\cos B_0 x$

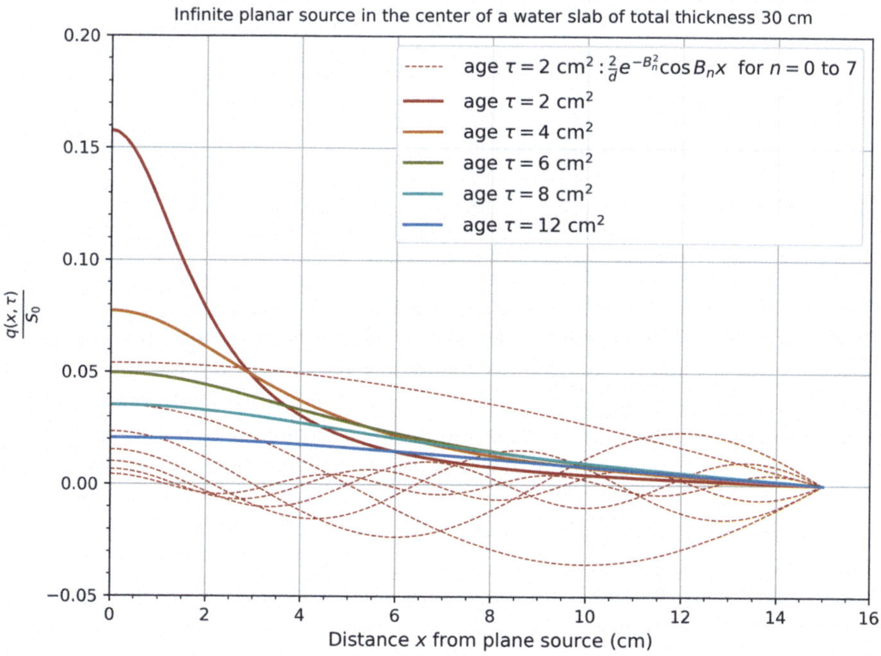

Fig. 9.1 Slowing down current as a function of the distance x to the plane source, for different age. Wall of 30 cm

function and the slowing down currents for this fixed age are harmoniously distributed throughout the space, according to the fundamental mode (see Sect. 10.1.4).

Planar Source in an Infinite Slab

We consider an infinite planar source which, emitting S_0 neutrons at the energy E_0 per second and per unit surface into an infinite homogeneous medium. The distance from the plane is x. The age equation is:

$$\frac{\partial q(x, \tau)}{\partial \tau} = \frac{\partial^2 q(x, \tau)}{\partial x^2}$$

After calculation, the final result is:

$$q(x, \tau) = \frac{S_0}{\sqrt{4\pi\tau}} \exp \frac{-x^2}{4\tau} \tag{9.11}$$

Point Source

In a spherical geometry: a source in the center of a sphere emits S_0 neutrons per second, with an energy E_0. By definition, the age of the neutrons produced by the source is zero. Due to spherical symmetry, the Laplacian can be written without azimuthal angle and zenith angle, and the age equation can be written as:

$$\frac{1}{r^2} \frac{\partial}{\partial r} \left(r^2 \frac{\partial q}{\partial r}(r, \tau) \right) = \frac{\partial q}{\partial \tau}(r, \tau) \tag{9.12}$$

It can be established that the solution of Eq. 9.12 is:

$$q(r, \tau) = \frac{S_0}{(4\pi\tau)^{3/2}} \cdot \exp \left(\frac{-r^2}{4\tau} \right) \tag{9.13}$$

Figure 9.2 represents the Eq. 9.13 (slowing down current as a function of the distance from the source) for different value of age. It illustrates that slightly slowed down neutrons are major contributors to slowing down current close to source of emission. Highly slowed down neutrons are spread by diffusion at great distances from the source. At a few cm from the source, the contribution to the slowing down current of neutrons of age $30\,\text{cm}^2$ (which are thermal neutrons if the medium is water at $20\,°\text{C}$ is not negligible: these neutrons have undergone shocks away from the source and have scattered back towards the source.

From Eq. 9.13, it is also possible to draw the slowing down current as a function of age, at a specific distance from the source. An example is given on Fig. 9.3 for $30\,\text{cm}$ from the point source.

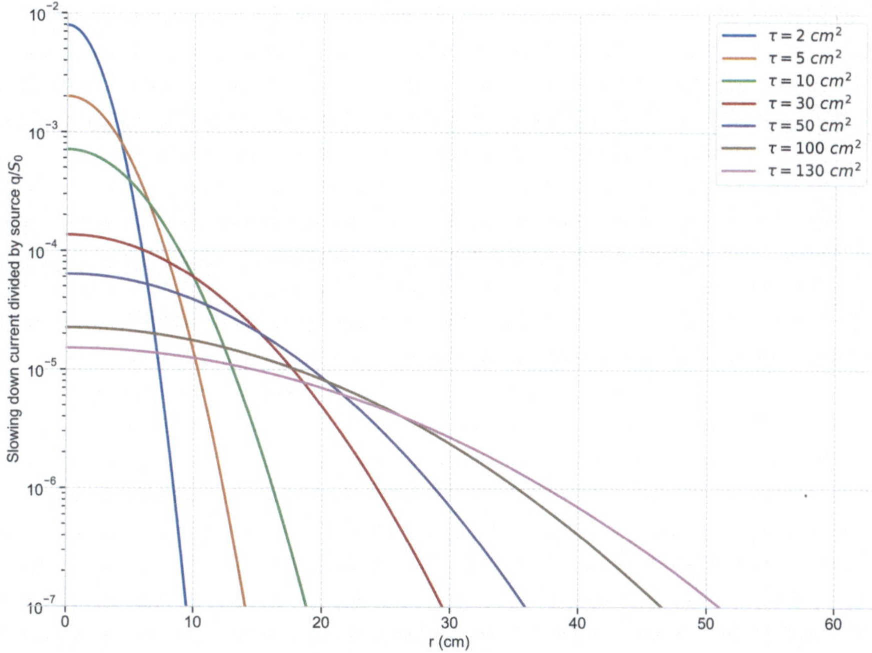

Fig. 9.2 Slowing down current as a function of the distance r to the point source, for different age and with a continuous slowing down process

9.1.4 Illustration with a Few Pratical Problems

Reminder About Reactor Operation During Start Up

Reactor reloading is monitored with source range neutron detectors installed outside the pressure vessel. These detectors are also into service during the pressure and temperature build up towards hot shutdown conditions (155 bar, 297.2 °C), and also during the approach to criticality.

These detectors are proportional counters for which each ionizing event is detected separately. Since they are inserted inside polyethylene, all the neutrons are thermalized before interacting with the detectors.

So as to reach the source range detectors, the neutrons emitted by fuel assemblies have to go through different media, as illustrated by Fig. 9.4: the baffle assembly enclosing the fuel assemblies, a water layer between the baffle assembly and the core barrel, the water inside the downcomer, and then the reactor vessel wall. The shortest straight line through water is approximately 40 cm (20 + 20).

During reloading and then during the pressure and temperature build up, the boron concentration BC is about 2500 ppm (this concentration is 3000 ppm for reactors loaded with MOX fuel assemblies). This BC and the rod assemblies inserted in their fuel assemblies maintain the effective multiplication factor of the core below 0.95.

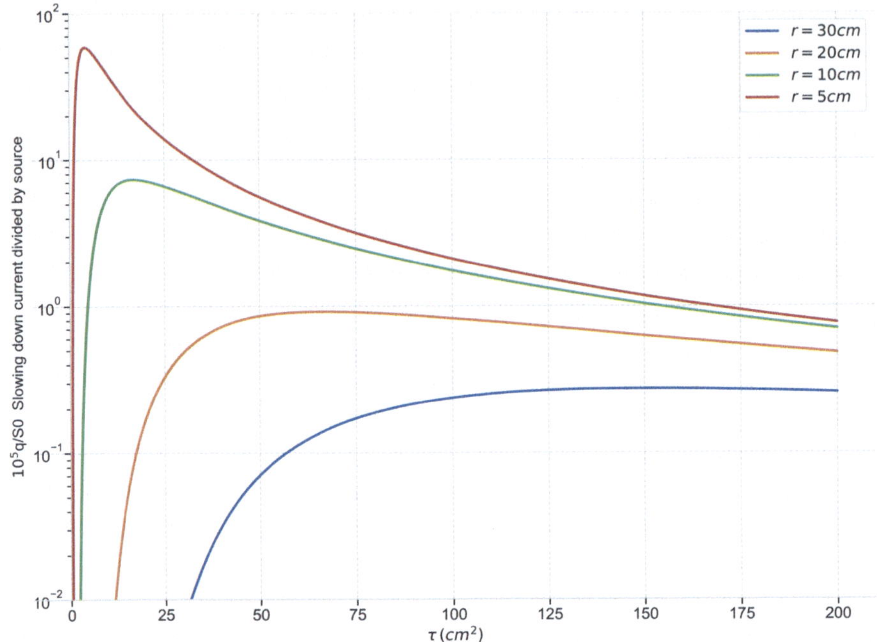

Fig. 9.3 Slowing down as a function of age, at 30, 20, 10, and 5 cm from the point source, with a continuous slowing down process

The fours fuel assemblies loading an Antimony-Beryllium Source are put first into the vessel (positions L1, R11, A5, E1). The loading of each of these 4 fuel assemblies triggers a count rate, about 15–40 hits/s (typical values) by the corresponding source range detector. At the end of the core loading, the count rate is higher, about 40–80 hits/s.

Probability of Leakage Out of a Sphere of Water and Application

A Monte Carlo calculation enables to estimate the probability of leakage out of a sphere of water, with a satisfactory accuracy even with a modest number of neutrons. The result is shown on Fig. 9.5. We can see that the probability of leakage is multiplied by a factor 5 if the temperature increases from 20 °C, 29 bar up to 297.2 °C, 155 bar.

And effectively, during the reactor pressure and temperature build up, the count rate of the source range neutron detectors increases significantly. Figure 19.2 highlights the evolution of neutron measurements (by source range detectors) between end of refueling (in the range [50 hits/s, 80 hits/s]) and end of heating ([200 hits/s, 300 hits/s]). This is due to the fact that the layer of water between the baffle and the vessel, which interposes itself between the detectors and the fuel assemblies, has a lower albedo if water density is lower.

We can deduce from these results that the increase of the count rate is mostly due the ability of neutrons to leak: the effect of the moderator temperature over

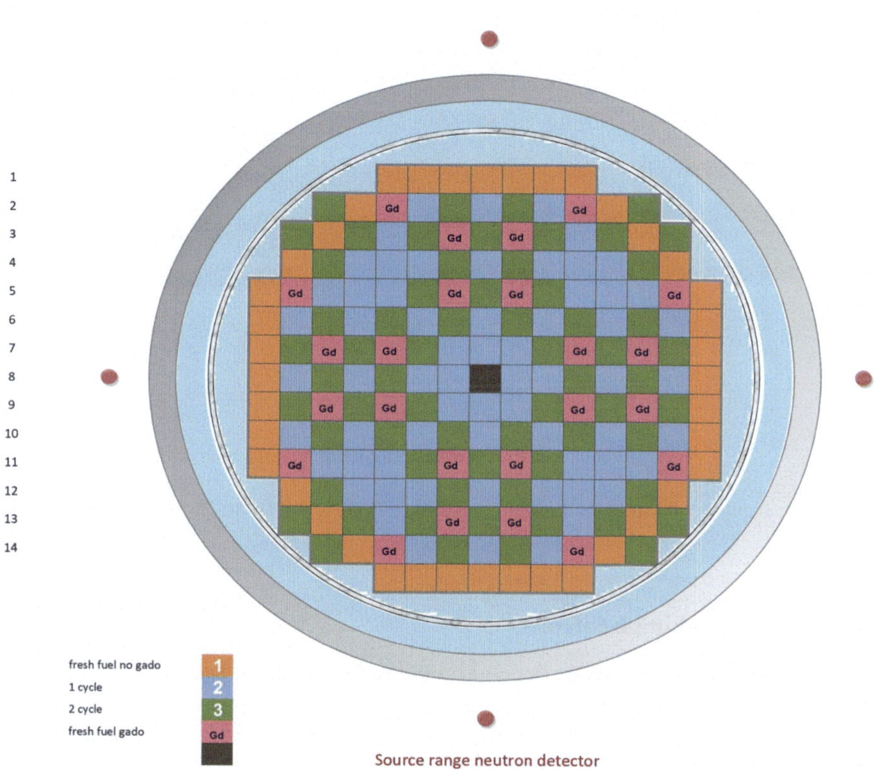

Fig. 9.4   The core (PWR 1300 MWe) and the media neutrons need to go through so as to reach the source range neutron detectors

the stationary flux (in a subcritical system, the flux always tends to a stationary state) can be neglected during pressure and temperature build up at 2500 ppm. Since the moderator temperature coefficient is slightly positive and balances the Doppler temperature coefficient, the isothermal coefficient is close to zero. Moreover, the reactor being deeply subcritical, it has a very little sensitivity to reactivity changes.

Power operation offers another example of how the probability of leakage affects the neutrons measurements. Due to the set point of the average coolant temperature, the temperature of cold water which flows inside the downcomer depends on the reactor power: 297.2 °C at 0%Pth and 289 °C at 100%Pth. When the power decreases, the leakage of neutrons that can be measured by the power range detectors increases. So as not to overestimate the nuclear power below rated power, the measurement of the power range detectors is corrected by a multiplier coefficient (<1).

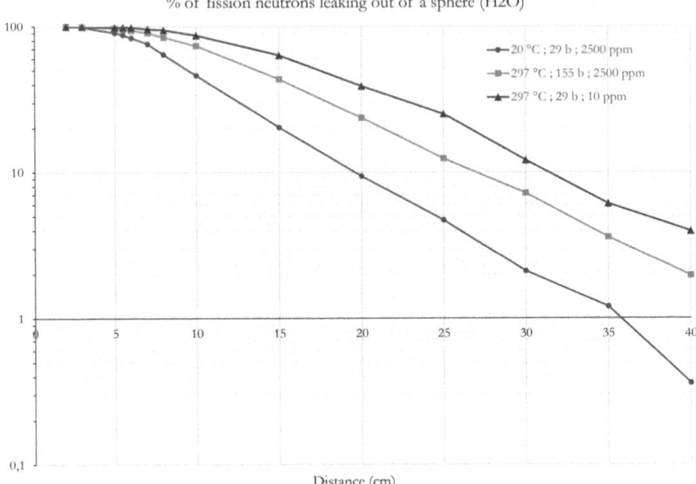

Fig. 9.5 Probability of leakage out of a sphere of water (radius from 2 to 40 cm) in different conditions

Energy of the Leaking Neutrons

The energy of leaking neutrons depend on the radius of the sphere and on the properties of water (density, boron concentration). Let's take the example of water at 297.2 °C, 155 bar, and 10 ppm of boron: 39% of leaking neutrons are fast neutrons, 56% are thermal neutrons and the other epithermal.

9.1.5   Age and Slowing Down Area

In the space between two spheres of radius r and $r + dr$, the number of neutrons of age τ in the process of slowing down is $4\pi r^2 q(r, \tau)dr$.

The probability that a neutron, emitted by a source, is in the place between the two spheres, r and $r + dr$, at the age τ is:

$$dP(r, \tau) = \frac{4\pi r^2 q(r, \tau)dr}{\int_0^\infty 4\pi r^2 q(r, \tau)dr}$$

We can deduce the mean squared crow flight distance of Fermi-aged τ neutrons:

$$\overline{r^2} = \int_0^\infty r^2 dp(r, \tau) = \frac{\int_0^\infty r^4 q(r, \tau)dr}{\int_0^\infty r^2 q(r, \tau)dr}$$

And we make the change of variable: $u^2 = \dfrac{r^2}{4\tau}$.

Then

$$\overline{r^2} = 4\tau \cdot \frac{\int_0^\infty u^4 e^{-u^2}\, du}{\int_0^\infty u^2 e^{-u^2}\, du}$$

We admit that the integrals $I_n = \int_0^\infty u^n e^{-u^2}\, du$ satisfy the following recurrence relation:

$$I_n = \frac{n-1}{2} I_{n-2}$$

And we can deduce that:

$$\overline{r^2} = 6\tau \tag{9.14}$$

This relationship is similar to the relationship linking the mean squared crow flight distance and the diffusion area L_D^2 in the case of thermal neutrons diffusion: $\overline{r^2} = 6L_D^2$.

For thermal neutrons, $L_D^2 = \dfrac{D}{\Sigma_a}$.

In a similar way, we can define a fictive absorption cross section $\Sigma_{\text{transfer}}(E)$ such as:

$$\tau(E) = \frac{D(E)}{\Sigma_{\text{transfer}}(E)}$$

The fictive absorption cross section Σ_{transfer} represents the removal of neutrons from their energy E due to the slowing down process.

The Fermi theory of age has been established by Enrico Fermi for the slowing down in graphite, for which some assumptions are justified. These assumptions do not apply for water. However the notion of age can be extended for the case of slowing down in water, if age is defined as the sixth of the squared value of the crow flight distance of a neutron.

Equation 9.14 enables to define the age of Fermi, regardless of how the neutrons slow down in a medium, that is, whether age theory is valid or not. Thus, for all moderators, age can be defined as:

$$\tau = \frac{1}{6}\overline{r^2}. \tag{9.15}$$

9.1.6   Thermal Age of Neutrons

Let us consider a point fission source emitting neutrons in an infinite moderating medium. The mean squared value of the distance from the source to the point where the neutron reaches thermal energy E_{th} is $6\tau_{th}$ where τ_{th} is the thermal age of neutrons.

The calculation of the thermal age of neutrons can be based on Eq. 9.4 with an energy decrease from 2 MeV down to 0.0253 eV, provided that the medium is not absorbing and that the slowing down process may be considered as continuous. Graphite satisfies these conditions, heavy water is acceptable.

The lethargy gain associated to a slowing down from 2 MeV down to 0.0253 eV is:

$$\Delta u = \ln \frac{2 \cdot 10^6}{0.0253} = 18.19$$

The age of thermal neutrons, for a continuous slowing down process, is defined by:

$$\tau_{th} = \int_0^{18.19} \frac{D(u)}{\xi \Sigma_s(u)} du \tag{9.16}$$

This Eq. 9.16 requires to preliminarily calculate $D(E)$. In the Sect. 9.4, we will estimate the age of thermal neutrons in heavy water with Eq. 9.16, assuming that the non continuous slowing down in heavy water does not overly distort the results. In order to take into account the anisotropy of scattering over heavy water during the slowing down, the correction transport presented in Sect. 9.3 will be applied.

9.1.7   *From the Equation of Age to Multi-group Diffusion Equations*

If absorption is introduced in the equation of age, this leads to establish the multi-group diffusion equations.

9.2   Migration Area in an Infinite Medium

9.2.1   *Thermal Scattering Area*

The thermal scattering area L_{th}^2 only applies to thermal neutrons and is not exactly the same than the squared diffusion length L_D^2.

L_{th}^2 is defined by:

$$L_{th}^2 = \frac{D_{th}}{\Sigma_{a,th}}$$

Where D_{th} and $\Sigma_{a,th}$ are averaged values over the thermal group.

9.2.2 Age of Thermal Neutrons

The mean squared value of the distance from the source to the point where the neutron reaches thermal energy E_{th} is $6\tau_{th}$ where τ_{th} is the thermal age of neutrons. It is relevant in an infinite medium.

9.2.3 Migration Area

The mean squared value of the distance from the source to the point where the neutron is absorbed after thermalization is six times the migration area M^2. According to the definition of τ given by Eq. 9.4, it should be noted that neutron ages are additive. In an infinite medium, the age of thermal neutrons adds up to thermal scattering area, and this addition defines the migration area:

$$M^2 = \tau_{th} + L_{th}^2. \tag{9.17}$$

9.3 Correction Transport in the Case of Anisotropic Diffusion

The rigorous migration area for anisotropic scattering mono kinetic neutron in an homogeneous and infinite medium medium is (cf. Chap. 1.1):

$$\frac{1}{6}\overline{r^2} = \frac{1}{3\,\Sigma_t\Sigma_a\,(1-\overline{\mu})} - \frac{\overline{\mu}\Sigma_a}{3\,\Sigma_t^2\,(1-\overline{\mu})^2}\left(\frac{1}{\Sigma_s} - \frac{\overline{\mu}}{\Sigma_t - \overline{\mu}\Sigma_s}\right) \tag{9.18}$$

If $\Sigma_a \ll \Sigma_s$, then Eq. 9.18 can be simplified, and becomes:

$$\overline{r^2} = \frac{2}{\Sigma_a\,[\Sigma_a + (1-\overline{\mu})\Sigma_s)]}$$

This enables to define the transport macroscopic cross section:

$$\Sigma_{tr} = \Sigma_a + (1-\overline{\mu})\Sigma_s$$

We can define a transport diffusion area:

$$\overline{r^2} = 6\,L_{Dtr}^2$$

And thus a transport diffusion coefficient, which is diffusion coefficient modified according to transport theory:

$$L^2_{Dtr} = \frac{D_{tr}}{\Sigma_a}$$

Since $\overline{r^2} = 6 \frac{D_{tr}}{\Sigma_a}$ and $\overline{r^2} = \frac{2}{\Sigma_a \left[\Sigma_a + (1 - \overline{\mu}) \Sigma_s\right]}$, we can write that:

$$D_{tr} = \frac{1}{3 \left[\Sigma_a + (1 - \overline{\mu}) \Sigma_s\right]} = \frac{1}{3\, \Sigma_{tr}} \tag{9.19}$$

Which is, from a formal point of view, the same equation than $D \approx \dfrac{1}{3\,(\Sigma_s + \Sigma_a)}$.

It is interesting to underline that Eq. 9.19 can be established from the transport equation:

$$D_{tr}(r, E, t) = \frac{1}{3 \left[\Sigma_t(r, E, t) - \overline{\mu} \Sigma_s(r, E, t)\right]} = \frac{1}{3\, \Sigma_{tr}(r, E, t)}$$

The correction transport consists in substituting $\Sigma_t = \Sigma_s + \Sigma_a$ for Σ_{tr}.

We recall that, without the transport correction, the scattering collisions are considered as isotropic even for light nuclei, and this simplifying assumption is not consistent with physics. Transport correction allows to preserve most of the effect of anisotropy, and migration of neutrons.

9.4 Calculation of the Age of Thermal Neutrons in Different Moderators

The Fermi age model is more appropriate for describing the slowing down of neutrons in heavy rather than in light moderators. However, this does not entirely rule out the use of this model, even for the lightest moderators.

Calculations of the age of thermalized neutrons are performed with a Monte Carlo method, and also by integration, with and without the transport correction. In this section, the neutron is considered thermalized from the moment its energy falls below 5 times the energy at the most probable velocity $E_0 = kT$ (arbitrary estimate of the upper limit of the Maxwell spectrum).

In order to be able to compare Monte Carlo calculations and calculations based on age and diffusion theory, the range of integration is $[5\,kT, 2\,\text{MeV}]$. Equations 9.20 and 9.21 have been used to calculate the age in a hydrogen medium having an atom density equal to that of the hydrogen in water, without transport correction, and with transport correction. Equations 9.22 and 9.23 have been used to calculate the age in water, without and with transport correction.

Equivalent calculation have been done with deuterium and heavy water.

Table 9.1 summarizes the values of age.

Age in a hydrogen medium without transport correction:

Table 9.1 Thermal age for different moderators

Moderator	Temperature (°C)	Density (g·cm^{-3})	τ_{th} from MC (cm^2)	τ_{th} with transport correction [2]	τ_{th} without transport correction [2]
Light water	20	0.9982	21.7	15.5	7.1
Hydrogen	Equiv. water 20 °C	$0.9982 \times \frac{2}{16+2}$	32.5	31.6	10.5
Light water	300	0.7265	42.2	28.9	13.2
Heavy water	20	1.1054	138.8	117	91.9
Deuterium	Equiv. heavy water 20 °C	$1.1054 \times \frac{4}{16+4}$	252	236	158

$$\tau(5\,kT) = \int_{5\,kT}^{2\mathrm{MeV}} \frac{\Sigma_s^H(E)}{3 \times (\Sigma_a^H(E) + \Sigma_s^H(E))^2} \frac{1}{\Sigma_s^H(E)} \frac{1}{E\, \underbrace{\xi_H}_{1}}\, dE \qquad (9.20)$$

Age in a hydrogen medium with transport correction:

$$\tau(5\,kT) = \int_{5\,kT}^{2\mathrm{MeV}} \frac{1}{3\,\Sigma_a^H(E) + (1 - \underbrace{\overline{\mu}_H}_{=\frac{2}{3A}=\frac{2}{3}})\Sigma_s^H(E)} \frac{1}{\Sigma_s^H(E)} \frac{1}{E\xi_H}\, dE \qquad (9.21)$$

Age in water without transport correction:

$$\tau(5\,kT) = \int_{5\,kT}^{2\mathrm{MeV}} \frac{\Sigma_s^H(E) + \Sigma_s^O(E)}{3 \times (\Sigma_a^H(E) + \Sigma_a^O(E) + \Sigma_s^H(E) + \Sigma_s^O(E))^2}$$
$$\times \frac{1}{\Sigma_s^H(E) + \Sigma_s^O(E)} \frac{1}{E\, \underbrace{\xi_{H_2O}}_{0.92}}\, dE \qquad (9.22)$$

Age in water with transport correction:

$$\tau(5\,kT) = \int_{5\,kT}^{2\mathrm{MeV}} \frac{1}{3\,\Sigma_a^H(E) + (1 - \underbrace{\overline{\mu}_{H_2O}(E)}_{\text{see Fig.\,4.17}})\Sigma_s^H(E)} \frac{1}{\Sigma_s^H(E)} \frac{1}{E\xi_{H_2O}}\, dE \qquad (9.23)$$

For light water at 20 °C, the age of thermal neutrons is determined by Monte Carlo: $\overline{r^2} = 130\,\mathrm{cm}^2$, and thus $\tau_{th} = 21.7\,\mathrm{cm}^2$.

For heavy water at 20 °C, the calculation with transport correction gives $\tau_{th} = 117\,\mathrm{cm}^2$. A Monte Carlo calculation returns $\overline{r^2} = 833\,\mathrm{cm}^2$ and thus $\tau_{th} = 138.8\,\mathrm{cm}^2$.

The differences between a slowing down in water, and in a hydrogen medium having an atom density equal to that of the hydrogen in water are interesting to notice. Although it is the hydrogen which is primarily responsible for the moderation of the neutrons in water, oxygen nuclei which are present play an important role in determining the age. This appears to be counterintuitive, since neutrons lose very little energy in elastic collisions compared with hydrogen, there are twice as many hydrogen atoms as oxygen atoms in water. Nevertheless, while the oxygen itself does not slow the neutrons clown appreciably, collisions with oxygen tend to prevent the neutrons from traveling very far from the source, particularly at the start of the slowdown phase (Lamarsh 1966, p. 201).

The calculation of thermal age by integration is subject to limitations: the slowing down process has to be continuous, that is true for graphite, it may be accepted in the case of heavy water, and far from being true for light water. In the case of a mixture of different nuclei (oxygen and hydrogen for example), the calculation by integration is much less accurate.

In graphite ($1.6\,\mathrm{g}\cdot\mathrm{cm}^3$), the age of thermal neutrons has been measured by Dlouhỳ (1962) ($355\,\mathrm{cm}^2$).

9.5 Migration Area in a Reactor Core

The slowingdown area of a lattice is a little larger than that of the pure moderator because slowing-down is negligible in the volume added for the fuel. On the other hand, the diffusion area of the lattice is far smaller than that of the moderator because of the great amount of absorption added by the presence of the fuel. (..) In water reactors, the migration area is very small and most of the leaks involve fast neutrons because the diffusion area is close to zero (a thermalized neutron in a water reactor is practically absorbed on the spot).

Reuss (2008, p. 540)

Typical values of the slowing down area (age of thermal neutrons), the diffusion area and the migration area are summarized on Table 9.2 for primary water and a PWR core. The values for PWR cores are from Reuss (2003, p. 409).

Table 9.2 Migration area in a PWR core compared to sole moderator

Medium	τ_{th} (^{2})	L^2_{th} (^{2})	$M^2 = \tau_{th} + L^2_{th}$ (^{2})
Light water $BC = 0$ ppm[a] at 300 °C	42.2	22	64.2
Light water $BC = 1200$ ppm[b] at 300 °C	44.2	6.5	50.7
PWR lattice at 300 °C	50	6	56

[a] The boron concentration at the end of the operating cycle and during stretch out is 10 ppm, and not 0 ppm

[b] Value at the beginning of the operating cycle, at rated power, xenon at equilibrium

References

Z. Dlouhỳ, A measurement of the neutron age in graphite by the pulsed source method. J. Nucl. Energy. Parts A/B. Reactor Sci. Technol. **16**(6), 311–316 (1962)

J.R. Lamarsh, *Introduction to Nuclear Reactor Theory*. Addison-Wesley Series in Nuclear Engineering (Addison-Wesley Publishing Company, 1966), ISBN 9780201041262, https://books.google.fr/books?id=by5RAAAAMAAJ

P. Reuss, *Précis de neutronique* (EDP Sciences, 2003)

P. Reuss, *Neutron Physics* (EDP Sciences, 2008)

Chapter 10
The Theory of Diffusion Applied to Fissile Systems with Monokinetic Neutrons

Abstract This chapter introduces fundamental concepts, in the simple case of monokinetic neutrons and homogeneous reactor. These concepts are criticality, the fundamental mode as an eigen function, the geometrical buckling, the effective multiplication factor. The last section presents calculations with the one group diffusion theory equation, which is from a formal point of view the same equation than for monokinetic neutrons, for a reactor with a fissile zone and a reflector. Chapter 12 will show that one group diffusion theory the model fails to represent the impact of the reflector on the flux.

10.1 Criticality of the Homogeneous Reactor, with Monokinetic Neutrons

10.1.1 Intuitive Approach to the Physical Concept of Criticality

A system is said to be subcritical if, for any nonzero initial neutron population, the expected population at late times, i.e., as $t \to \infty$, will die out unless it is sustained by a neutron source, internal or external. Similarly, a system is described as supercritical when the expected neutron population diverges at late times, starting from any nonzero population or with a source. Finally, a critical system is defined as one in which a steady, time-independent expected neutron population can be maintained in the absence of a source.

Bell and Glasstone (1970, p. 38)

10.1.2 The Steady State Diffusion Equation

In the case of an homogeneous reactor, we can write the Eq. 7.3 established in Chap. 7:

$$\frac{\partial n}{\partial t} = \nu \Sigma_f \Phi - \Sigma_a \Phi + D \Delta \Phi \tag{10.1}$$

In a steady state, Eq. 7.3 becomes Eq. 10.2.

$$\nu \Sigma_f \Phi - \Sigma_a \Phi + D \Delta \Phi = 0 \tag{10.2}$$

k_∞ is the ratio between the production rate of neutrons (n/s) in the reactor and the absorption rate. For an homogeneous reactor, we can write:

$$k_\infty = \frac{\nu \Sigma_f}{\Sigma_a}$$

Then

$$\frac{D \Delta \Phi}{\Sigma_a} + (k_\infty - 1)\Phi = 0$$

We define L^2 as $L^2 = \dfrac{D}{\Sigma_a}$

$$\Delta \Phi + \frac{(k_\infty - 1)}{L^2} \cdot \Phi = 0. \tag{10.3}$$

10.1.3 Curvature of the Flux

If the medium is characterized by $k_\infty > 1$, then the Laplacian of the flux is negative. The curvature of the flux is convex if the production rate of neutrons $\nu \Sigma_f \Phi$ is greater than the absorption rate $\Sigma_a \Phi$. This is a also true if the system is non homogeneous: the local curvature of the flux is convex if $\nu \Sigma_f(r)\Phi(r) > \Sigma_a(r)\Phi(r)$. When production rate is greater than absorption rate, there is a net source of neutrons characterized by a negative Laplacian, as illustrated by Fig. A.5.

10.1.4 Eigen Values of the Laplacian, with Dirichlet Boundary Conditions

The steady-state diffusion Eq. 10.3 has an infinite number of solutions. Boundary conditions, that complement the steady-state diffusion equation enable to make the problem definite and specified, as explained in Chap. 11.

We consider the following general eigenvalue problem for the Laplacian:

$$\begin{cases} \Delta f(r) = \lambda f(r) \\ f(r) \text{ satisfies boundary conditions (BC)} \end{cases} \tag{10.4}$$

The most common boundary condition is the Dirichlet which consists in specifying only values along the boundary. A special case of Dirichlet condition is $f = 0$ on the boundary.

There are an infinity of eigenvalues λ_i. All eigenvalues are negative in the Dirichlet case. The eigenfunctions corresponding to distinct eigenvalues are orthogonal.

The eigen values can be ranked in decreasing order:

$$0 > \lambda_0 > \lambda_1 > \lambda_2 > (..) > \lambda_n > (..)$$

The eigenfunctions corresponding to distinct eigenvalues are orthogonal and form a basis.

The eigenfunction f_0 associated to the less negative eigenvalue λ_0 has the specificity to remain positive within the domain, except on the boundary where it is null in accordance with the Dirichlet BC. This function is the fundamental mode.

For $n \neq 0$, the functions $f_n(r)$ change of sign over the domain. These functions are transitory terms (harmonics) appearing after a modification in the core and fade over time, leaving the flux according to the fundamental mode.

10.1.5 Application to the Diffusion Equation

The "buckling" B_g^2 is defined by $B_g^2 \overset{\text{def}}{=} -\lambda_0$.

B_g^2 is related to the Laplacian, and thus it depends on the dimensions of the reactor, and on the geometrical shape, notably the curvature of the boundary. It determines the amount of neutron leakage. B_g^2 is known today in the nuclear engineering community as the "geometrical buckling" or "buckling."

From Eq. 10.3, we can define $B_m^2 \overset{\text{def}}{=} \dfrac{(k_\infty - 1)}{L^2}$.

Equation 10.3 is satisfied if $-B_m^2$ is equal to an eigenvalue of the Laplacian over the considered domain.

if $k_\infty < 1$ then $B_m^2 < 0$ and Eq. 10.3 cannot be solved. Criticality cannot be achieved and a steady stade is only possible in the case of a neutron source (production of neutrons independently from the chain reaction).

10.1.6 Critical Condition

The critical condition is $-B_m^2$ is equal to the less negative eigenvalue of the Laplacian, for which the eigenfunction, positive over the considered domain, has a physical meaning.

$$B_m^2 = B_g^2 \quad \text{Critical condition}$$

$$\frac{(k_\infty - 1)}{L^2} = B_g^2$$

$$\frac{k_\infty}{1 + L^2 B_g^2} = 1$$

$$\frac{\nu \Sigma_f}{\Sigma_a (1 + L^2 B_g^2)} = 1$$

At criticality, we know that:

$$\frac{\text{production rate of neutrons in the reactor (n/s)}}{\text{absorption + leakage rates (n/s)}} = 1$$

Thus we can write that, in a reactor (homogeneous or no):

- production rate of neutrons in the reactor (n/s) $= \iiint\limits_V \nu \Sigma_f(r) \Phi(r)\, d\tau$

- absorption rate (n/s) $= \iiint\limits_V \Sigma_a(r) \Phi(r)\, d\tau$

- leakage rate (n/s) $= \iiint\limits_V D(r) B_g^2 \Phi(r)\, d\tau.$

10.1.7 Effective Multiplication Factor

The effective multiplication factor is defined as the ratio between the production rate of neutrons in the reactor (n/s) and the loss rate of neutrons (n/s), by both absorption and leakage.

$$k_{\text{eff}} = \frac{\nu \Sigma_f}{\Sigma_a (1 + L^2 B_g^2)}$$

At criticality, $k_{\text{eff}} = 1$.

$$k_{\text{eff}} = k_\infty \times \frac{1}{1 + L^2 B_g^2}$$

And $\frac{1}{1 + L^2 B_g^2}$ is the non-leakage probability.

10.1.8 Subcriticality of the Harmonics

The critical condition has be defined as the criticality of the fundamental mode:

$$\frac{(k_\infty - 1)}{L^2} = B_g^2 = -\lambda_0$$

If the reactor is critical, it is easy to check that the harmonics are strongly subcritical:

$$\frac{(k_\infty - 1)}{L^2} = -\lambda_0 < -\lambda_1 < -\lambda_2(..).$$

10.2 Application to Different Geometries

10.2.1 Parallel Wall

If the faces are infinite, the case turns into 1D problem:

$$\Delta\Phi = \frac{d^2}{dx^2}\Phi(x)$$

Then, the Eq. 10.5 we have to solve:

$$\frac{d^2}{dx^2}\Phi(x) + B_g^2\Phi = 0 \quad \text{Within the wall} \tag{10.5}$$

We now have to determinate the "bulking" as a function of the geometry (wall of thickness e).

The solutions of the second order differential Eq. 10.5 are:

$$\Phi(x) = A_1 \cos B_g x + \underbrace{A_2 \sin B_g x}_{\substack{A_2 = 0 \\ \text{symmetry requirement}}}$$

The Dirichlet condition can be written:

$$A_1 \cos B_g\left(\frac{e}{2}\right) = 0$$

This enables us to determinate the buckling: $B_g = \dfrac{\pi}{e}$

The first eigenvalues and eigenfunctions of the equation $\dfrac{d^2}{dx^2}\Phi(x) = \lambda_n \Phi(x)$ are:

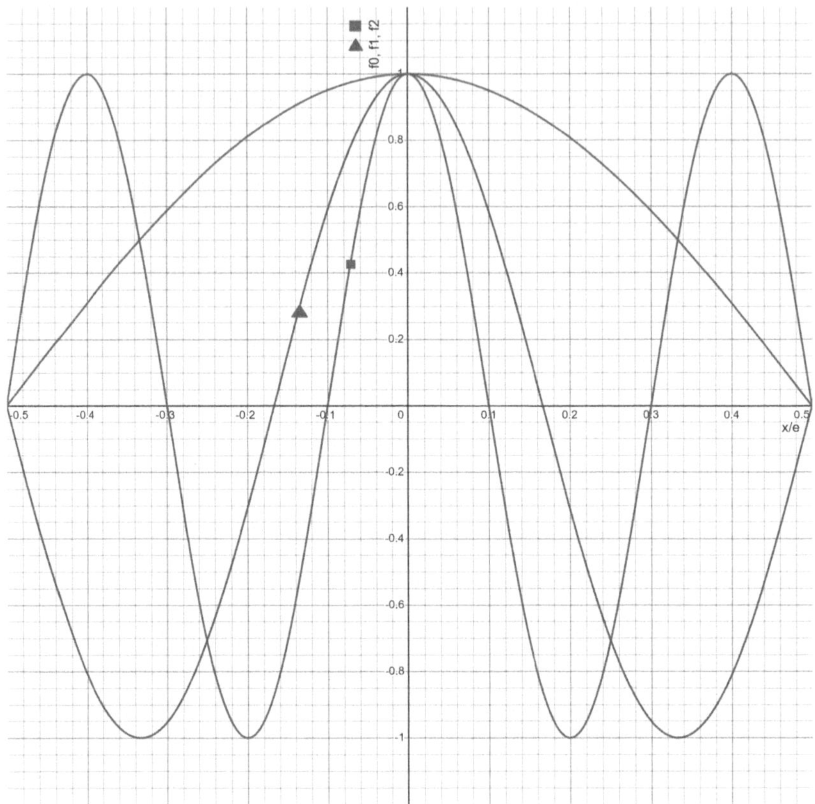

Fig. 10.1 First eigenfunctions of $\dfrac{d^2}{dx^2}\Phi(x) = \lambda_n \Phi(x)$, with a Dirichlet condition on $x \pm \frac{e}{2}$

$$\lambda_0 = -B_g^2 \qquad\qquad\qquad f_0(x) = \cos B_g x$$

$$\lambda_1 = -\left(\frac{3\pi}{e}\right)^2 \qquad\qquad\qquad f_1(x) = \cos \frac{3\pi x}{e}$$

$$\lambda_2 = -\left(\frac{5\pi}{e}\right)^2 \qquad\qquad\qquad f_2(x) = \cos \frac{5\pi x}{e}$$

$$(..)$$

Figure 10.1 illustrates that the eigenfunctions f_n associated to λ_n for $n > 1$ change sign over the domain and cannot be physical solutions of the steady state. The eigenfunction f_0 (fundamental mode) is the only one to remain positive in the domain.

10.2.2 Sphere

We have to solve Eq. 10.6:

$$\Delta \Phi(r) + B_g^2 \Phi(r) = 0 \tag{10.6}$$

Where r is a scalar and not a vector. In spherical geometry, Eq. 10.6 turns into Eq. 10.7:

$$\frac{d^2}{dr^2}\Phi(r) + \frac{2}{r} \cdot \frac{d}{dr}\Phi(r) + B_g^2 \Phi(r) = 0 \tag{10.7}$$

Equation 10.7 is equivalent to Eq. 10.8:

$$\frac{d^2}{dr^2}(r\Phi(r)) + B_g^2 \Phi(r) = 0 \tag{10.8}$$

It can be deduced that:

$$r\Phi(r) = A\cos(B_g r) + C\sin(B_g r)$$

$$\Phi(r) = \frac{A}{r}\cos(B_g r) + \frac{C}{r}\sin(B_g r)$$

Φ shall not tend to be infinite on $r = 0$, therefore $A = 0$.
The Dirichlet condition at the boundary of the sphere $(r = R)$ is $\Phi(R) = 0$. Thus:

$$\frac{C}{R}\sin(B_g R) = 0$$

$$B_g R = \pi$$

$$B_g = \frac{\pi}{R}.$$

10.2.3 Cylinder

We have to solve the Eq. 10.9

$$\frac{\partial^2}{\partial r^2}\Phi(r, z) + \frac{1}{r} \cdot \frac{\partial}{\partial r}\Phi(r, z) + \underbrace{\frac{\partial^2}{\partial z^2}\Phi(r, z)}_{\text{This term can be dropped if the cylinder is infinite}} + B_g^2 \Phi(r, z) = 0 \tag{10.9}$$

We separate the two variables r and z: $\Phi(r, z) = H(z)\theta(r)$. This enables to transform equation 10.9 and write:

$$\underbrace{\frac{1}{\theta}(r) \cdot \left(\frac{d^2}{dr^2}\theta(r) + \frac{1}{r} \cdot \frac{d}{dr}\theta(r) \right)}_{-\alpha^2} + \underbrace{\frac{1}{H(z)} \cdot \frac{d^2}{dz^2}H(z)}_{-\beta^2} + B_g^2 = 0 \qquad (10.10)$$

The buckling can be divided in two parts: $-\alpha^2$ for the radial part and $-\beta^2$ for the axial part.

$$B_g^2 = \alpha^2 + \beta^2$$

The radial part of the Eq. 10.10 can be written as:

$$\frac{d^2}{dr^2}\theta(r) + \frac{1}{r} \cdot \frac{d}{dr}\theta(r) + \alpha^2\theta(r) = 0$$

Then Eq. 10.10 is equivalent to the system:

$$\begin{cases} \dfrac{d^2\theta(\alpha r)}{d(\alpha r)^2} + \dfrac{1}{\alpha r} \cdot \dfrac{d\theta(\alpha r)}{d(\alpha r)} + \theta(\alpha r) = 0 \\ \dfrac{1}{H(z)} \cdot \dfrac{d^2}{dz^2}H(z) + \beta^2 = 0 \end{cases} \qquad (10.11)$$

The first equation of the system (10.11) is the zero order Bessel Differential equation (see Appendix B):

$$y'' + \frac{y'}{x} + y = 0$$

This equation has been solved by Bernouilli in 1732 and its general solution is:

$$y = C_1 J_0(x) + C_2 Y_0(x)$$

Where $J_0(x)$ and $Y_0(x)$ are two order 0 Bessel functions, please refer to Appendix B (Fig. B.1 for J_0 and Fig. B.2 for Y_0). These two functions constitute a basis and any linear combination of these two functions is a solution of Bessel equation of zero order.

The function J_0 has a physical meaning, but Y_0 has no physical meaning because $\lim_{x \to 0} Y(x) = -\infty$.

Then the general solution for the first equation of system (10.11) (radial equation) is $C_1 J_0(\alpha r)$. The boundary condition does impose that $C_1 J_0(\alpha R) = 0$, where R is the radius of the cylinder.

The first root of J_0 is around 2.4048. Therefore $\alpha = \frac{2.4048}{R}$ and we can write:

$$\theta(r) = C_1 J_0 \left(\frac{2.4048}{R} r \right)$$

The second equation of the system (10.11) is:

$$\frac{d^2}{dz^2} H(z) + \beta^2 H(z) = 0 \tag{10.12}$$

Equation 10.12 is similar to Eq. 10.5.
The solution of Eq. 10.12 is $H(z) = A_1 \cos(\beta z)$.
The Dirichlet boundary condition for the two faces of the cylinder is:
$A_1 \cos\left(\beta \frac{h}{2}\right) = 0$ where h is the height of the cylinder. Therefore $\beta = \frac{h}{\pi}$ and:

$$H(z) = A_1 \cos\left(\frac{h}{\pi} z\right)$$

The solution of Eq. 10.9 is:

$$\Phi(r, z) = J_0\left(\frac{2.4048}{R} r\right) \cos\left(\frac{h}{\pi} z\right)$$

Then, from Eq. 10.10, the buckling of the cylinder is $B_g^2 = \alpha^2 + \beta^2$, and:

$$B_g^2 = \left(\frac{2.4048}{R}\right)^2 + \left(\frac{\pi}{h}\right)^2 \tag{10.13}$$

For an infinite cylinder, the critical condition is:

$$\left(\frac{2.4048}{R}\right)^2 = B_m^2. \tag{10.14}$$

10.2.4   Optimal Cylinder Shape

We assume that our homogeneous cylinder is characterized by B_m^2.
   The at criticality the cylinder radius doe satisfy the equation:

$$R^2 = \frac{2.4048^2 \cdot h^2}{B_m^2 \cdot h^2 - \pi^2}$$

And the critical volume is:

$$V_c = \frac{\pi \cdot 2.4048^2 \cdot h^3}{B_m^2 \cdot h^2 - \pi^2}$$

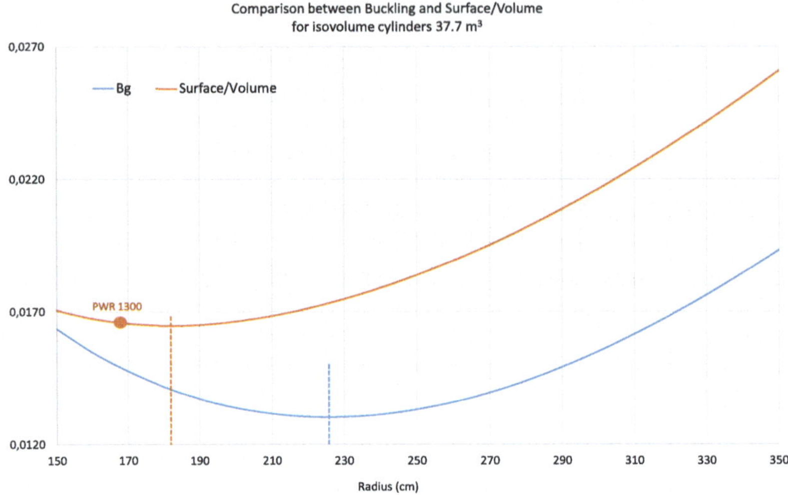

Fig. 10.2 Comparison between buckling B_g (cm^{-1}) and the ratio surface/volume for isovolume cylinders of $3.7\,\text{m}^3$

The minimal value of V_c is obtained with $B_m^2 \cdot h^2 = 3\pi^2$.

Then $R^2 = \dfrac{2.4048^2 \cdot h^2}{3\pi^2 - \pi^2}$.

The optimal cylinder shape for an homogeneous reactor, minimizing the critical volume is:

$$\frac{R}{h} = \frac{2.4048}{\pi \cdot \sqrt{2}} \approx 0.54$$

This optimal cyclinder shape, previously calculated in order to minimize buckling, assumes that the flux has the shape of the fundamental mode calculated for an homogeneous system, for both axial and radial dimensions. The radial and axial shapes of the flux in the case of industrial PWRs are quite flattened, thus it can be expected that the designed ratios $\frac{R}{h}$ are different from the optimal shape calculated in the fundamental mode. Intuitively, we can say that if the flux distribution were homogeneous, which is not feasible in practice, then the most favourable ratio $\frac{R}{h}$ for the neutron balance would be obtained by minimising surface area (leakage) over volume (neutron production).

The example of PWR 1300 MWe is a good illustration. We consider an isovolume of 37.7 m^3, approximatively the volume of the core and we calculate for various radii the ratio surface over volume and the buckling. Figure 10.2 shows that the design of PWR 1300 MWe is close to the minimal ratio surface over volume and significantly different from the cylinder shape minimizing the buckling.

10.3 Application: Criticality of a Plutonium Sphere

> Lady Godiva (she was unclad, like the lady[1]), began operation in 1951. This essentially bare sphere of enriched uranium, simpler than the two-component Topsy, had not been planned earlier because of undetermined sensitivity to outside influence–facetiously, even the effect of a fly alighting on the surface. The critical size was established earlier by means of a near-critical pseudosphere set up on the Comet machine. The Lady Godiva sphere, on the Elsie machine, was subdivided into a stationary central section with retractable upper and lower caps. The central section contained a diametral glory hole with fillers, and channels for two highly enriched uranium control rods. Compensation for openings in the glory hole was provided by mass-adjustment buttons of enriched uranium that fit into recesses in the spherical caps. Lightweight component mounts resulted in small extraneous reflection. By late 1962, the usual critical characteristics of Lady Godiva had been thoroughly established.
>
> Paxton (1983, p. 11)

The Jezebel experiment of 1954–1955 was a very small, nearly-spherical, nearly-bare (unreflected), nearly-homogeneous assembly of plutonium alloyed with gallium. This experiment was used to determine the critical mass of spherical, bare, homogeneous Pu-alloy.

> In 1954, Jezebel was set up in Kiva 2[2] as an unreflected spherical critical assembly of delta-phase plutonium. This assembly was to satisfy the need for neutronic information about plutonium that would parallel the information about enriched uranium provided by Lady Godiva.
>
> Paxton (1983, p. 16)

10.3.1 Jezebel Critical Mass

The characteristics of Jezebel are summarized in Favorite (2017). Jezebel is an homogeneous sphere of bare plutonium alloyed with gallium, its plutonium vector is described on Table 10.1. The radius is 6.3849 cm and the of the plutonium alloy is 15.61 g/cm^3. In 1969, a Los Alamos report specified the full material and gave the critical mass as 17.02 kg Pu-alloy.

In the Jezebel sphere, due to the small size of the fissile mass (taking as scale dimension the mean free path of neutrons), the diffusion does not apply. However, it is interesting to calculate the critical radius of the plutonium alloyed sphere and

---

[1] Lady Godiva (died between 1066 and 1086) is an Anglo-Saxon gentlewoman famous for her legendary ride while nude through Coventry, Warwickshire. Godiva was the wife of Leofric. Her husband, in exasperation over her ceaseless imploring that he reduce Coventry's heavy taxes, declared he would do so if she rode naked through the crowded marketplace. She did so, her hair covering all of her body except her legs.

[2] These experiments at Pajarito Site (Los Alamos National Laboratory) were conducted in remote-control laboratories, known as kivas, which were located at some distance from the main laboratory building housing an individual control room for each kiva. Among the modern Hopi and most other Pueblo peoples, "kiva" means a large room that is circular and underground, and used for spiritual ceremonies.

Table 10.1 Plutonium vector composition of Jezebel

Nuclide	^{239}Pu	^{240}Pu	^{241}Pu
Atom fraction in plutonium (%)	95.2	4.5	0.3

Table 10.2 Cross sections of plutonium isotopes

Nuclide (Fiss)	^{2}39Pu (barn)	^{2}40Pu (barn)	^{2}41Pu (barn)
σ_f	1.7916	1.3493	1.65
σ_a	0.055	0.0833	0.12
σ_{tr}	5.984	6.464	6.1687

to compare it with the Jezebel critical benchmark. The neutrons produced by fission in the Jezebel Sphere are not slowed down buy the heavy nuclei of plutonium, thus they may be considered as fast monokinetic neutrons. Then the critical condition is:

$$\frac{\nu \Sigma_f}{\Sigma_a} = B_g^2 L_D^2 + 1$$

Where L_D is the diffusion length defined by $L_D^2 = \frac{D_{tr}}{\Sigma_a}$ and $D_{tr} = \frac{1}{\Sigma_{tr}}$ The buckling for a spherical geometry is $B_g^2 = \frac{\pi^2}{R^2}$ Hence, the critical radius of the sphere is:

$$R_c = \sqrt{\frac{\pi^2 L_D^2}{\frac{\nu \Sigma_f}{\Sigma_a} - 1}} \tag{10.15}$$

For plutonium, we use cross sections avereged over a fast fission spectrum, shown on Table 10.2.

We finally calculate the following values for the plutonium alloy:

$$\Sigma_f = 0.069653 \, \text{cm}, \, \Sigma_a = 2.2 \times 10^{-3} \, \text{cm}, \, \Sigma_{tr} = 0.23633 \, \text{cm}$$

For all the three isotopes, we assume that $\nu = 2.98$. The critical radius with Eq. 10.15 is $R_c = 8.23$ cm. Equation 10.15 assumes that boundary condition is Dirichlet, with a null value of flux on the boundary. Since the realistic condition at the boundary is vacuum (cf. Chap. 11, we need to calculate a better estimation of the radius. This cannot be done by subtracting the extrapolation distance, presented in Sect. 11.3.2. It has to be noted thas the formula of the extrapolation distance, $2D$ or its enhanced formulation $0.7104 \times \lambda_{tr}$, cannot be used validely for several reasons: it has been established with the assumption that the layer of the system at the boundary is a scattering and absorbing medium, and not fissile ; moreover it is relevant for plane geometries and not for a sphere, even less considering the high curvature (the radius is not far above λ_{tr}). Indeed, $0.7104 \times \lambda_{tr} = 3$ cm and substracting this value to $R_c = 8.23$ cm would underestimate the critical radius.

So as to enhance our estimation of the critical radius, a numerical calculation with vacuum boundary conditions is performed, and gives $R_c = 7.73$ cm. This value is significantely higher than the critical radius of Jezebel: 6.3849 cm.

This raises awareness on limits to the validity of the diffusion theory. For the diffusion approximation to be applicable, it is necessary that the neutron mean free path is much smaller than the dimensions of the system and that the angular neutron flux is approximately isotropic. Such conditions are satisfied in nuclear reactors (homogenized), but not in Jezebel sphere nor in nuclear explosives. In addition, in a nuclear explosion, due to the hydrodynamic expansion of the fissile mass, there are continuous variations of the dimensions, density and composition of the fissile material during the fission chain reaction propagation. Thus, the criticality and the neutron flux distribution in space and energy also vary continuously.

10.3.2 Bare Plutonium Critical Mass as a Function of Density

The bare plutonium (with the vector defined in Table 10.1) critical mass can be calculated as a function of density.

10.4 From Monokinetic Neutrons to One Group Diffusion Theory

The one group theory deals with neutrons *as if* they were monokinetic, although their velocities range between about $2000 \, \text{ms}^{-1}$ and $2.0000 \times 10^4 \cdot \text{s}^{-1}$. The one group theory provides acceptable results insofar the infinite multiplication factor (k_∞ is the ratio between the production rate of neutrons (n/s) in the reactor and the absorption rate) and the migration area (see Eq. 9.17) are close to their values in the two groups therory (Bussac and Reuss 1985, p. 198).

Let's take as an example k_∞ and M^2 calculated for an homogeneous infinite cylinder in the Chap. 13, Sect. 13.1. In the one group theory, critical reactors are characterized by the couples of values (B_g^2, k_∞) on the line defined by the equation:

$$k_\infty = 1 + M^2 \, B_g^2 \tag{10.16}$$

Chapter 13 establishes that in the two groups theory, the critical reactors are characterized by a parabola (see Fig. 13.1) of equation:

$$k_\infty = \tau_{1 \to 2} L_2^2 \, B_g^4 + M^2 \, B_g^2 + 1$$

These two function are asymptotic at the point (0, 1) and k_∞ determines almost the same bucklings. With $k_\infty = 1.0115$ and $M^2 = 56\,\text{cm}^2$, the bucklings are respectively, for one group and two groups, $2.0512 \times 10^{-4}\text{cm}^{-2}$ and $2.0490 \times 10^{-4}\text{cm}^{-2}$.

And the corresponding critical fissile radius are 167.91 and 168 cm.

This example illustrates that the one group reactor, derived from the two groups inputs, and defined in the homogeneous area by $k_{\infty,1G} = k_{\infty,2G}$ (Eqs. 10.16 and 13.14) and $M^2 = \tau_{1\to2} + L_2^2$ results in almost the same bucklings and can be considered as a one group approximation of two groups.

10.5   Reactor with Several Homogeneous Zones: in One Group Diffusion Theory

In each area, the reactor in a stationary state is described by the following on group diffusion equation:

$$\Delta\Phi(r) + \frac{k_\infty - 1}{M^2}\,\Phi(r) = 0$$

We study a system with two zones: a central zone, characterized by $k_\infty > 1$ and an external zone characterized by $k_\infty < 1$, which may be a water reflector or a low enrichment area. The migration areas in the central and external regions are M_c and M_r, and the diffusion coefficients are D_c and D_r.

In the central area, the diffusion equation can be written as:

$$\Delta\Phi(r) + \chi^2\,\Phi(r) = 0 \tag{10.17}$$

With:

$$\chi^2 = \frac{k_\infty - 1}{M_c^2}$$

And in the external area:

$$\Delta\Phi(r) - K^2\,\Phi(r) = 0 \tag{10.18}$$

With:

$$K^2 = \frac{k_\infty - 1}{M_r^2}$$

The solutions of Eqs. 10.17 and 10.18 are summarized on table 10.3. Some functions have to be eliminated, for symmetry reasons and to prevent infinite limits: for example, $\lim\limits_{x\to0} Y(x) = -\infty$.

We remind that:

$$\cosh x = \frac{e^x + e^{-x}}{2} \quad \text{and} \quad \sinh x = \frac{e^x - e^{-x}}{2}$$

Table 10.3 General solutions for one group diffusion Eqs. 10.17 and 10.18

Region	Slab geometry	Cylindric geometry
χ^2	$A \cos(\chi x) + B \sin(\chi x)$	$A\, J_0(\chi r) + B\, Y_0(\chi r)$
$-K^2$	$E \cosh(K x) + F \sinh(K x)$	$E\, I_0(K r) + F\, K_0(K r)$

$$\frac{d^2}{dx^2} \cosh x = \cosh x \quad \text{and} \quad \frac{d^2}{dx^2} \sinh x = \sinh x.$$

10.5.1   Slab Geometry

The core is an infinite slab of thickness d_c wrapped by two reflective slabs of water, of thickness $\frac{d_r}{2}$. The distance from the center is x.

In the core area, the term $B \sin(\chi x)$ has to be removed for symmetry reasons. Thus the flux in the core area is: $\Phi(x) = A \cos(\chi x)$.

In the reflector zone, the function $C \cosh$ has to be removed because for all x, $C \cosh x > 1$ and cannot vanish at the boundary of the reactor (Dirichlet boundary condition). So that the flux vanishes at the boundary, the hyperbolic sine function for the flux has to be:

$$\sinh\left(K \left(\frac{d_c + d_r}{2} - x \right) \right) = 0 \text{ if } x = \frac{d_c + d_r}{2}$$

The Dirichlet boundary condition being taken into account, there only remains two variables: A and F.

Two conditions have to be satisfied: the continuity of flux at the interface between the core area and the reflector $x = \frac{d_c}{2}$, and continuity of current.

$$\begin{cases} A \cos(\chi \frac{d_c}{2}) - F \sinh(K \frac{d_r}{2}) \\ A\, D_c\, \chi\, \sin(\chi \frac{d_c}{2}) + F\, D_r\, K\, \cosh(K \frac{d_r}{2}) \end{cases} \tag{10.19}$$

The critical condition is that the determinant of the system is null, which leads to a single relationship between A and F.

Table 10.4 Input data for one group PWR core modelization (homogeneous cylinder)

$\Sigma_{a,1G}$ (1 Group)	D_{1G} (1 Group)	$(\nu\Sigma_f)_{1G}$ (1 Group)
$= \Sigma_{r2}$ (Th)	$= (L_2^2 + \tau_{1\to2}) \times \Sigma_{a,1G}$	$= k_{\infty,2G} \times \Sigma_{a,1G}$
0.179 54 cm	10.054 cm	0.18161 cm

Table 10.5 Input data for one group: water reflector 300 °C, 155 bar, 1200 ppm

M^2	$(\Sigma_{a,w})_{1G}$ (1 Group)	$D_{1G,w}$ (1 Group)
50.655 cm^2	0.037037 cm	1.8761 cm

10.5.2   Infinite Cylinder with Two Zones: Homogeneous Cylinder and a Water Reflector

Critical Condition and Influence of the Reflector's Thickness

The values from Table 10.4 define a radially homogeneous infinite cylinder for two groups, which is critical with a Dirichlet boundary condition for a radius of 168 cm. From Sect. 10.4 and assuming that the absorption in the group is the thermal absorption, we define the properties of a one group medium in the Table 10.4.

The cylinder is surrounded by a water reflector of 40 cm. The water is supposed to be at 300 °C, 155 bar with a boron concentration 1200 ppm. For one group modelization, the water is caracterized by the absorption of thermal neutrons, and a migration area (see Table 9.2) $M^2 = \tau_{th} + L_{th}^2 = 50.655$ cm^2. The absorbing cross section of the considered water for thermal neutrons is 0.037 037 037 037 037 035 cm, and thus the one group diffusion coefficient of this water reflector is: $D_{1G,w} = 50.655 \times 0.037037 = 1.8761$ cm. The values are summarized on Table 10.5.

The general solution in the fissile cylinder is $A\,J_0(\chi r) + B\,Y_0(\chi r)$ (Table 10.3), and B has to be set to zero in order to eliminate the function $Y_0(\chi r)$, which is not regular on $r = 0$.

In the reflector, the general solution is $E\,I_0(Kr) + F\,K_0(Kr)$.

The couple of diffusion equations becomes:

$$\begin{cases} \Phi(r) = A\,J_0(\chi r) & r \le R_f \\ \Phi(r) = E\,I_0(Kr) + F\,K_0(Kr) & R_f < r \le R_{tot} \end{cases}$$

Where R_f is the fissile radius and R_{tot} the external radius of reflector.

Criticality can be presented as a relationship between the variables A, E, and F. The critical reactor satisfies an equation for the continuity of current at the interface between the two zones, an equation for the continuity of flux at the interface, and an equation for the respect of the Dirichlet boundary condition at the external interface of the reflector. These are three equations for three variables A, E, and F.

$$\begin{cases} D_{1G} \, A \, \chi(- J_1(\chi R_f)) = D_{1G,w} \, E \, K \, I_1(K R_f) - D_{1G,w} \, F \, K \, K_1(K R_f) \\ A \, J_0(\chi R_f) = E \, I_0(K R_f) + F \, K_0(K R_f) \\ 0 = E \, I_0(K R_{tot}) + F \, K_0(K R_{tot}) \end{cases}$$

$$(10.20)$$

$$\begin{cases} A\,(-D_{1G}\,\chi\,J_1(\chi R_f)) & -E\,(D_{1G,w}\,K\,I_1(K R_f)) & +F\,(D_{1G,w}\,K\,K_1(K R_f)) & = 0 \\ A\,J_0(\chi R_f) & -E\,I_0(K R_f) & -F\,K_0(K R_f) & = 0 \\ A \times 0 & +E\,I_0(K R_{tot}) & +F\,K_0(K R_{tot}) & = 0 \end{cases}$$

$$(10.21)$$

Since the flux in a critical reactor is a multiplication factor (that represents the level of power) times a shape function f_0, there cannot be a unique solution to the system of three equations. Thus it's necessary to consider that the equations are not independant and one equation has to be a linear combination of the two other equations: the determinant of the coefficient matrix (system of Eq. 10.21) has to be null.

The critical condition is therefore:

$$\begin{vmatrix} -D_{1G}\,\chi\,J_1(\chi R_f) & -D_{1G,w}\,K\,I_1(K R_f) & D_{1G,w}\,K\,K_1(K R_f) \\ J_0(\chi R_f) & -I_0(K R_f) & -K_0(K R_f) \\ 0 & I_0(K R_{tot}) & K_0(K R_{tot}) \end{vmatrix} = 0$$

If $R_f = R_{tot}$ (which means there is no reflector), the determinant is null at the condition that $J_0(\chi R_f)$. In this case, χR_f has to be the first root of the Bessel function J_0 (around 2.4048). This is the result previously established in Eq. 10.14.

This determinant can easily be numerically calculated with the inputs of Tables 10.4 and 10.5. The fissile radius and the external radius of the water reflector are two variables for the calculation. The fissile radius is lowered with an increasing reflector thickness, as shown by Fig. 10.3.

Calculation of the Functions Describing the Flux in a Critical Reactor

We can write F as a function of E:

$$F = -E \, \frac{I_0(K R_{tot})}{K_0(K R_{tot})}$$

The system of Eq. 10.21 becomes:

$$\begin{cases} A\,(-D_{1G}\,\chi\,J_1(\chi R_f)) \;+E\left[-D_{1G,w}\,K\,I_1(K R_f) - \dfrac{I_0(K R_{tot})}{K_0(K R_{tot})}\,D_{1G,w}\,K\,K_1(K R_f)\right] = 0 \\[2em] A\,J_0(\chi R_f) \qquad\qquad +E\left[-I_0(K R_f) + K_0(K R_f)\,\dfrac{I_0(K R_{tot})}{K_0(K R_{tot})}\right] = 0 \end{cases}$$

$$(10.22)$$

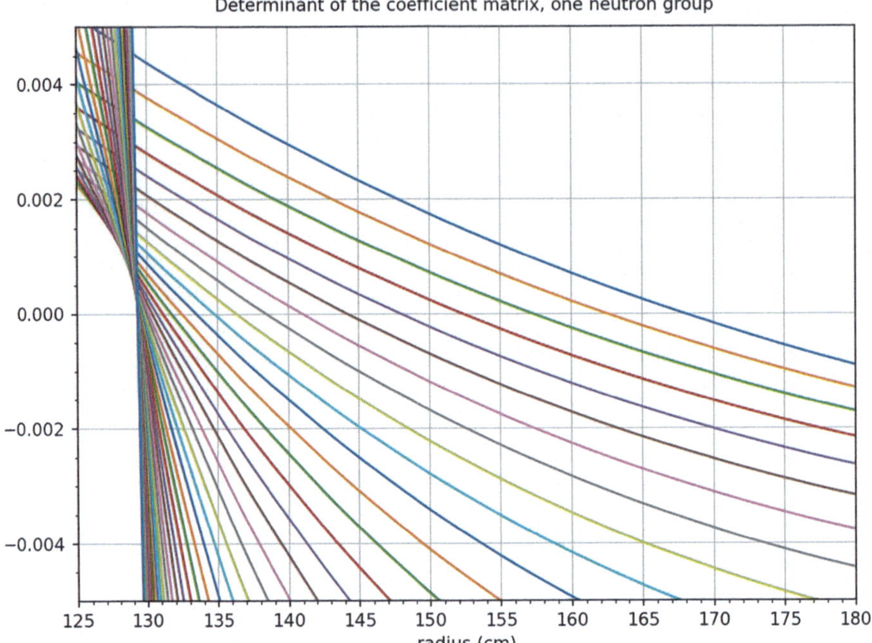

Fig. 10.3 Determination of the coefficient matrix (1 group) with reflector thicknesses rangins from 0 to 40 cm

We consider the infinite cylinder at criticality with a reflector's thickness of 40 cm. In this case, the critical condition calculated with the determinant is $R_f = 129.34$ cm. Without reflector, the critical radius for one group is 167.9 cm. At criticality, the determinant of the coefficient matrix (system of Eq. 10.21) is null, and therefore the determinant of the system (10.22) is also null. We can calculate that $\frac{F}{E} = -1.4935$. With the considered numerical values, the determinant of (10.22) is:

$$\begin{vmatrix} -0.083786 & -1.5123 \times 10^{11} \\ 0.30944 & 5.5852 \times 10^{11} \end{vmatrix} = 0$$

The two equations of (10.22) establish equivalent relationship between A and E. If we choose arbitrarily $A = 1$, then $E = -5.5404 \times 10^{-13}$ and $F = 8.2744 \times 10^{-7}$. The calculation of χ and K gives: $\chi = 0.014322$ and $K = 0.14050$.

The two diffusion equations, solution of the critical problem with a reflector of 40 cm are given in (10.23) and are drawn on Fig. 10.4:

$$\begin{cases} \Phi(r) = J_0(\chi \times r) & r \leq R_f \\ \Phi(r) = -5.5404 \times 10^{-13} \times \ I_0(K \times r) + 8.2744 \times 10^7 \times K_0(K \times r) & R_f < r \leq R_{tot} \end{cases}$$

$$(10.23)$$

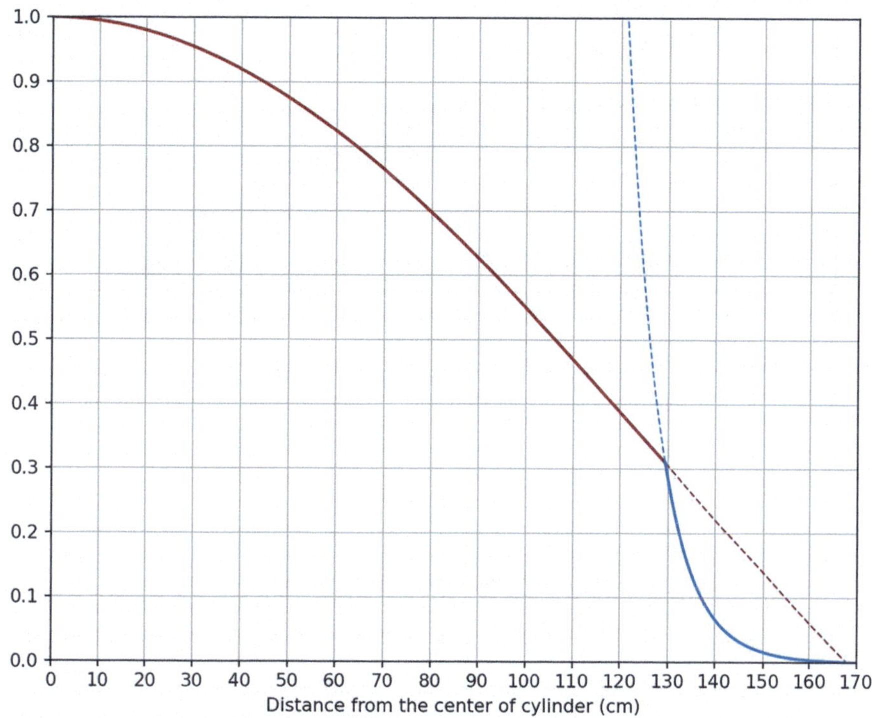

Fig. 10.4 Analytical solutions for a critical reactor with a reflector of 40 cm

Table 10.6 Reactivity and leaks as a function of the reflector's thickness

Reflector thickness (cm)	Reactivity ρ (pcm)	Net leak out of fissile area (pcm)	Reduction of net leak (pcm)
0	0	1136	0
4	227	909	227
10	355	781	355
20	386	750	386
38	388	748	388

Shape of the Flux as a Function of Reflector's Thickness

A numerical calculation of the diffusion equation (1 group) with a fissile radius of 167.9 cm and reflector thicknesses ranging from 0 to 40 cm is performed. The flux shapes or shown on Fig. 10.5. Due to the reflector, flux tends to be flatened since it is pulled up at the boundary between the fissile area and the reflector. Somes results from the numerical calculation are presented on Table 10.6 and highlight that the reactivity increase goes along the reduction of the net leak out of the fissile area.

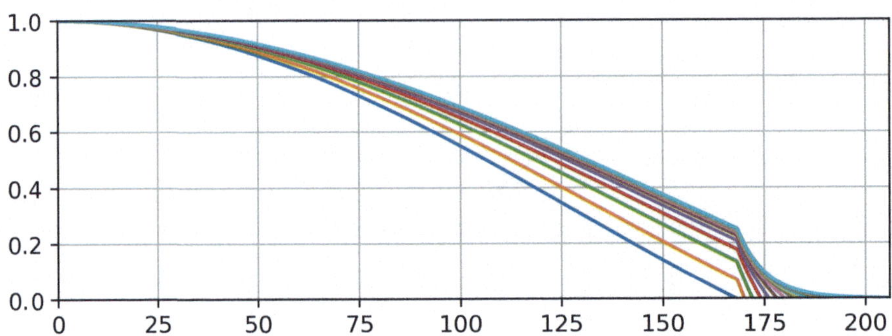

Fig. 10.5 Numerical calculation of the diffusion equation (1 group) with reflector thicknesses ranging from 0 to 40 cm

References

G.I. Bell, S. Glasstone, *Nuclear Reactor Theory* (Van Nostrand Reinhold Company, 1970), https:// books.google.fr/books?id=RNQmAQAAMAAJ

J. Bussac, P. Reuss, *Traité de neutronique: physique et calcul des réacteurs nucléaires avec application aux réacteurs à eau pressurisée et aux réacteurs à neutrons rapides*, vol. 25 (Hermann, 1985)

J.A. Favorite, Jezebel: reconstructing a critical experiment from 60 years ago. Technical report, Los Alamos National Lab. (LANL), Los Alamos, NM (United States), 2017

H.C. Paxton, History of critical experiments at Pajarito site. Technical report, Los Alamos National Lab., NM (USA), 1983

Chapter 11
Boundary Conditions

Abstract Dirichlet boundary condition has been previously used in Chap. 10. This chapter gives an overview of the main boundary conditions and the mathematical expression of linear boudary conditions in 1D geometry. The notion of extrapolation distance and the behavior of flux and currents at the boundary of the system are also presented at the end of the chapter. We only consider convex reactors: thus any neutron escaping from the system will not be able to re-enter.

11.1 Most Common Boundary Conditions

11.1.1 General

A problem is not uniquely specified if we give the differential equation which the solution must satisfy, for there are an infinite number of solutions (..). In order to make the problem a definite one, with a unique answer, we must pick, out of the mass of possible solutions, the one which has certain definite properties along definite boudary surfaces. Any physical problem must state not only the differential equation which is to be solved but also the boundary conditions which the solution must satisfy. (..) Usually they take the form of the specification of the behavior of the solution on or near some boundary line (or surface, in three dimensions).

Morse and Feshbach (1953, p. 676, 677)

Boundary condition is a relationship (usually linear but it may also be non-linear) between a function and its first derivatives at the boundary of the system.

In 1D geometry, and with the fist derivative, the boundary condition (with $r = R$) can be written as:

$$A(R)\Phi(R) + B(R) \frac{d\Phi}{dr}(R) = C(R) \tag{11.1}$$

Using the general boundary condition, we can enforce a variety of boundary conditions.

11.1.2   *Dirichlet*

> The specifying only of values along the boundary is called Dirichlet conditions, and the specifying only of slopes is called Neumannconditions. (..) we may, at times, need to specify the value of some linear combination of value and slope, a single boundary condition which is intermediate between Dirichlet and Neumann conditions.
>
> Morse and Feshbach (1953, p. 679)

A Dirichlet boundary condition of the form $\Phi(R) = c$ can be enforced by setting A = 1, B = 0, and C = c.

The most common boundary condition is the Dirichlet with $f = 0$ on the boundary.

11.1.3   *Neumann*

> The other usual boundary condition is the Neumann condition where the gradient of the density is fixed at the boundary. This arises in several models when symmetry requires the normal grandient to vanish on some surface, e.g. the centre plane of a symmetric reactor
>
> Lewins (1965, p. 3)

11.1.4   *Albedo*

An albedo boundary condition is used for the case where some fraction of the neutrons that leave the system are reflected back. In general, the albedo is the same in all energy groups.

If we call this fraction that is reflected α, which depends on the neutron properties of the reflector, the albedo boundary condition can be written as:

$$\alpha = \frac{-\dfrac{\Phi(R)}{4} - \dfrac{D(R)}{2}\dfrac{d\Phi}{dr}(R)}{\left(\dfrac{\Phi(R)}{4} - \dfrac{D(R)}{2}\dfrac{d\Phi}{dr}(R)\right) \times (-1)}$$

$$-\alpha\frac{\Phi(R)}{4} + \alpha\frac{D(R)}{2}\frac{d\Phi}{dr}(R) = -\frac{\Phi(R)}{4} - \frac{D(R)}{2}\frac{d\Phi}{dr}(R)$$

$$(1-\alpha)\frac{\Phi(R)}{4} + (1+\alpha)\frac{D(R)}{2}\frac{d\Phi}{dr}(R) = 0$$

$$\frac{(1-\alpha)}{4\,(1+\alpha)}\,\Phi(R) + \frac{D(R)}{2}\,\frac{d\Phi}{dr}(R) = 0$$

Where $D(R)$ is the diffusion coefficient of the system at the boundary R of the system, where albedo condition applies.

This enables us to write that $A(R) = \dfrac{(1-\alpha)}{4\,(1+\alpha)}$; $B(R) = \dfrac{D(R)}{2}$ and $C(R) = 0$.

11.1.5 Reflective

The reflective boundary condition is a special case of albedo with $\alpha = 1$. Then $A(R) = 0$, $C(R) = 0$ and $B(R)$ is undetermined.

As an example, we consider the following homogeneous system: a source of neutrons is dispersed (Q n/cm$^{3/s}$) within a scattering and absorbing material, in a slab geometry. The faces of the slab are at the abscisses $x = R$ and $x = -R$. The equation to solve is:

$$\frac{d^2}{dx^2}\Phi(x) - \frac{\Sigma_a}{D}\Phi(x) = -\frac{Q}{D} \tag{11.2}$$

A flat flux inside the slab $\Phi(x) = \frac{Q}{\Sigma_a}$ satisfies both Eq. 11.2 and the reflective boundary condition, characterized by $\frac{d^2}{dx^2}\Phi(R) = 0$.

If the system is not homogeneous, the solution with a reflective boundary condition is not anymore flat.

When performing the transport calculation of an assembly or cell, it is usually assumed that the system consists in an infinite lattice of identical symmetric assemblies or cells. This is equivalent to the reflection of neutrons on the boundaries of a single cell.

11.1.6 Marshak and Vacuum

11.1.6.1 Marshak

The final boundary condition that we consider is the partial current boundary condition, also called a Marshak boundary condition. These allow us to specify the amount of neutrons that enter the system from the edge, rather than specifying the scalar flux on the boundary as in the Dirichlet condition.

At $r = R$, the incoming partial current is given by:

$$|J| = \frac{\Phi(R)}{4} + \frac{D(R)}{2}\,\frac{d\Phi}{dr}(R)$$

Therefore, if we wish to set the incoming partial current into the system at a particular value J_{in}, we set our boundary constants to be:

$$A(R) = \frac{1}{4}, \ B(R) = \frac{D(R)}{2}, \ C(R) = J_{in}.$$

11.1.6.2 Vacuum

Vacuum boundary condition is a special case of Marshak condition, obtained by setting $J_{in} = 0$. It means that the reactor is surrounded by vacuum, and a neutron leaving the system has no probability of returning. This boundary condition is called free-surface boundary condition by Bell and Glasstone (1970, p. 17). Although free-surface boundary conditions are an idealization, they are very useful because for many systems the probability of neutron return is negligible, so it can be asumed that no neutrons are reflected back into the reactor.

> Although the surrounding atmosphere is not truly a vacuum, the mean free path of neutrons in air is much larger than in non gaseous materials, and it is usually possible to treat it as a vacuum in reactor calculations. It will be recalled that diffusion theory is not valid near such a surface, and it is necessary to handle such problems in a special way.
>
> Lamarsh (1966, p. 133)

The incoming diffusion current is defined by Eq. 6.6 in the Chap. 6. The boundary condition at the surface of the medium can be written as:

$$\frac{\Phi(R)}{4} + \frac{D(R)}{2} \frac{d\Phi}{dr}(R) = 0$$

$$\frac{1}{\Phi(R)} \frac{d\Phi}{dr}(R) = -\frac{1}{2\,D(R)}$$

In two or three dimensions, the normal derivative takes the place of the 1D derivative $\frac{d\Phi}{dr}(R)$:

$$\frac{1}{\Phi(r)} \frac{\partial\Phi}{\partial n}(r) = -\frac{1}{2\,D(r)}$$

with $\frac{\partial\Phi}{\partial n}(r) = \boldsymbol{grad}\,\Phi(r).\boldsymbol{n}$.

Vacuum condition is also a special case of albedo condition, for which $\alpha = 0$.

11.2 Calculation of a 1D Albedo for One Group Theory

11.2.1 Calculation of the Reflected Fraction α

The albedo α is the fraction of neutrons that is reflected, thus we can write:

$$\frac{J_{1,\,\text{incoming}}}{J_{1,\,\text{outgoing}}} = \frac{\dfrac{\Phi_0}{4} + \dfrac{D_1}{2}\left(\dfrac{d\Phi}{dx}\right)_{0,1}}{\dfrac{\Phi_0}{4} - \dfrac{D_1}{2}\left(\dfrac{d\Phi}{dx}\right)_{0,1}} = \frac{J_{2,\,\text{outgoing}}}{J_{2,\,\text{incoming}}} = \frac{\dfrac{\Phi_0}{4} + \dfrac{D_2}{2}\left(\dfrac{d\Phi}{dx}\right)_{0,2}}{\dfrac{\Phi_0}{4} - \dfrac{D_2}{2}\left(\dfrac{d\Phi}{dx}\right)_{0,2}}$$

where Φ_0 is the flux at the interface between the neutron multiplying system (indice 1) and the reflector (indice 2).

$$\alpha = \frac{J_{2,\,\text{outgoing}}}{J_{2,\,\text{incoming}}} = \frac{1 + \dfrac{2D_2}{\Phi_0}\left(\dfrac{d\Phi}{dx}\right)_{0,2}}{1 - \dfrac{2D_2}{\Phi_0}\left(\dfrac{d\Phi}{dx}\right)_{0,2}}. \tag{11.3}$$

11.2.2 Reflected Fraction α in a Plane Geometry

We consider a scattering and absorbing wall of infinite thickness (reflector, index 2), at the interface of a neutron multiplying system (index 1). This is a 1D configuration. And we remind Eq. 8.5, which is a result from the Chap. 9:

$$\Phi(x) = \frac{Q\, L_{Dtr}}{2D}\, e^{-\frac{x}{L_{Dtr}}}$$

L_{Dtr} is the depth inside the wall at which the flux is divided by e.

In a plane geometry, we can write:

$$\alpha = \frac{\dfrac{\Phi_0}{4} + \dfrac{D_2}{2}\left(\dfrac{-1}{L_{Dtr}}\right)\Phi_0}{\dfrac{\Phi_0}{4} - \dfrac{D_2}{2}\left(\dfrac{-1}{L_{Dtr}}\right)\Phi_0}$$

$$\alpha = \frac{1 - 2\dfrac{D_2}{L_{Dtr}}}{1 + 2\dfrac{D_2}{L_{Dtr}}} \tag{11.4}$$

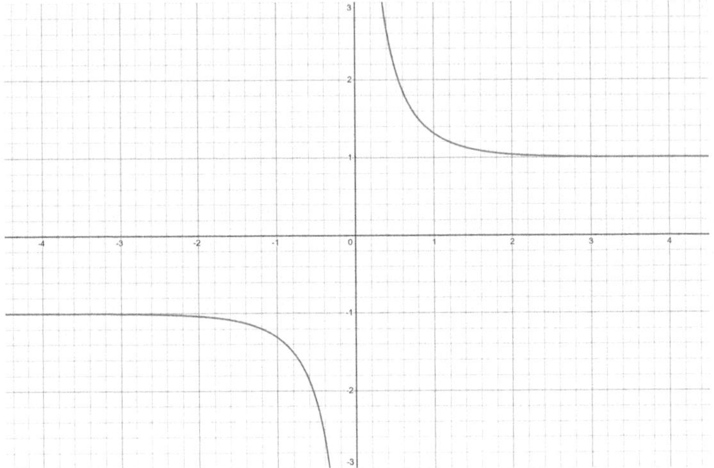

Fig. 11.1 Function coth

This albedo does not depend on the flux at $x = 0$; it tends to be purely reflective (mirror) when $\frac{D_2}{L_{Dtr}}$ tends to 0.

If the thickness of the wall is finite (noted a), it can be demonstrated that:

$$\alpha = \frac{1 - 2\dfrac{D_2}{L_{Dtr}} \coth \dfrac{a}{L_{Dtr}}}{1 + 2\dfrac{D_2}{L_{Dtr}} \coth \dfrac{a}{L_{Dtr}}} \tag{11.5}$$

We remind that coth is the hyperbolic cotangent function defined by:

$$\coth x = \frac{\cosh x}{\sinh x}$$

Then

$$\coth x = \frac{e^x + e^{-x}}{e^x - e^{-x}}$$

The graph of the function coth is drawn on Fig. 11.1. It show that if the thickness of the slab is at least twice the diffusion length, i.e. $\dfrac{a}{L_{Dtr}} > 2$, the value of coth is almost down to 1 and then the albedo of the finite slab tends to the infinite slab value.

11.2.3 Example of a Water Wall for Thermal and Fast Neutrons

Separated albedo coefficients are calculated from fast and thermal neutron properties of water, as a function of the thickness of water wall. This calculation does not take into accound that part of fast neutrons thermalized in the water reflector are reflected back into the system, and add up the reflected thermal neutrons. These albedo coefficients are shown on Fig. 11.2.

11.2.4 Reflected Fraction α in a Spherical Geometry

We consider the case of an infinite scattering and absorbing medium surrounding a sphere of radius R. In this medium ($r > R$), taking the origin at the center of the sphere, the flux is:

$$\Phi(r) = C \, \frac{\exp\left(\frac{-r}{L_{Dtr}}\right)}{r}$$

At the interface, where $r = R$:

$$\frac{1}{\Phi}\frac{d\Phi}{dr} = -\left(\frac{r}{L_{Dtr}} + \frac{1}{R}\right)$$

From Eq. 11.3, we can write:

$$\alpha = \frac{1 - 2D_2\left(\dfrac{1}{L_{Dtr}} + \dfrac{1}{L_R}\right)}{1 + 2D_2\left(\dfrac{1}{L_{Dtr}} + \dfrac{1}{L_R}\right)} \tag{11.6}$$

The albedo at an interior spherical surface is less than for an infinite flat surface, since the probability of a neutron being scattered back depends on the solid angle that intercepts the core region (Glasstone and Edlund 1952, p. 133).

11.3 Equivalence Within the Reactor Between Dirichlet with Extrapolation Distance and Vacuum Conditions

It should be noted that the true flux value is not zero at any point. Furthermore, it should also be pointed out that if there were a non-zero flux in vacuum, it would not be possible to represent it by the diffusion equation without a diffusion coefficient or any cross section. Only the flux in transport theory has a physical meaning. In a 1D medium and with simple

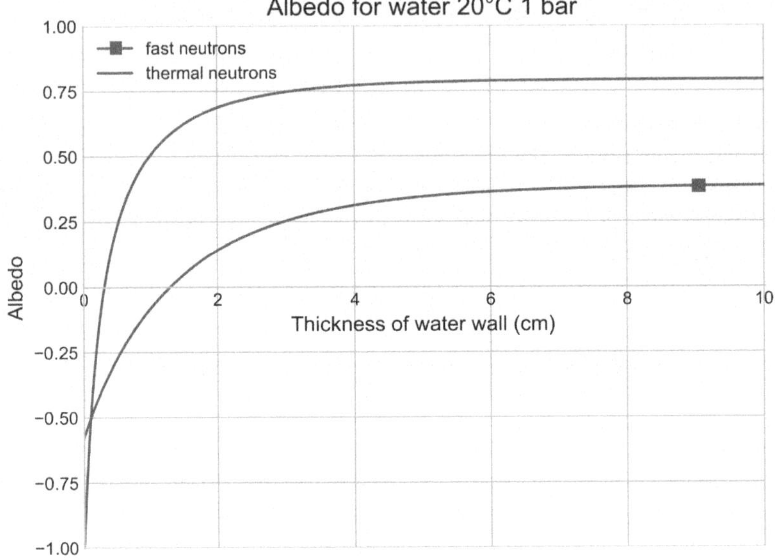

(a) Cold water 20 °C

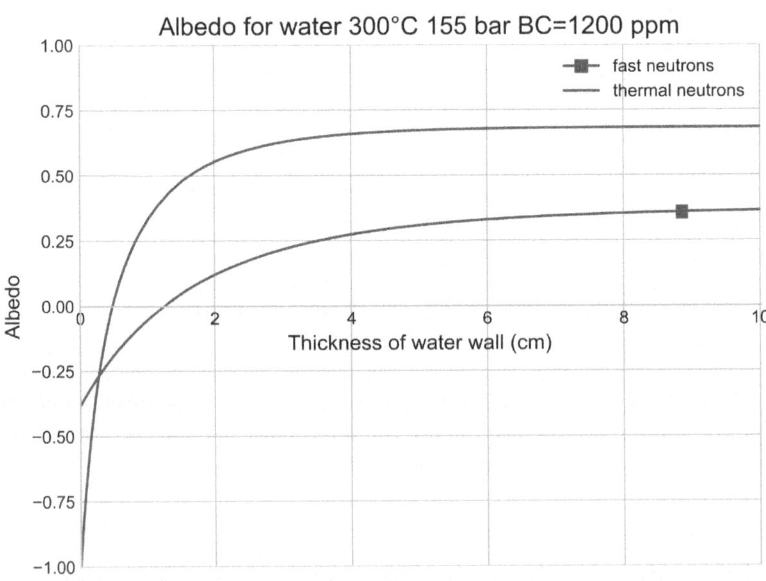

(b) Hot water 300 °C 1200 ppm

Fig. 11.2 Albedo of water for fast and thermal neutrons

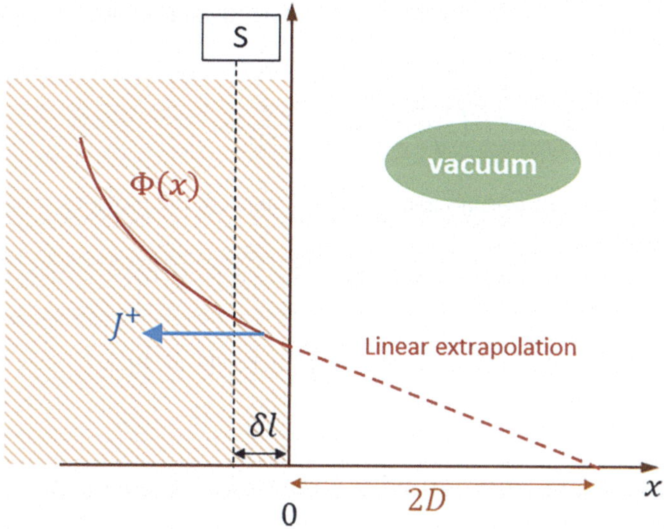

Fig. 11.3 Interface with vacuum

ideas, it can be shown that transport flux is constant in a vacuum. The extrapolation distance depends solely on the diffusion coefficient (..)

Marguet (2018, p. 744)

11.3.1 Calculation of the Flux in a Thin Layer at Interface with Vacuum

It is instructive to see what behavior for the neutron flux at the boundary follows from a naive application of elementary diffusion theory.

Beckurts and Wirtz (1964, p. 95)

We consider a plane geometry, having a interface with vacuum, represented on Fig. 11.3.

δl is the thickness of a thin layer at the interface with vacuum. The order of magnitude of δl is a few mean paths. This layer is almost transparent to neutrons, thus the neutrons crossing the surface S move out the system towards the vacuum.

J^+ is the current of neutrons coming back inside the system, considered as positive.

$$J^+(x = 0) = \frac{\Phi_0}{4} + \frac{D}{2}\left(\frac{d\Phi}{dx}\right)_{x=0^-}$$

No neutrons coming from the vacuum cross the boundary, thus $J^+(x = 0) = 0$. This enables to write that:

$$\left(\frac{d\Phi}{dx}\right)_{x=0^-} = -\frac{1}{2D}\Phi_0$$

Since, within the layer δl, no neutrons are scaterred, the previous relation is true within the layer:

$$\frac{d\Phi}{dx}(x) = -\frac{1}{2D}\Phi(x) \text{ for } x \in [-\delta l, 0]$$

$$\Phi(x) = K \exp\left(-\frac{x}{2D}\right) \text{ for } x \in [-\delta l, 0].$$

11.3.2　Dirichlet Condition with Extrapolation Distance

If we linearly extrapolate the flux after interface, then its equal to zero at a distance $2 \times D$ from interface. This distance is called extrapolation distance, and the term zero-flux distance is also sometimes used.

In the case of a power reactor, extrapolation distance is small compared with the dimensions of the reactor.

We consider that a layer of $2 \times D$ thickness is added to the system. This new system, expanded with extrapolation distance, and with Dirichlet condition, is almost equivalent to the non expanded system with vacuum condition, both in diffusion theory.

As a example, we consider a plane source emitting monokinetic thermal neutrons within a water wall of half-thickness 4 cm. A numerical calculation enables to compare the flux in the vicinity of the boundary with two boundary conditions: vacuum and Dirichlet, with an expansion of water wall. The expansion is the extrapolation distance $2 \times D = 0.316$ cm. The Fig. 11.4 shows that the flux calculated with the two boundary conditions vacuum and extrapolated Dirichlet are almost equal.

11.4　Extrapolation Distance and Transport

We assume a boundary condition at the surface of the medium of the form:

$$\frac{1}{\Phi(r)}\frac{\partial\Phi}{\partial n}(r) = -\frac{1}{2d} \tag{11.7}$$

Where d is the extrapolation distance, the distance at which the extrapolated flux is equal to zero.

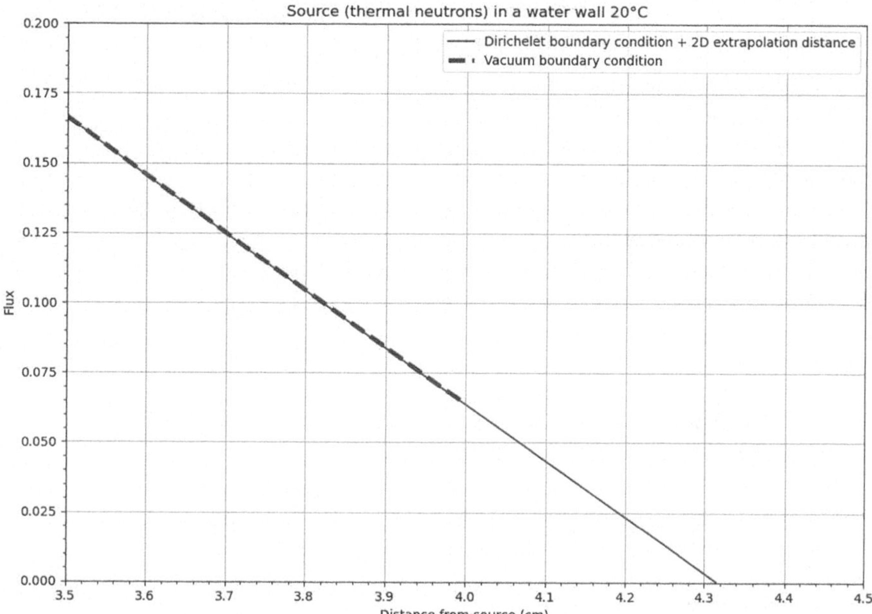

Fig. 11.4 Comparison of extrapolated Dirichlet and vacuum boundary conditions

A value of d is calculated in such a way that the solution to the diffusion equation satisfying the boundary condition (11.7) matches as nearly as possible the rigorous solution given by transport theory in the interior of the medium.

Extrapolation distance d could first be estimated with transport corrected diffusion coefficient $D_{1,tr} = \dfrac{1}{3\Sigma_{tr}}$, then $2D_{1,tr} = \dfrac{2}{3}\lambda_{tr}$.

Diffusion theory is not relevant at a distance below a few scattering mean free paths from a boundary. Transport theory enables to calculate the extrapolation of the asymptotic transport flux, in the case of a plane surface forming the boundary of a weakly absorbing medium, with a vacuum or black absorber. The plot of the neutron flux in transport theory against the distance undergoes a pronounced change of slope within a few mean free paths from the interface, this is why the extrapolation is related to the asymptotic flux and not the derivative of the flux at the interface. This extrapolation is:

$$d = 0.7104\,\lambda_{tr}$$

The extrapolation of the asymptotic transport flux is also called extrapolated endpoint (Beckurts and Wirtz 1964, p. 94).

In the application of the diffusion theory, the boundary condition at the surface of a vacuum or back absorber is that the neutron flux vanished on the extrapolated boundary $0.7104\,\lambda_{tr}$.

d approaches the planar value for large radius of curvature R ($d \ll R$).

The flux obtained by solving the diffusion equation, subject to boundary condition (11.7), provides a good approximation to the actual flux everywhere within the medium except near the surface.

The neutron flux does not, in fact, vanish at the extrapolation distance, and solutions obtained by this method do not give the correct flux near the boundary. Neutron leakage through the boundary to a vacuum surrounding the system cannot be represented by a diffusion equation, since there is no diffusion coefficient nor any cross section. But transport theory can apply in vacuum. In a 1D configuration, flux is just constant in vacuum.

11.5 Flux and Currents at the Boundary of the System in the Theory of Diffusion

11.5.1 Flux Resulting from a Point Source in a Water Sphere

Thermal neutrons are emitted in the center of a water sphere (20°, 1bar) of radius 15 cm, surrounded by vacuum.

The analytical formulation has been established in the Chap. 8, Eq. 8.9. This analytical formulation assumes that the medium in infinite and thus does not take into account the boundary wity the vacuum, at R= 15 cm.

Close to the edge, flux calculated by Monte Carlo and with a numerical 1D diffusion model differ, as shown on Fig. 11.5. Both calculations result in the same leak (number of neutrons leaking over 100000×10^5 neutrons emitted): 4400 pcm. This may seem counter-intuitive since the gradient of flux calculated in diffusion theory is higher, therefore we could expect a higher leak.

It is important to remember that, in the theory of diffusion, the flux is isotropic and the leak result from the diffusion current, which is a consequence of the flux gradient.

In the Monte Carlo calculation, the anisotropy of the flux is growing when we get closer to the edge: the thin external layer of water (the thickness is a few λ_s) at the interface with the vacuum can be considered as transparent to neutrons and all the neutrons motion vectors are oriented towards the vacuum. In Monte Carlo calculation, the currents should not be interpreted as flux gradients.

11.5.2 Incoming and Outcoming Currents within the Same Water Sphere

Figure 11.6 represents the incoming or outcoming currents derived from the diffusion numerical calculation, and the Monte Carlo calculation. At a distance r from the

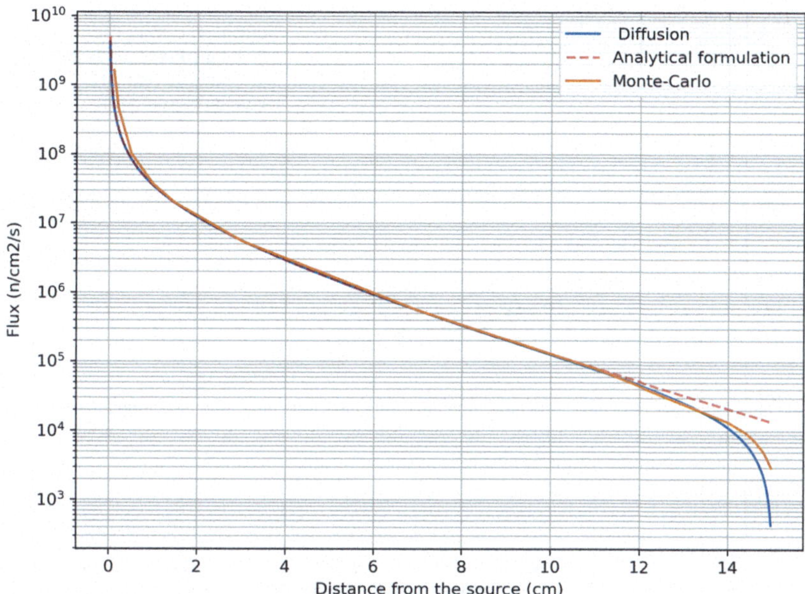

Fig. 11.5 Flux as a function of the distance from a point source of thermal neutrons in a sphere of water (15 cm) surrounded by vacuum

source, the incoming current is defined as the inward current integrated through the surface of a sphere of radius r. The outcoming current is directed from the outside to the inside. In both calculations (diffusion and Monte Carlo), the outcoming current is higher than the incoming current, reflecting the net flow of neutrons directed towards the vacuum.

Close to the edge, the incoming current calculated in Monte Carlo drops sharply. Close to the edge, the angular flux varies fast and the diffusion theory is not valid.

Because of vacuum boudary conditions, the incoming current is null at the boudary. As a result of the phenomenon of diffusion, a decrease of the incoming current spreads within the sphere.

Since the flux calculated with Monte Carlo and diffusion are different closed to the edge, reaction rates are also different. If we now consider a reactor, it is easy to seet that the reaction rates calculated with diffusion and Monte Carlo differ close the boundary of the reactor, notably the fission reaction rate.

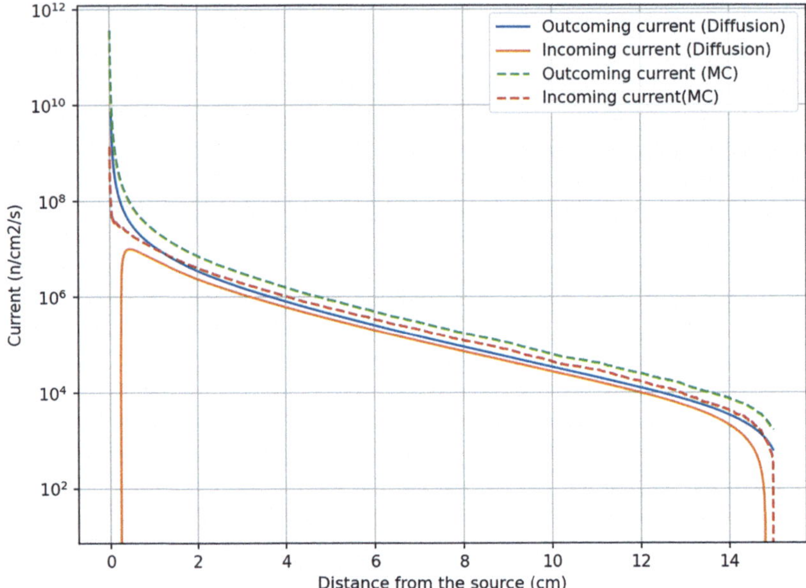

Fig. 11.6 Currents as a function of the distance from a point source of thermal neutrons in a sphere of water (15 cm) surrounded by vacuum

11.6 Boundary Conditions for a Two Groups Model

In the case of a two groups modelization (see Chap. 12), the boundary condition expressed by Eq. 11.1 has to be written for each group:

$$\begin{cases} A_1(R)\Phi_1(R) + B_1(R)\,\frac{d\Phi_1}{dr}(R) = C_1(R) \\ A_2(R)\Phi_2(R) + B_2(R)\,\frac{d\Phi_2}{dr}(R) = C_2(R). \end{cases} \tag{11.8}$$

References

K.-H. Beckurts, K. Wirtz, *Neutron Physics* (Springer, Berlin Heidelberg GmbH, 1964), https://link.springer.com/book/10.1007/978-3-642-87614-1#bibliographic-information

G.I. Bell, S. Glasstone, *Nuclear Reactor Theory* (Van Nostrand Reinhold Company, 1970), https://books.google.fr/books?id=RNQmAQAAMAAJ

S. Glasstone, M.C. Edlund, *The Elements of Nuclear Reactor Theory* (D. Van Nostrand Company, 1952), ISBN 9780598464224, https://books.google.fr/books?id=MSRRAAAAMAAJ

J.R. Lamarsh, *Introduction to Nuclear Reactor Theory*. Addison-Wesley Series in Nuclear Engineering (Addison-Wesley Publishing Company, 1966), ISBN 9780201041262, https://books.google.fr/books?id=by5RAAAAMAAJ

J. Lewins, *Importance: The Adjoint Function. The Physical Basis of Variational and Perturbation Theory in Transport and Diffusion Problems* (Pergamon Press, 1965)

S. Marguet, *The Physics of Nuclear Reactors* (Springer International Publishing, 2018), ISBN 9783319595603, https://books.google.fr/books?id=9DlODwAAQBAJ

P.M. Morse, H. Feshbach, *Methods of Theoretical Physics* (1953)

Chapter 12
Two Groups Diffusion

Abstract The Chap. 13 highlights that the one group modelization cannot properly calculate a reflected reactor, whether in terms of critical condition or shape of flux. This justifies the need to take into account of changes in neutron energy during the slowing down process. Thus, the energy-dependent diffusion equation is commonly transformed into a set of coupled equations, in the energy groups resulting from the discretization of the neutron energy domain. We provide a simplified approach to the neutron transfer from one group to the underneath group. Most of industrial codes for core calculation in the world rely on two neutron energy group model for the calculation of PWRs (However in the case of BWRs more groups are needed). Thus, in this chapter, we establish the two neutron energy group set of diffusion equations.

12.1 Simplified Neutron Balance in Multi Group Diffusion Equations

The very complex preliminary calculation of the energy condensed and space homogeneized cross sections at the scale of a fuel assembly is not part of this book. The reader can refer to Marguet (2018). The resulting condensed on homogeneized data are the inputs of coupled diffusion equations that are solved at the scale of the core.

12.1.1 Neutron Balance of a Group

An interval of energy $[E_2, E_1]$ is considered. Its lethargy equivalent is $[u - \Delta u, u]$. This interval of energy is a group of energy.

The neutron balance of the hypervolume $\Delta V \, \Delta u$ can be written as:

$$- \overline{D} \cdot \Delta \Phi_u + \overline{\Sigma_a} \cdot \Phi_u + q(u) - q(u - \Delta u) = \text{Sources}$$

where:

© The Author(s), under exclusive license to Springer Nature Switzerland AG 2026

H. Grard, *Diffusion of Neutrons in Nuclear Reactors*,

https://doi.org/10.1007/978-3-032-05088-5_12

$q(u)$ is the slowing down current, which means the number of neutrons /cm^3/s crossing the lethargy u.

Sources take into account the neutrons created within the interval of energy, by fission, or nuclear reactions like spontaneous fission and (α, n).

Φ_u is the flux of the energy interval (or group), it is an integral parameter which has the dimension of a flux:

$$\Phi_u = \int_{u-\Delta u}^{u} \Phi(u)du$$

$\overline{D}$ is a mean diffusion coefficient:

$$\overline{D} \int_{u-\Delta u}^{u} \Delta\Phi_u du = \int_{u-\Delta u}^{u} D(u)\Phi(u)du$$

Reaction cross sections are averaged over the energy interval in such a way that the integral reaction rate is preserved. The mean cross section of absorption $\overline{\Sigma_a}$ is defined by:

$$\overline{\Sigma_a} \cdot \Phi_u = \int_{u-\Delta u}^{u} \Sigma_a(u)\Phi(u)du$$

12.1.2   The Transfer of Neutrons to Lower Energy Groups

In the asymptotic energy area, $\phi(u) = \dfrac{q(u)}{\Sigma_s(u)\xi}$

Then:

$$q(u) = \frac{\xi\Sigma_s(u)}{\Delta u} \cdot \Phi(u)\Delta u$$

We define a cross section of transfer towards lower energy groups Σ_{tr}:

$$\Sigma_{tr}(u) = \frac{\xi\Sigma_s(u)}{\Delta u}$$

$$q(u) = \Sigma_{tr}(u) \cdot \Phi(u)\Delta u \approx \Sigma_{tr}(u) \cdot \Phi_u$$

The significance of Σ_{tr} is the probability (/cm path) that a neutron in the energy band $[u - \Delta u, u]$ is transferred to lower energy groups.

Along with absorption, the transfer to lower energies contributes to the removal of neutrons. Thus, the diffusion equation becomes

$$-\overline{D} \cdot \Delta\Phi_u + (\overline{\Sigma_a} + \Sigma_{tr}(u)) \cdot \Phi_u - q(u - \Delta u) = \text{Sources}$$

The removal cross section can be introduced: $\Sigma_r = \overline{\Sigma_a} + \Sigma_{tr}(u)$

$$\overline{D} \cdot \Delta \Phi_u - \Sigma_r \cdot \Phi_u + q(u - \Delta u) + \text{Sources} = 0 \qquad (12.1)$$

12.1.3 Coupled Multi Group Equations

The total energy area can be divided into bands, or groups of energy. For each group, an equation similar to (12.1) can be written. In Eq. (12.1), $q(u - \Delta u)$ represents the arrival of neutrons in the lethargy band $[u - \Delta u, u]$ coming from higher energies. In the multi group equations, this is written as $\sum_{j=1}^{i-1} \Sigma_{j \to i} \cdot \Phi_j$. The equations are coupled by the transfer of neutrons between groups.

$$D_i \cdot \Delta \Phi_i - \Sigma_{r,i} \cdot \Phi_i + \sum_{j=1}^{i-1} \Sigma_{j \to i} \cdot \Phi_j + P_i S = 0 \qquad (12.2)$$

with

$$\Sigma_{r,i} = \Sigma_{a,i} + \Sigma_{i \to i+1} + \Sigma_{i \to i+2} + \cdots$$

$\Sigma_{j \to i} \cdot \Phi_j$ is the transfer of neutrons from higher energy group j to group i. Due to inelastic collisions, neutrons can be transferred to i from groups at higher energy than the immediate neighbor group $i - 1$.

P_i is the probability that the new neutrons are produced in the energy interval i.

12.2 Two Energy Groups

12.2.1 Macroscopic Cross Section for Energy Transfer

Since fission neutrons appear with an energy which can reach $10\,\text{MeV}$ and then are slowed down below electronvolt, this energy domain spans at least 7 decades. The PWR are often calculated with two energy groups (thermal cut-off at $0.625\,\text{eV}$[1]). Two energy groups of neutrons are considered:

- Group 2 in the range $[0, E_{\text{cut off}}]$
- Group 1 in the range $[E_{\text{cut off}}, 10\,\text{MeV}]$

We assume that neutrons of the first group are not susceptible to upscatter to the fast group, which means that the cut-off energy is such that upscattering from the thermal group into the fast group does not occur. However, physically, upscattering may occur up to $5\,\text{eV}$ (see Sect. 4.1.6).

---

[1] The isotope of cadmium ^{113}Cd exhibits a giant resonance connected to a high capture cross section at lower energies. As a consequence, neutrons of kinetic energy below the cadmium cut-off energy (by convention, 0.5 or 0.625 eV) are strongly absorbed by ^{113}Cd.

A mean number of shocks, n, is necessary to decrease energy from fission energy 2 MeV down to thermal energy.

The associated lethargy gain is $\Delta u = u_2 - u_1 = 18.19$.

$$n = \frac{\Delta u}{\xi}$$

The macroscopic cross for transfer from fast to thermal group is defined by:

$$\Sigma_{tr} = \frac{\Sigma_s}{n} = \frac{\xi \Sigma_s}{\Delta u}$$

where Σ_s is the scattering cross section of the medium. The definition of Σ_{tr} means that at each scattering shock, the probability of transfer from groups 1 to 2 is $\frac{1}{n}$. Transfer means that the neutron is removed from group 1 due to slowing down.

From Eq. (9.4), we can write that the increase of the age of Fermi is:

$$\tau_{1 \to 2} = \int_{u_1}^{u_2} \frac{D(u)}{\xi \Sigma_s} du \approx \frac{1}{\xi \Sigma_s} \Delta u \cdot \overline{D_1}$$

where $\overline{D_1}$ is a mean value between u_1 and u_2. We can define the macroscopic cross section for energy transfer from groups 1 to 2, $\Sigma_{tr1 \to 2}$:

$$\overline{r^2} = 6\tau_{1 \to 2} \approx 6 \frac{\overline{D_1}}{\Sigma_{tr1 \to 2}}$$

$$\Sigma_{tr1 \to 2} = \frac{\overline{D_1}}{\tau_{1 \to 2}} \tag{12.3}$$

For reasons of brevity, henceforth $\Sigma_{tr1 \to 2}$ will be written Σ_{tr}.

12.2.2 Equations for Two Groups

$$\begin{cases} D_1(r)\Delta\Phi_1(r) - (\Sigma_{a1}(r) + \Sigma_{tr}(r))\Phi_1(r) + \nu_2\Sigma_{f2}(r)\Phi_2(r) + \nu_1\Sigma_{f1}(r)\Phi_1(r) = 0 \\ D_2(r)\Delta\Phi_2(r) - \Sigma_{a2}(r)\Phi_2(r) + p \cdot (\Sigma_{a1}(r) + \Sigma_{tr}(r))\Phi_1(r) = 0 \end{cases} \tag{12.4}$$

$\nu_1\Sigma_{f1}\Phi_1$ represents the fast fission. The process of fast fission occurs at higher than thermal energies, at energies higher than 1 MeV. **The fast fission occurs mainly with** $^{238}_{92}$U but also with other fissile isotopes ($^{235}_{92}$U and $^{239}_{94}$Pu). The fission cross-section of $^{238}_{92}$U is similar to the other fissile isotopes. They are relatively low (of the order of barns). But there is a significantly larger amount of the $^{238}_{92}$U isotope in the reactor core (in thermal, low-enriched uranium reactors).

The removal of neutrons from group 1 times p, the resonance escape probability, is the cross section of arrivals from group 1 into group 2:

$$\Sigma_{s1\to 2} = p \cdot (\Sigma_{a1} + \Sigma_{tr})$$

We can define a cross section of removal from group 1:

$$\Sigma_{r1} = \Sigma_{a1} + \Sigma_{tr}$$

and we change the notation of thermal absorption, Σ_{r2} instead of Σ_{a2}

Thus the coupled Eq. (12.4) for two groups become:

$$\begin{cases} D_1(r)\Delta\Phi_1(r) - \Sigma_{r1}(r)\Phi_1(r) + \nu_2\Sigma_{f2}(r)\Phi_2(r) + \nu_1\Sigma_{f1}(r)\Phi_1(r) = 0 \\ D_2(r)\Delta\Phi_2(r) - \Sigma_{r2}(r)\Phi_2(r) + \Sigma_{s1\to 2}(r)\Phi_1(r) = 0 \end{cases}$$

$$(12.5)$$

The equations are said to be coupled since the function $\Phi_2(r)$ appears in the first equation and $\Phi_1(r)$ appears in the second equation.

12.2.2.1 Another Form for the Coupled Equations

From Eq. (13.8), and $\Sigma_{s1\to 2} = p\Sigma_{r1}$, it can be written that:

$$\frac{k_\infty}{p}\,\Sigma_{r2}\,\Phi_2 = \frac{1}{p} \cdot \frac{\nu_2\Sigma_{f2} + \nu_1\Sigma_{f1}\cdot\frac{\Phi_1}{\Phi_2}}{\Sigma_{r2}} \cdot p \cdot \Sigma_{r2}\Phi_2$$

$$\frac{k_\infty}{p}\,\Sigma_{r2}\,\Phi_2 = \nu_1\Sigma_{f1}\Phi_1 + \nu_2\Sigma_{f2}\Phi_2 \qquad (12.6)$$

Equation (13.8) does not require a critical reactor to be valid, thus this last Eq. (12.6) neither.

The coupled equations for two groups (12.5) can be now written with a new form:

$$\begin{cases} D_1(r)\Delta\Phi_1(r) - \Sigma_{r1}(r)\Phi_1(r) + \frac{k_\infty}{p}\Sigma_{r2}(r)\Phi_2 = 0 \\ D_2(r)\Delta\Phi_2(r) - \Sigma_{r2}(r)\Phi_2(r) + p\Sigma_{r1}(r)\Phi_1(r) = 0. \end{cases} \qquad (12.7)$$

Reference

S. Marguet, *The Physics of Nuclear Reactors* (Springer International Publishing, 2018), ISBN 9783319595603, https://books.google.fr/books?id=9DlODwAAQBAJ

Chapter 13
The Critical Reactor with Two Neutron Groups

Abstract We will study in this chapter the criticality of an infinite cylinder considered from a radial point of view, which is therefore a one dimension problem. Our starting point will be an homogeneous reactor with Dirichlet conditions, which enables hand calculations. From this basis, we will go forward by introducing zones with various fuel enrichments, and a water reflector surrounding the fissile system. A particular focus will be brought on the following topics: the influence of the boundary conditions, the influence of the reflector and of the loading pattern over the radial shape of the flux, the thermal and fast neutron radial leakage. Through this approach, we will attempt to approach some characteristics of a PWR 1300 MWe.

13.1 The Bare Homogeneous Reactor

Since we consider an homogeneous reactor, cross sections and diffusion coefficient become constant values within the volume of the reactor: D_1, D_2, Σ_{f1}, Σ_{f2} etc.

In the case of a bare homogeneous reactor, the values from Table 13.1 will be used both for hand calculations and numerical calculations with centered finite difference solver, solving the diffusion equation by power iteration method, for two neutron groups, and various boundary conditions.

13.1.1 Reflective Boundary Conditions

In the Sect. 11.1.5, the example of an homogeneous system with monokinetic neutrons emitted by a dispersed source in a scattering and absorbing slab illustrated the fact that an homogenous system has a flat flux in the case of a reflective boundary condition. This conclusion remains the same with a fissile system, and two groups of neutrons.

A numerical calculation is performed, with a cancellation of fast fission and a reflective boundary condition: $k_{\text{eff}} = 0.945\,76$.

© The Author(s), under exclusive license to Springer Nature Switzerland AG 2026

H. Grard, *Diffusion of Neutrons in Nuclear Reactors*,

https://doi.org/10.1007/978-3-032-05088-5_13

Table 13.1 Input data for PWR core modelization (homogeneous cylinder)

D_1 (Fast)	$\nu_1 \Sigma_{f1}$ (Fast)	Σ_{r1} (Fast)	Σ_{s12} (Fast)
1.9753 cm	2.5935×10^{-3} cm	0.039506 cm	0.028571 cm
D_2 (Th)	$\nu_2 \Sigma_{f2}$ (Th)	Σ_{r2} (Th)	f thermal utilization factor
1.0773 cm	0.234.40 cm	0.17954 cm	0.851 67

Without fast fissions, from Eq. 13.14 we can write:

$$k_\infty = p\,f\,\eta = \frac{\nu_2 \Sigma_{f2}}{\Sigma_{r2}} \cdot \frac{\Sigma_{s1\to 2}}{\Sigma_{r1}}$$

A calculation of k_∞ with the values from Table 13.1 returns 0.945 76 illustrating the fact that due to reflective conditions, there is no leakage of neutrons. And thus:

$$k_\infty = k_{\text{eff}}$$

13.1.2   Fundamental Mode. The Critical Reactor with Dirichlet Boundary Conditions

The reader can refer to Sect. 20.2.2. The Eq. 12.5 are critical equations, and the distributions of $\Phi_1(r)$ and $\Phi_2(r)$ are the fundamental mode. In an homogeneous system, the fundamental mode is an eigenfunction f_0 of the Laplacian operator (the eigenfunction that does not change of sign) in the studied geometry, equal to zero at the boundary of the system.

Since the system is homogeneous, cross sections and diffusion coefficients do not depend on r, and are constants.

Thus $\Phi_1(r) = \Phi_1 \cdot f_0(r)$ and $\Phi_2(r) = \Phi_2 \cdot f_0(r)$, where Φ_1 and Φ_2 are just multiplication factors of the function $f_0(r)$

Then we can write, from the fast neutrons equation (first equation of 12.5):

$$D_1 \Phi_1 \Delta f_0(r) + (-\Sigma_{r1}\Phi_1 + \nu_2 \Sigma_{f2}\Phi_2 + \nu_1 \Sigma_{f1}\Phi_1) \cdot f_0(r) = 0 \qquad (13.1)$$

The Eq. 13.1 is an eigenvalue equation, thus we can introduce $B_g^2 = -\lambda_0$, where λ_0 is the less negative of the eigenvalues.

$$\Delta f_0(r) + B_g^2 f_0(r) = 0 \qquad (13.2)$$

13.1.3 Determination of the Six Factors

13.1.3.1 Critical Condition

By combining the Eqs. 13.1 and the 13.2, we can write the critical condition:

$$D_1 B_g^2 = -\Sigma_{r1} + \nu_2 \Sigma_{f2} \frac{\Phi_2}{\Phi_1} + \nu_1 \Sigma_{f1}$$

$$\frac{\Phi_1}{\Phi_2} = \frac{\nu_2 \Sigma_{f2}}{D_1 B_g^2 + \Sigma_{r1} - \nu_1 \Sigma_{f1}}$$

It should be recalled that the age of thermal neutrons, that can be written $\tau_{1 \to 2}$ or τ_{th} is linked to the diffusion coefficient of fast neutrons:

$$\tau_{1 \to 2} = \frac{D_1}{\Sigma_{r1}} \tag{13.3}$$

This enables to write:

$$\frac{\Phi_1}{\Phi_2} = \frac{\nu_2 \Sigma_{f2}}{\Sigma_{r1}(1 + \tau_{th} B_g^2) - \nu_1 \Sigma_{f1}} \tag{13.4}$$

And similarly, the thermal scattering area L_2^2 is linked to the diffusion coefficient of thermal neutrons:

$$L_2^2 = \frac{D_2}{\Sigma_{r2}} \tag{13.5}$$

From the thermal neutrons equation (second equation of 12.5), we can write:

$$\frac{\Phi_2}{\Phi_1} = \frac{\Sigma_{s1 \to 2}}{\Sigma_{r2}(1 + L_2^2 B_g^2)} \tag{13.6}$$

At criticality, the ratio $\dfrac{\Phi_2}{\Phi_1}$ can be calculated indifferently from the fast Eq. 13.4 or the thermal equation 13.6. The critical condition can be written independently from Φ_2 and Φ_1:

$$\frac{\nu_2 \Sigma_{f2}}{\Sigma_{r1}(1 + \tau_{th} B_g^2) - \nu_1 \Sigma_{f1}} = \frac{\Sigma_{r2}(1 + L_2^2 B_g^2)}{\Sigma_{s1 \to 2}} \quad \text{(critical condition)} \tag{13.7}$$

Now we calculate the infinite multiplication factor k_∞. In the one group theory, k_∞ is the ratio of production by fission over absorption. In the second group theory, the thermal absorption does not take into account resonant absorption nor absorption by moderator during the slowing down. k_∞ can be considered as the ratio between production by fission (neutrons produced in the fast group) and removal in the thermal group, times the probability that a neutron removed from fast group appears in thermal group.

$$k_\infty = \frac{\nu_2 \Sigma_{f2} \Phi_2 + \nu_1 \Sigma_{f1} \Phi_2 \cdot \frac{\Phi_1}{\Phi_2}}{\Sigma_{r2} \Phi_2} \cdot \frac{\Sigma_{s1\to2} \Phi_1}{\Sigma_{r1} \Phi_1} \text{ (Not a critical condition)} \qquad (13.8)$$

By introducing k_∞, we can write the Eq. 13.7 in a new form:

$$k_\infty = (1 + \tau_{th} B_g^2)(1 + L_2^2 B_g^2) \text{ (critical condition)} \qquad (13.9)$$

If the reactor is critical, then $k_{\text{eff}} = 1$. Then we can write an estimation of k_{eff}:

$$k_{\text{eff}} \approx \frac{k_\infty}{(1 + \tau_{th} B_g^2)(1 + L_2^2 B_g^2)} \qquad (13.10)$$

Equation 13.10 is an exact equality at criticality.

13.1.3.2   Fast and Thermal Non-leakage Probabilities

The fast non-leakage factor, $P_{NL,f}$, is defined as the ratio of the number of fast neutrons that do not leak from the reactor core during the slowing down process to the number of fast neutrons produced by fissions at all energies. The thermal non-leakage factor, $P_{NL,t}$, is defined as the ratio of the number of thermal neutrons that do not leak from the reactor core during the neutron diffusion process to the number of neutrons that reach thermal energies.

$$P_{NL,f} \approx \frac{1}{1 + \tau_{th} B_g^2} \qquad (13.11)$$

Equation 13.11 is a valid equality if the reactor is critical, and is an approximation if the reactor is non critical.

$$P_{NL,t} = \frac{1}{1 + L_2^2 B_g^2} \qquad (13.12)$$

The fast non-leakage probability $P_{NL,f}$ and the thermal non-leakage probability $P_{NL,t}$ may be combined into one term that gives the fraction of all neutrons that do not leak out of the reactor core. This term is called the total non-leakage probability and is given the symbol PNL, and may be expressed by the following equation:

$$P_{NL} \approx \frac{1}{(1 + \tau_{th} B_g^2)(1 + L_2^2 B_g^2)} \qquad (13.13)$$

13.1.3.3 Fast Fission Factor

The fast fission factor is defined as the ratio of the fast neutrons produced by fission at all energies to the number of fast neutrons produced in thermal fission. From Eq. 13.7, which defines criticality, another expression of k_∞ can be written:

$$k_\infty = \frac{\nu_2 \Sigma_{f2}}{\Sigma_{r2}} \cdot \frac{\Sigma_{s1\to 2}}{\Sigma_{r1}} + \frac{\nu_1 \Sigma_{f1}}{\Sigma_{r2}} \cdot \frac{\Sigma_{s1\to 2}}{\Sigma_{r1}} \cdot \frac{1 + L_2^2 B_g^2}{\Sigma_{s1\to 2}} \cdot \Sigma_{r2}$$

$$k_\infty = \frac{\nu_2 \Sigma_{f2}}{\Sigma_{r2}} \cdot \frac{\Sigma_{s1\to 2}}{\Sigma_{r1}} + (1 + L_2^2 B_g^2) \cdot \frac{\nu_1 \Sigma_{f1}}{\Sigma_{r1}}$$

$$k_\infty = \underbrace{\frac{\nu_2 \Sigma_{f2}}{\Sigma_{r2}} \cdot \frac{\Sigma_{s1\to 2}}{\Sigma_{r1}}}_{k_\infty \text{ without fast fission}} + \underbrace{(1 + L_2^2 B_g^2) \cdot \frac{\nu_1 \Sigma_{f1}}{\Sigma_{r1}}}_{\text{supplementary for fast fission}} \tag{13.14}$$

This enables us to estimate, at criticality, the fast fission factor ϵ, as the ratio between k_∞ with fast fissions and k_∞ without fast fission.

$$\epsilon = 1 + \frac{\nu_1 \Sigma_{f1}}{\nu_2 \Sigma_{f2}} \cdot \frac{\Sigma_{r2}}{\Sigma_{s1\to 2}} \cdot (1 + L_2^2 B_g^2) \tag{13.15}$$

13.1.3.4 Thermal Utilization Factor

The neutrons that escape the resonance absorption and remain in the core will be thermalized. In thermal reactors, these neutrons continue to diffuse throughout the reactor until they are absorbed. But there are many materials in the reactor core in which these neutrons may be absorbed. The thermal utilization factor, f, is the fraction of the thermal neutrons absorbed in the nuclear fuel in all isotopes of the nuclear fuel.

Of course, in the frame of the two groups diffusion theory applied to an homogeneous reactor, there is no way to estimate the thermal utilization factor f.

By definition,

$$f = \frac{\Sigma_{a,\,\text{fuel}}}{\Sigma_{r2}}$$

13.1.3.5 Reproduction Factor

The reproduction factor, η, is defined as the ratio of the number of fast neutrons produced by thermal fission to the number of thermal neutrons absorbed in the fuel.

$$\eta = \frac{\nu_2 \Sigma_{f2}}{\Sigma_{a,\,\text{fuel}}}$$

$$\eta = \frac{\nu_2 \Sigma_{f2}}{f \Sigma_{r2}} \tag{13.16}$$

13.1.3.6   Resonance Escape Probability

During the thermalization, neutrons may collide not only with moderator nuclei but also with fuel nuclei. Unfortunately, especially $^{238}_{92}\mathrm{U}$ exhibits resonance behavior between the fast region and the thermal region. While the neutrons are slowing down through the resonance region of $^{235}_{92}\mathrm{U}$, which extends from about 6–200 eV, there is a chance that some neutrons will be captured. The probability that a neutron will not be captured in the entire resonance region is called the resonance escape probability .

The definitions of the fours factors (Resonance escape probability , Reproduction factor, Thermal utilization factor and Fast fission factor) enable to write that:

$$k_\infty = \epsilon p f \eta \tag{13.17}$$

The product $f\eta$ times ϵ is:

$$\epsilon f \eta = \frac{\nu_2 \Sigma_{f2}}{\Sigma_{r2}} + \frac{\nu_1 \Sigma_{f1}}{\Sigma_{s1\to2}}(1 + L_2^2 B_g^2) \tag{13.18}$$

Then, from Eqs. 13.14, 13.18 and 13.17 we can conclude that:

$$p = \frac{\Sigma_{s1\to2}}{\Sigma_{r1}} \tag{13.19}$$

As for the thermal utilization factor, the antitrap factor these factors cannot be deduced from knowledge of the properties of a homogeneous system and the resonance escape probability depends on the arrangement and the geometry of the reactor core (see Sect. 5.3).

13.1.4   *Bare Homogeneous Reactor: Application with Numerical Values*

Equation 13.9 is a critical condition for an homogeneous reactor with two groups:

$$k_\infty = (1 + \tau_{1\to2} B_g^2)(1 + L_2^2 B_g^2)$$

This is a second degree equation with one variable B_g^2:

$$\tau_{1\to2} L_2^2 B_g^4 + (\tau_{1\to2} + L_2^2) B_g^2 + (1 - k_\infty) = 0$$

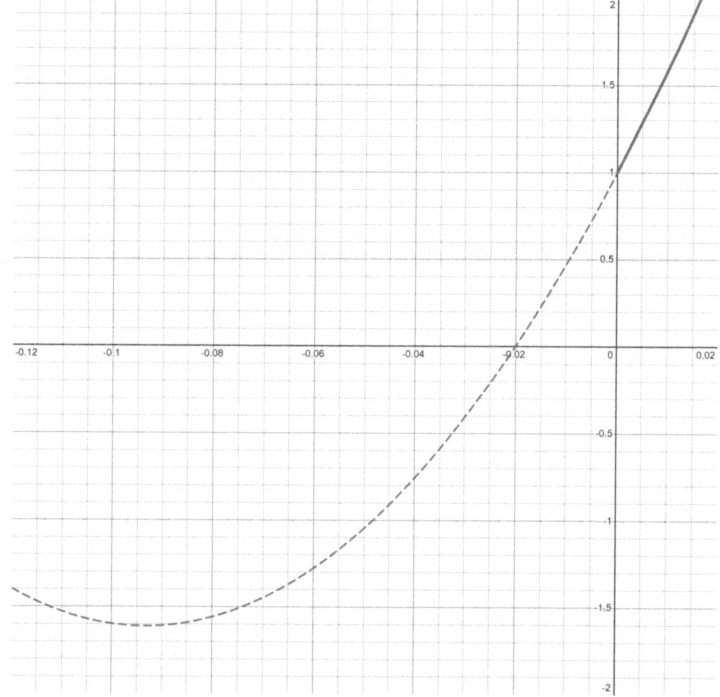

Fig. 13.1 Parabola of critical reactors characterized by buckling and k_∞. On this example, $\tau_{1\to 2} = 50\,\mathrm{cm}^2$ and $L_2^2 = 6\,\mathrm{cm}^2$

$$\tau_{1\to 2} L_2^2\, B_g^4 + M^2\, B_g^2 + 1 = k_\infty$$

This second degree polynomial function has the shape drawn on Fig. 13.1 and passes systematically through the point (0 and 1).

Slowing down area and thermal scattering area were given in Eqs. 13.3 and 13.5 as a function of mean diffusion coefficients (overline is omitted):

$$\tau_{1\to 2} = \frac{D_1}{\Sigma_{r1}}$$

$$L_2^2 = \frac{D_2}{\Sigma_{r2}}$$

With the values from Table 13.1, we can calculate $\tau_{1\to 2} = 50\,\mathrm{cm}^2$ and $L_2^2 = 6\,\mathrm{cm}^2$.

If $k_\infty < 1$, criticality is not achievable since the two roots of the secondary equation are negative and cannot fit with B_g^2. For a specific value $k_\infty > 1$, the critical condition becomes:

$$B_g^2 = \frac{-M^2 + \sqrt{M^4 - 4\,\tau_{1\to 2}\,L_2^2\,(1 - k_\infty)}}{2\,\tau_{1\to 2}\,L_2^2} \tag{13.20}$$

If $k_\infty > 1$, it is therefore possible to calculate the buckling at criticality and then the critical fissile radius, as a function of the effective multiplication factor.

13.1.4.1   Calculation of B_g^2 Without Previous Knowledge of k_∞

We can also solve the Eq. 13.21 so as to calculate B_g^2 at criticality:

$$\frac{\nu_2 \Sigma_{f2}}{\Sigma_{r2}} \cdot \frac{\Sigma_{s1\to 2}}{\Sigma_{r1}} + (1 + L_2^2 B_g^2) \cdot \frac{\nu_1 \Sigma_{f1}}{\Sigma_{r1}} = (1 + \tau_{1\to 2} B_g^2)(1 + L_2^2 B_g^2) \tag{13.21}$$

The calculated value is $B_g^2 = 2.0490 \times 10^{-4}$, and the critical fissile radius is then: $R_{f,c} = 168.00$ cm. k_∞ associated to $B_g^2 = 2.0490 \times 10^{-4}$ is: $k_\infty = 1.0115$.

13.1.4.2   Probability of Non-leakage

Equations 13.11 and 13.12 enable to calculate the fast non-leakage factor and the thermal non-leakage factor for an homogeneous reactor, with Dirichlet boundary conditions:

$$P_{NL,f} = \frac{1}{1 + \tau_{1\to 2} B_g^2}$$

$$P_{NL,t} = \frac{1}{1 + L_2^2 B_g^2}$$

The fast non leakage factor and the thermal non leakage factor can now be calculated, and then the non leakage probability $P_{NL} = P_{NL,f} \times P_{NL,t}$.

$$P_{NL,f} = \frac{1}{1 + 50 \times 2.0490} = 0.98986$$

$$P_{NL,t} = \frac{1}{1 + 6 \times 2.0490} = 0.99877$$

$$P_{NL} = 0.98986 \times 0.99877 = 0.98864$$

13.1.4.3 Worth of Leakage in pcm

The non leakage probabilitie enables to calculate a reactivity difference (Marguet 2018, p. 841) between an hypothetical non leaking reactor, and the leaking reactor:

$$
\begin{aligned}
\rho_{\text{leakage}} &= \frac{k_\infty - k_{\text{eff}}}{k_\infty \times k_{\text{eff}}} \\
&= \frac{k_\infty \times (1 - P_{NL})}{k_\infty \times k_{\text{eff}}} \\
&= \frac{1 - P_{NL}}{k_{\text{eff}}}
\end{aligned}
$$

A numerical calculation, with a 1D solver, enables to check if k_{eff} with values from Table 13.1, $R_{f,c} = 168\,\text{cm}$, for an infinite cylinder, Dirichlet boundary conditions and fast fissions enabled, is almost equal to 1: $k_{\text{eff}} = 1.0000$. The criticality of the system is confirmed.

The reactivity loss due to leakage can then be calculated: $\rho_{\text{leakage}} = 1135.7\,\text{pcm}$. This value can be splited in two contributions: fast leakage and thermal leakage.

In a finite cylinder, axial leakage also occurs. Equation 10.13 enables to establish that for a radius $R = 168\,\text{cm}$ and an active height $h = 426\,\text{cm}$, the contributions of radial and axial leakage to total leakage as respectively 76 and 24%. The leakage of the finite cylinder, with an height of $h = 426\,\text{cm}$, can be estimed to 1494 pcm.

13.1.5 The Ratio Between Fast and Thermal Flux

Figure 13.2, resulting from a numerical calculation, shows that thermal and fast flux are both proportional to the fundamental mode (see Sect. 13.1.2) which is the Bessel function J_0. The numerically calculated ratio, with 1D solver, is $\frac{\Phi_1}{\Phi_2} = 6.2918$.

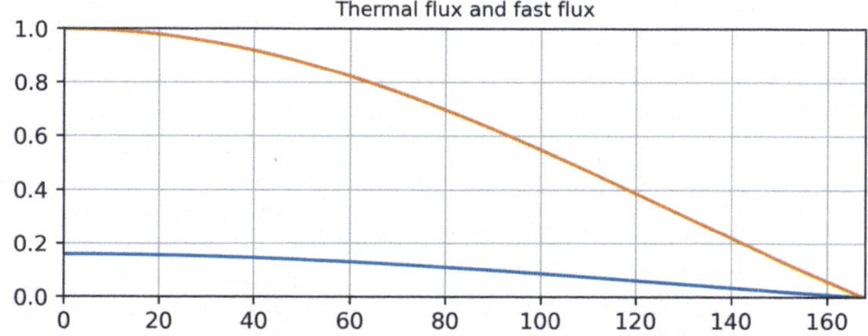

Fig. 13.2 Fast and thermal radial flux for an homogeneous cylinder (fast flux is set to 1 in the center of the cylinder)

The ratio $\frac{\Phi_1}{\Phi_2}$ can be hand-calculated with Eq. 13.4, which is only relevant at criticality:

$$\frac{\Phi_1}{\Phi_2} = \frac{\nu_2 \Sigma_{f2}}{\Sigma_{r1}(1 + \tau_{th} B_g^2) - \nu_1 \Sigma_{f1}}$$

The resulting value is $\frac{\Phi_1}{\Phi_2} = 6.2918$.

The ratio of thermal to fast flux (also called epithermal flux) is termed spectrum index $\gamma = \frac{\Phi_1}{\Phi_2}$. A spectrum hardening can be associated with a decrease of the spectrum index (Stammler and Abbate 1983, p. 351).

13.1.6   Calculation of the Four Factors

The fast fission factor ϵ for an homogeneous reactor with two groups can be calculated with Eq. 13.15:

$$\epsilon = 1 + \frac{\nu_1 \Sigma_{f1}}{\nu_2 \Sigma_{f2}} \cdot \frac{\Sigma_{r2}}{\Sigma_{s1 \to 2}} \cdot (1 + L_2^2 B_g^2)$$

$$\epsilon = 1.0695$$

The reproduction factor, η, defined as the ratio of the number of fast neutrons produced by thermal fission to the number of thermal neutrons absorbed in the fuel, can be calculated with Eq. 13.16:

$$\eta = \frac{\nu_2 \Sigma_{f2}}{f \Sigma_{r2}}$$

$$\eta = 1.5355$$

The resonance escape probability can be calculated with Eq. 13.19:

$$p = \frac{\Sigma_{s1 \to 2}}{\Sigma_{r1}}$$

$$p = 0.72322$$

The thermal utilization factor is an input data, with an accepted value 0.851 67.

From hand-calculated four factors and probability of non-leakage, it is possible to calculate the effective multiplication factor:

$$k_{\text{eff}} = \epsilon \, \eta \, p \, f \, P_{NL} = 1.0000$$

13.2 Coupling Within the Reflector

13.2.1 Study of Fast Neutron Source in a Water Wall

In order to study the coupling of fast and thermal fluxes in the reflector, we consider a water wall, infinite in the plane (x, y). This water wall is delimitated on one side of x axis by an infinite plane source of fast neutrons, and on the other side the water wall is infinite. This problem has already been studied in the Sect. 8.2 of the Chap. 9.

$$\begin{cases} D_{1,w} \Delta \Phi_1(x) - \Sigma_{r1,w} \Phi_1(x) = 0 \\[2mm] D_{2,w} \Delta \Phi_2(x) - \Sigma_{r2,w} \Phi_2(x) + \Sigma_{s1 \to 2,w} \Phi_1(x) = 0 \end{cases} \qquad \text{for } x \neq 0 \qquad (13.22)$$

For the fast group, the solution is analog to the solution established in the Sect. 8.2. We can write:

$$\Phi_1(x) = \frac{Q L_1}{2 D_{1,w}} e^{-\frac{x}{L_1}}$$

where $L_1 = \sqrt{\frac{D_{1,w}}{\Sigma_{r1,w}}} = \sqrt{\tau_{1 \to 2}}$.

The solution for thermal equation is:

$$\Phi_2(x) = \frac{\Phi_1(x)}{\gamma_w} + G e^{-\nu_w x}$$

where:

$$\nu_w^2 = \frac{1}{L_2^2} = \frac{\Sigma_{r2,w}}{D_{2,w}}$$

The second group equation can be written with the solutions of the system:

$$\frac{D_{2,w}}{\gamma_w} \frac{1}{L_1^2} \Phi_1(x) + D_{2,w} G \nu_w^2 e^{-\nu_w x} - \Sigma_{r2,w} \frac{\Phi_1(x)}{\gamma_w} - \Sigma_{r2,w} G e^{-\nu_w x} + \Sigma_{s1 \to 2,w} \Phi_1(x) = 0$$

$$\left(\frac{D_{2,w}}{\gamma_w} \frac{1}{L_1^2} - \frac{\Sigma_{r2,w}}{\gamma_w} + \Sigma_{s1 \to 2,w} \right) \Phi_1(x) = 0$$

This requires that:

$$\gamma_w = \frac{\Sigma_{r2,w} - \frac{D_{2,w}}{L_1^2}}{\Sigma_{s1 \to 2,w}}$$

And then:

$$\gamma_w = \frac{D_{2,w}\left(\frac{1}{L_2^2} - \frac{1}{L_1^2}\right)}{\Sigma_{s1\to2,w}}$$

In order to calculate G, we need to write that:

$$\frac{d\Phi_2}{dx}(0) = 0$$

$$\frac{-1}{L_1}\frac{QL_1}{2D_1\gamma_w} - G\nu_w e^{-\nu x} = 0$$

$$G = \frac{-1}{L_1}\frac{QL_1}{2D_1\gamma_w\nu_w}$$

This enables to write the flux for the thermal group:

$$\Phi_2(x) = \frac{QL_1}{2D_1\gamma_w}\left(e^{\frac{-x}{L_1}} - \frac{L_2}{L_1}e^{\frac{-x}{L_2}}\right)$$

It is interesting to note that, in the water wall:

$$\frac{\Phi_2(0)}{\Phi_1(0)} = \frac{1}{\gamma_w}\left(1 - \frac{L_2}{L_1}\right)$$

and

$$\lim_{x\to\infty}\frac{\Phi_2(x)}{\Phi_1(x)} = \frac{1}{\gamma_w}$$

In this example, all the thermal neutrons within the water wall come from the slowing down of fast neutrons, which is not true for a reactor with a fissile region and a reflector. Indeed, the calculation of albedos for fast and thermal neutrons from $\Phi_1(x)$ and $\Phi_2(x)$ give a valid result for the fast group, and 1 (reflective) for the thermal group.

With two neutron groups, the albedo of the fast group can be calculated from the Eq. 11.5 (Chap. 11), since $\Phi_1(x)$ is the same equation than the Eq. 11.4 established fot the flux with one group of neutrons.

$$\alpha_f = \frac{1 - 2\dfrac{D_{1,w}}{L_1}}{1 + 2\dfrac{D_{1,w}}{L_1}}$$

Concerning the thermal group, since $\frac{d\Phi_2}{dx}(0) = 0$, it comes immediately that $\alpha_{th} = 1$: this is a reflective condition for thermal neutrons.

13.2.2 *Water Wall with Boundary Condition $\Phi_1(0)$ and $\Phi_2(0)$*

The system (13.22) is now studied with a boundary condition $\Phi_1(0)$ and $\Phi_2(0)$. It is easy to establish that:

$$\Phi_1(x) = \Phi_1(0)e^{\frac{-x}{L_1}}$$

and

$$\Phi_2(x) = \frac{\Phi_1(0)}{\gamma_w}\left(e^{\frac{-x}{L_1}} - e^{\frac{-x}{L_2}}\right) + \Phi_2(0)\,e^{\frac{-x}{L_2}}$$

With these functions $\Phi_1(x)$ and $\Phi_2(x)$, the albedo coefficient for thermal neutrons depends from $\frac{\Phi_1(0)}{\Phi_2(0)}$.

13.3 The Reactor with Several Homogeneous Zones, Including a Reflector

13.3.1 *Complexity of the Real Reflector*

The radial reflector of a PWR is composed of successive materials with very different properties: steel of the baffle, water in the by-pass, steel of the core barrel, possibly the thermal shield that protects the reactor vessel depending on the reactor model, water in the downcomer, the steel of the vessel and its shaft, which is of concrete (..) We can consider that only the first 40 centimeters really influence the neutron properties of the reflector given that it is highly improbable for a neutron at that distance to be reflected back into the core. Nevertheless, the succession of highly absorbing material (steel) and a scattering moderator (water) leads to difficult homogenization calculation in diffusion theory with a homogeneous reflector material. Furthermore, given the geometrical complexity of the circumference of the core that includes square assemblies with a cylindrical reflector, calculation of the reflector is very sensitive to the industrial calculation scheme.

Marguet (2018, p. 751)

13.3.2 *Characterization of Each Zone*

We consider an infinite cylindric system, with several concentric rings. It should be noted that in each zone, the system is described by the coupled equations (see Sect. 12.5), with constant values for D_1, Σ_{r1}, Σ_{f2}, Σ_{f1}, D_2, Σ_{r2}, $\Sigma_{s1\to2}$:

$$\begin{cases} D_1\Delta\Phi_1(r) - \Sigma_{r1}\Phi_1(r) + \nu_2\Sigma_{f2}\Phi_2(r) + \nu_1\Sigma_{f1}\Phi_1(r) = 0 \\[2mm] D_2\Delta\Phi_2(r) - \Sigma_{r2}\Phi_2(r) + \Sigma_{s1\to2}\Phi_1(r) = 0 \end{cases}$$

The equations are said to be coupled since the function $\Phi_2(r)$ appears in the first equation and $\Phi_1(r)$ appears in the second equation.

Let's write that f_λ, λ are an eigen function and the associated eigen value of the Laplacian in a cylindric geometry: $\Delta f_\lambda + \lambda f_\lambda = 0$ (in this section, the eigen value equation is written $\Delta f_\lambda + \lambda f_\lambda = 0$ and not $\Delta f_\lambda = \lambda f_\lambda$). We admit that $\Phi_1(r) = \Phi_1 f_\lambda(r)$ and $\Phi_2(r) = \Phi_2 f_\lambda(r)$, where Φ_1 and Φ_2 are two constants, are some solutions of the coupled equations. And we define:

$$\Gamma = \frac{\Phi_1}{\Phi_2}$$

The solutions are introduced in the coupled equations:

$$\begin{cases} D_1 \Delta\Phi_1 f_\lambda(r) - \Sigma_{r1}\Phi_1 f_\lambda(r) + \nu_2\Sigma_{f2}\Phi_2 f_\lambda(r) + \nu_1\Sigma_{f1}\Phi_1 f_\lambda(r) = 0 \\[2mm] D_2 \Delta\Phi_2 f_\lambda(r) - \Sigma_{r2}\Phi_2 f_\lambda(r) + \Sigma_{s1\to2}\Phi_1 f_\lambda(r) = 0 \end{cases}$$

$$\begin{cases} -D_1\lambda\Phi_1 f_\lambda(r) + (\nu_1\Sigma_{f1} + \dfrac{\nu_2\Sigma_{f2}\Phi_2}{\Gamma} - \Sigma_{r1})\Phi_1 f_\lambda(r) = 0 \\[3mm] -D_2\lambda\Phi_2 f_\lambda(r) + (\Sigma_{s1\to2}\Gamma - \Sigma_{r2})\Phi_2 f_\lambda(r) = 0 \end{cases}$$

$$\begin{cases} -D_1\lambda + (\nu_1\Sigma_{f1} + \dfrac{\nu_2\Sigma_{f2}\Phi_2}{\Gamma} - \Sigma_{r1}) = 0 \\[3mm] -D_2\lambda + (\Sigma_{s1\to2}\Gamma - \Sigma_{r2}) = 0 \end{cases} \qquad (13.23)$$

The determinant of the matrix coefficient of the system of Eq. 13.23 has to be null at criticallity; the two equations are equivalent in this case and:

$$D_1(\Sigma_{s1\to2}\Gamma - \Sigma_{r2}) = D_2(\nu_1\Sigma_{f1} + \frac{\nu_2\Sigma_{f2}\Phi_2}{\Gamma} - \Sigma_{r1})$$

This enables to write that:

$$\Gamma = \frac{\Sigma_{r2} + \lambda D_2}{\Sigma_{s1\to2}} = \frac{\nu_2\Sigma_{f2}}{\Sigma_{r1} - \nu_1\Sigma_{f1} + \lambda D_1}$$

The condition to have a null determinant is therefore:

$$\Sigma_{s1\to2}\nu_2\Sigma_{f2} = (\Sigma_{r2} + \lambda D_2)(\Sigma_{r1} - \nu_1\Sigma_{f1} + \lambda D_1)$$

We remind that:

$$\Sigma_{r1} = \frac{D_1}{\tau_{1\to2}} \quad \text{and} \quad L_2^2 = \frac{D_2}{\Sigma_{r2}}$$

Then comes:

$$\underbrace{\frac{\nu_2 \Sigma_{f2} \Sigma_{s1\to 2}}{\Sigma_{r1}\Sigma_{r2}} + \left(1 + L_2^2 \lambda\right)\frac{\nu_1 \Sigma_{f1}}{\Sigma_{r1}}}_{k_\infty \text{ of the zone}} = (1 + L_2^2 \lambda)(1 + \tau_{1\to 2}\lambda) \qquad (13.24)$$

Equation 13.24, established within a zone of a reactor having several zones is the same than Eq. 13.21 previously established for an homogeneous reactor with a single zone. Equation 13.24 enables to calculate the eigen values as roots of a second degree equation, whose coefficients depend on the parameters of the zone. Since there are tow roots λ, there are also two values for Γ: the two coupling factors between the two equations depends on the value of the eigen value λ, and thus on the corresponding eigen function f_λ.

Table 13.2 gives an overview of the whole problem for a reactor with several zones and two groups of neutrons.

If the multiplicative region $k_\infty > 1$ is the central cylinder, then the functions Y_0 and K_0 have to be eliminated, because they are not regular on zero, the center of the cylinder (Bussac and Reuss 1985, p. 268). It has to be noted that the curvature of J_0 is characteristic of a multiplicative medium.

The reflector has to be considered in a different way, since $\nu_1 \Sigma_{f1}$ and $\nu_2 \Sigma_{f2}$ are null in this zone. And the system of equation in the reflector:

$$\begin{cases} D_{1,w}\Delta\Phi_1(\mathbf{r}) - \Sigma_{r1,w}\Phi_1(\mathbf{r}) = 0 \\[2mm] D_{2,w}\Delta\Phi_2(\mathbf{r}) - \Sigma_{r2,w}\Phi_2(\mathbf{r}) + \Sigma_{s1\to 2,w}\Phi_1(\mathbf{r}) = 0 \end{cases}$$

Considering that the two roots in the reflector are:

$$-\mu_w^2 = \frac{-1}{\tau_{1\to 2}} = -\frac{\Sigma_{r1,w}}{D_{1,w}}$$

Table 13.2 Overview: reactor with several zones and two groups of neutrons

k_∞ of a zone	$0 < k_\infty < 1$	$k_\infty \geq 1$
Signs of roots	Two negative roots $-\nu^2$ and $-\mu^2$	One negative and one positive ≥ 0 root $-\nu^2$ and $+\mu^2$
Coupling factors	$\gamma_{[-\nu^2]}$ and $\gamma_{[-\mu^2]}$	$\gamma_{[-\nu^2]}$ and $\gamma_{[\mu^2]}$
Eigen values Equ.	$\Delta f - \mu^2 f = 0$ $\Delta f - \nu^2 f = 0$	$\Delta f + \mu^2 f = 0$ $\Delta f - \nu^2 f = 0$
Flux group 1	$A\,I_0(\mu r) + B\,K_0(\mu r)$ $+C\,I_0(\nu r) + D\,K_0(\nu r)$	$A\,J_0(\mu r) + B\,Y_0(\mu r)$ $+C\,I_0(\nu r) + D\,K_0(\nu r)$
Flux group 2	$\dfrac{1}{\gamma_{[-\mu^2]}}(A\,I_0(\mu r) + B\,K_0(\mu r))$ $+\dfrac{1}{\gamma_{[-\nu^2]}}(C\,I_0(\nu r) + D\,K_0(\nu r))$	$\dfrac{1}{\gamma_{[\mu^2]}}(A\,J_0(\mu r) + B\,Y_0(\mu r))$ $+\dfrac{1}{\gamma_{[-\nu^2]}}(C\,I_0(\nu r) + D\,K_0(\nu r))$

and

$$- \nu_w^2 = \frac{-1}{L_2^2} = -\frac{\Sigma_{r2,w}}{D_{2,w}}$$

This leads to the following functions:

$$\begin{cases} \Phi_1(r) = E \, I_0(\mu_w r) + F \, K_0(\mu_w r) \\[2mm] \Phi_2(r) = \dfrac{E}{\gamma_w} \, I_0(\mu_w r) + \dfrac{F}{\gamma_w} \, K_0(\mu_w r) + G \, I_0(\nu_w r) + H \, K_0(\nu_w r) \end{cases}$$

13.3.3   Infinite Cylinder with Two Regions: A Fissile Region and a Water Reflector

13.3.3.1   Input Data

We consider an infinite cylindric system, with a fissile region (radius R_f) and a water reflector (external radius R_t), and two neutron groups. The properties of water are given in Table 13.3. In this case of a reflected reactor and an homogeneous fissile region, the values from Table 13.4 will be used: they are the same than from Table 13.1, except for $\nu_2 \Sigma_{f2}$ which is lower, so as to keep the same critical radius (168 cm) after introducing a reflector of 40 cm.

13.3.3.2   Calculation of the Critical Condition

The roots of Eq. 13.24, calculated with the input data of Tables 13.3 and 13.4, are:

$$\text{for the core (fissile region): } - \nu_c^2 = -0.18553 \,/\text{cm}^2 \quad \text{and } \mu_c^2 = 1.766 \times \text{cm}^2$$

$$\text{for the reflector: } - \nu_w^2 = -0.15412 \, \text{cm}^2 \quad \text{and } - \mu_w^2 = -0.022642 \, \text{cm}^2$$

Table 13.3   Input data for two groups: water reflector 300°, 155 bar, 1200 ppm

$\tau_{1\rightarrow2}$	$\Sigma_{r1,w}$ (Group 1)	$D_{1,w}$ (Group 1)	$\Sigma_{s12,w}$	L_2^2	$\Sigma_{r2,w}$ (Group 2)	$D_{2,w}$ (Group 2)
44.167 cm^2	0.034223 cm	1.5115 cm	0.032683 cm	6.4883 cm^2	0.037 037 cm	0.240 31 cm

Table 13.4   Input data for two groups: homogeneous cylinder with reflector

$D_{1,c}$ (Fast)	$\nu_1 \Sigma_{f1}$ (Fast)	$\Sigma_{r1,c}$ (Fast)	$\Sigma_{s12,c}$ (Fast)
1.9753 cm	0.00259348 cm	0.039 506 cm	0.028 571 cm
$D_{2,c}$ (Th)	$\nu_2 \Sigma_{f2}$ (Th)	$\Sigma_{r2,c}$ (Th)	f thermal utilization factor
1.0773 cm	0.234 40 cm	0.179 54 cm	0.851 67

The J_0 function, associated to the root μ_c^2 vanishes at 180.95 cm, inside the reflector and outside the fissile region.

The coupling factors between fast and thermal flux can now be calculated:

$$\gamma_{[\mu_c^2]} = \frac{\Sigma_{r2,c} + \mu_c^2 D_{2,c}}{\Sigma_{s1\to2,c}} = \frac{\nu_2 \Sigma_{f2}}{\Sigma_{r1,c} - \nu_1 \Sigma_{f1} + \mu_c^2 D_{1,c}} \tag{13.25}$$

$$\gamma_{[-\nu_c^2]} = \frac{\Sigma_{r2,c} - \nu_c^2 D_{2,c}}{\Sigma_{s1\to2,c}} = \frac{\nu_2 \Sigma_{f2}}{\Sigma_{r1,c} - \nu_1 \Sigma_{f1} - \nu_c^2 D_{1,c}}$$

$$\gamma_w = \frac{\Sigma_{r2,w} - \mu_w^2 D_{2,w}}{\Sigma_{s1\to2,w}} \tag{13.26}$$

The calculation of coupling factors gives:

$$\gamma_{[\mu_c^2]} = 6.2907; \; \gamma_{[-\nu_c^2]} = -0.71124; \gamma_w = 0.96674$$

13.3.3.3 Analytical Determination of Flux and Critical Condition

The infinite cylinder with a water reflector has two regions: a multiplicative one characterized by $k_\infty > 1$ and a non-multiplicative one, the reflector, characterized by $k_\infty = 0$. In the multiplicative region, the fast and thermal flux are:

$$\begin{cases} \Phi_{1,c}(r) = A\,J_0(\mu_c r) + C\,I_0(\nu_c r) \\[2ex] \Phi_{2,c}(r) = \dfrac{A}{\gamma_{[\mu_c^2]}}\,J_0(\mu_c r) + \dfrac{C}{\gamma_{[-\nu_c^2]}}\,I_0(\nu_c r) \end{cases}$$

In the water reflector region, the equations are:

$$\begin{cases} \Phi_{1,w}(r) = E\,I_0(\mu_w r) + F\,K_0(\mu_w r) \\[2ex] \Phi_{2,w}(r) = \dfrac{E}{\gamma_w}\,I_0(\mu_w r) + \dfrac{F}{\gamma_w}\,K_0(\mu_w r) + G\,I_0(\nu_w r) + H\,K_0(\nu_w r) \end{cases}$$

Thus there are 6 variables (A, C, E, F, G, H). The flux in the critical reactor is characterized by continuity of the flux at the interface between the fissile region and the reflector, continuity of current, and flux vanishing at the boundary according to Dirichlet condition. The critical condition is that the determinant of the coefficient matrix is null.

Let's call $[M]$ the coefficient matrix:

$$[M] \cdot \begin{bmatrix} A \\ C \\ E \\ F \\ G \\ H \end{bmatrix} = [0]$$

The coefficient matrix is constructed with the continuity of flux (first and then second group), then the continuity of current and the Dirichlet conditions. The critical fissile radius is noted below as R_f.

$$[M] = \begin{bmatrix} m_{11} & m_{12} & m_{13} & m_{14} & m_{15} & m_{16} \\ m_{21} & m_{22} & m_{23} & m_{24} & m_{25} & m_{26} \\ m_{31} & m_{32} & m_{33} & m_{34} & m_{35} & m_{36} \\ m_{41} & m_{42} & m_{43} & m_{44} & m_{45} & m_{46} \\ m_{51} & m_{52} & m_{53} & m_{54} & m_{55} & m_{56} \\ m_{61} & m_{62} & m_{63} & m_{64} & m_{65} & m_{66} \end{bmatrix}$$

The coefficients are:

$$m_{11} = J_0(\mu_c\, R_f) \qquad m_{12} = I_0(\nu_c\, R_f) \qquad m_{13} = -I_0(\mu_w\, R_f)$$

$$m_{14} = -K_0(\mu_w\, R_f) \qquad m_{15} = 0 \qquad m_{16} = 0$$

$$m_{21} = \frac{1}{\gamma_{[\mu_c^2]}} J_0(\mu_c\, R_f) \qquad m_{22} = \frac{1}{\gamma_{[-\nu_c^2]}} I_0(\nu_c\, R_f) \qquad m_{23} = \frac{-1}{\gamma_w} I_0(\mu_w\, R_f)$$

$$m_{24} = \frac{-1}{\gamma_w} K_0(\mu_w\, R_f) \qquad m_{25} = -I_0(\nu_w\, R_f) \qquad m_{26} = -K_0(\nu_w\, R_f)$$

$$m_{31} = -D_{1,c}\mu_c J_1(\mu_c\, R_f) \qquad m_{32} = D_{1,c}\nu_c I_1(\nu_c\, R_f) \qquad m_{33} = -D_{1,w}\mu_w I_1(\mu_w\, R_f)$$

$$m_{34} = D_{1,w}\mu_w K_1(\mu_w\, R_f) \qquad m_{35} = 0 \qquad m_{36} = 0$$

$$m_{41} = -D_{2,c}\frac{\mu_c}{\gamma_{[\mu_c^2]}} J_1(\mu_c\, R_f) \qquad m_{42} = D_{2,c}\frac{\nu_c}{\gamma_{[-\nu_c^2]}} I_1(\nu_c\, R_f) \qquad m_{43} = -D_{2,w}\frac{\mu_w}{\gamma_w} I_1(\mu_w\, R_f)$$

$$m_{44} = D_{2,w}\frac{\mu_w}{\gamma_w} K_1(\mu_w\, R_f) \qquad m_{45} = -D_{2,w}\nu_w I_1(\nu_w\, R_f) \qquad m_{46} = D_{2,w}\nu_w K_1(\nu_w\, R_f)$$

$$m_{51} = 0 \qquad m_{52} = 0 \qquad m_{53} = I_0(\mu_w\, R_t)$$

$$m_{54} = K_0(\mu_w\, R_t) \qquad m_{55} = 0 \qquad m_{56} = 0$$

$$m_{61} = 0 \qquad m_{62} = 0 \qquad m_{63} = \frac{-1}{\gamma_w} I_0(\mu_w\, R_t)$$

$$m_{64} = \frac{-1}{\gamma_w} K_0(\mu_w\, R_t) \qquad m_{65} = -I_0(\nu_w\, R_t) \qquad m_{66} = -K_0(\nu_w\, R_t)$$

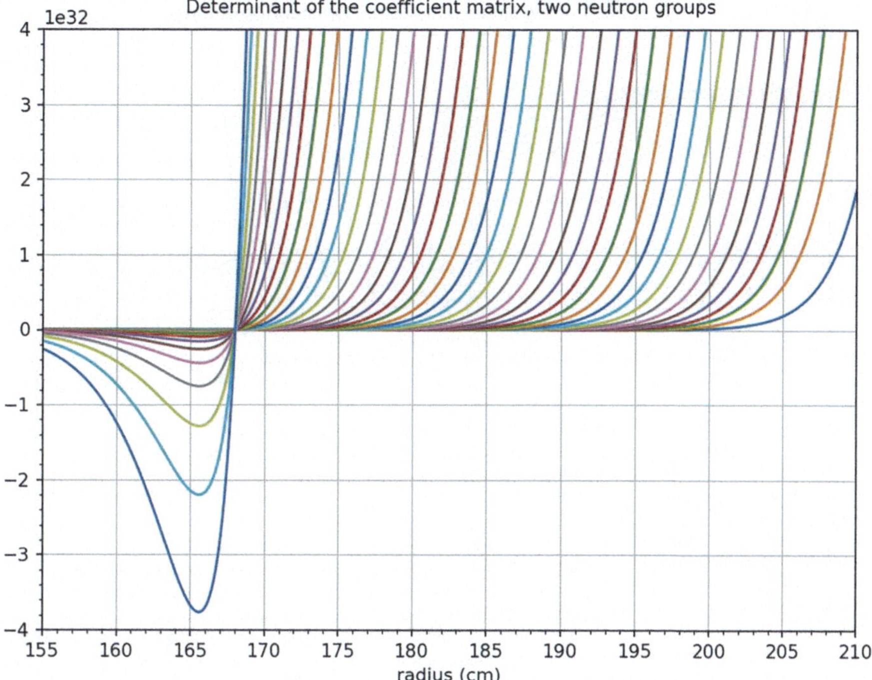

Fig. 13.3 Determinant of the coefficient matrix, for an homogeneous cylinder surrounded by a water reflector, with thickness ranging from 0 to 40 cm. Two neutron groups

The determinant is now calculated as a function of the fissile radius R_f, as shown on Fig. 13.3. The Fig. 13.4 shows that, with the input data of Tables 13.3 and 13.4, the critical radius with a reflector of thickness 40 cm is $R_f = 168$ cm, as estimated with our 1D Diffusion solver. The increasing of the critical fissile radius when the thickness of the reflector is reduced is visible on Fig. 13.4.

The determinant of the coefficient matrix is null at critiality: this means that one equation is a linear combination of the other equations. One of the coefficients may be chosen arbitrarily and set of the other coefficients comes with it. Let's choose $A = 1$. The other coefficients are the solution of the following system:

$$
\begin{bmatrix}
m_{22} & m_{23} & m_{24} & m_{25} & m_{26} \\
m_{32} & m_{33} & m_{34} & m_{35} & m_{36} \\
m_{42} & m_{43} & m_{44} & m_{45} & m_{46} \\
m_{52} & m_{53} & m_{54} & m_{55} & m_{56} \\
m_{62} & m_{63} & m_{64} & m_{65} & m_{66}
\end{bmatrix}
\begin{bmatrix}
C \\ E \\ F \\ G \\ H
\end{bmatrix}
=
\begin{bmatrix}
-m_{21} \\ -m_{31} \\ -m_{41} \\ -m_{51} \\ -m_{61}
\end{bmatrix}
$$

Thus, the system to solve is;

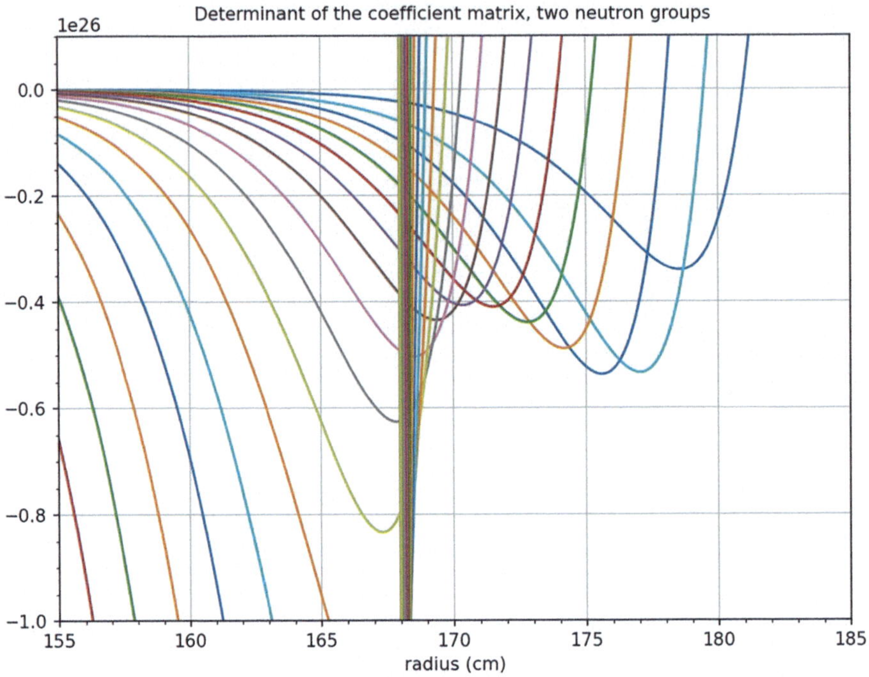

Fig. 13.4 Determinant of the coefficient matrix, for an homogeneous cylinder surrounded by a water reflector, with thickness ranging from 0 to 40 cm. Two neutron groups

$$
\begin{bmatrix}
-1.7659 \times 10^{30} & -7.8535 \times 10^{09} & -2.6953 \times 10^{-12} \\
1.0612 \times 10^{30} & -1.6923 \times 10^{9} & 6.0424 \times 10^{-13} \\
-8.1371 \times 10^{29} & -2.7830 \times 10^{08} & 9.9370 \times 10^{-14} \\
0.0 & 2.802 \times 10^{12} & 5.7015 \times 10^{-15} \\
0.0 & -2.8987 \times 10^{12} & -5.8977 \times 10^{-15}
\end{bmatrix}
$$

$$
\begin{bmatrix}
-2.1670 \times 10^{27} & -3.4984 \times 10^{30} \\
0.0 & 0.0 \\
-2.0289 \times 10^{26} & 3.3254 \times 10^{-31} \\
0.0 & 0.0 \\
-1.2860 \times 10^{34} & -4.7616 \times 10^{37}
\end{bmatrix}
\begin{bmatrix} C \\ E \\ F \\ G \\ H \end{bmatrix}
=
\begin{bmatrix} 0 \\ 0 \\ 0 \\ 0 \\ 0 \end{bmatrix}
$$

The solution is:

$$
A = 1.0; \quad C = -5.1199 \times 10^{-33}; \quad E = -6.7008 \times 10^{-17}; \quad F = 3.2935 \times 10^{10}
$$

$$
G = 6.8865 \times 10^{-43}; \quad H = -1.8598 \times 10^{28}
$$

We remind the previously calculated values:

$$\mu_c = 0.013290; \; \nu_c = 0.43073; \; \mu_w = 0.15047; \; \nu_w = 0.39258 \text{ and}$$

$$\gamma_{[\mu_c^2]} = 6.2907; \; \gamma_{[-\nu_c^2]} = -0.71124; \; \gamma_w = 0.96674$$

These values enable to calculate the functions describing $\Phi_1(r)$ and $\Phi_2(r)$ in the fissile region, and in the reflector, which are drawn on Figs. 13.5 and 13.6.

In the core of the fissile cylinder, $\Phi_{1,c}(r)$ and $\Phi_{2,c}(r)$ are respectively dominated by $A(=1) \times J_0(\mu_c r)$ and $\dfrac{1}{\gamma_{[\mu_c^2]}} J_0(\mu_c r)$. Thus, at a sufficient distance away from the reflector, we can write that:

$$\frac{\Phi_{1,c}(r)}{\Phi_{2,c}(r)} \approx \gamma_{[\mu_c^2]} \quad \text{for } r < R_f$$

The spectrum index $\gamma(r) = \dfrac{\Phi_1(r)}{\Phi(r)}$ varies in the vincinity of the reflector, and tends to an asymptotic value away from the reflector. In the fissile zone at a sufficient

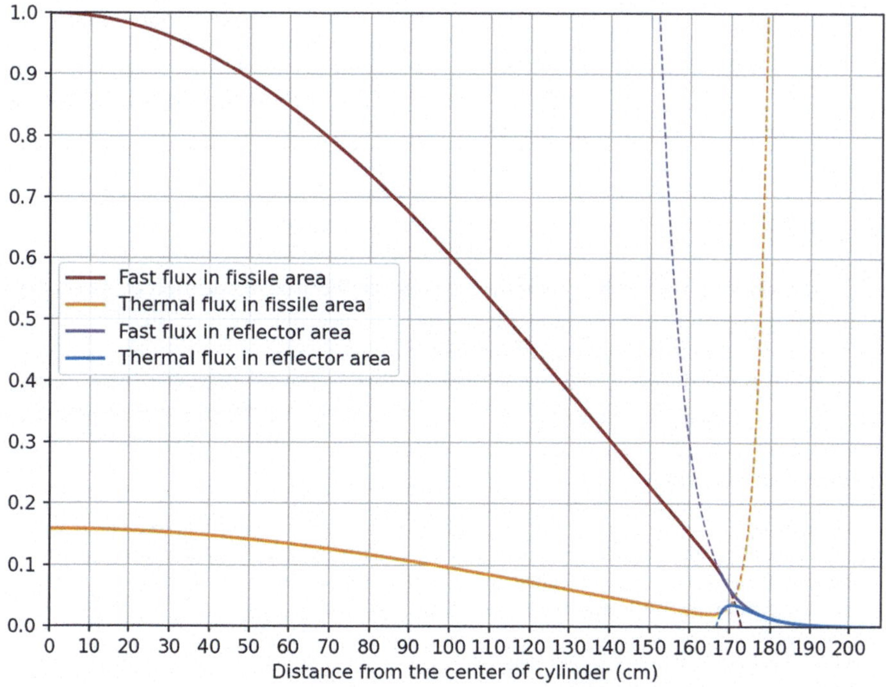

Fig. 13.5 Functions for thermal and fast flux in the fissile region of an homogeneous cylinder, and in the reflector (40 cm)

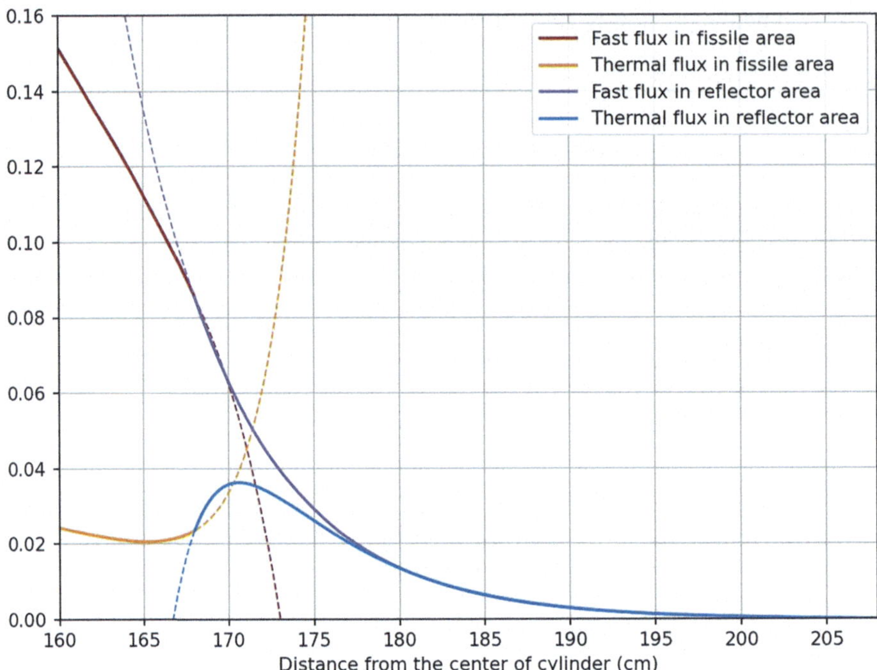

Fig. 13.6 Functions for thermal and fast flux in the fissile region of an homogeneous cylinder, and in the reflector (40 cm). Focus on the reflector

distance from the reflector, the spectrum index is equal to the positive coupling factor $\gamma_{[\mu_c^2]}$.

The reason why $\Phi_{2,c}$ is less than $\Phi_{1,c}$ is that the thermal absorption cross section in the core is greater than the slowing-down cross section. As a result, the thermal flux is depressed below the value of the fast flux.

Figure 13.6 highlights a rise of the thermal flux in the fissile region close to the reflector, and a peak of flux in the reflector. This is due to the transfer of neutrons from the fast to the thermal group in the reflector. This behavior is not predicted by the one-group theory.

> The thermalized neutrons are not absorbed as quickly in the reflector as neutrons thermalized in the core since the reflector, being unfueled, has a much smaller absorption cross-section. The thermal neutrons tend to accumulate in the reflector until they lay back into the core, escape from the outer surface of the reflector, or are absorbed. An importance consequence of the rise in the thermal flux in the reflector is that it tends to flatten the thermal flux distribution in the core. (..) The flux in the core near the core-reflector interface is built up by the flow of thermal neutrons out of the reflector.
>
> Lamarsh (2001, p. 305)

13.3.3.4 Fast and Thermal Leaks

The neutron leaks are considered here from the fissile region point of view. The discussed case is the critical reactor, with a critical fissile radius of 168 cm and a reflector of 40 cm. Fast and thermal leaks have to be distinguished, since some fast neutrons leaking into the reflector are thermalized and reflected back to the fissile region. This explains why the fast leak is positive and the thermal leak negative: in this example, the values are respectively 1200 and −175 pcm.

13.3.3.5 Comparison Between Two Groups and One Group Calculation

The two groups calculation is compared to the equivalent one group calculation; the inputs for one group are summarized in Table 10.5 for the reflector, and in Table 13.5.

Figure 13.7 shows the comparison between the critical fissile radius, as a function of the water reflector thickness, calculated with two groups or with one group. Thus a reflector not only reduces the maximum-to-average flux ratio, it also reduces the critical size and mass of a reactor. This highlights that the one group modelization cannot properly calculate a reflected reactor, whether in terms of critical condition or shape of flux.

13.4 Flattening the Radial Flux

13.4.1 Reactor Partioned Into Zones

Reactors are not homogeneous and realistic reactor models may consist of hundreds or thousands of different homogenized regions, or zones, after performing spatial homogenization (fuelmoderator), usually at the scale of a fuel assembly. In this chapter, we will consider a radial zoning, by partitionning the cylinder into a few concentric homogeneous rings. Beyond two groups and two regions, finding an analytical solution becomes excessively laborious and a numerical solution has to be calculated.

An example of numerical equation is the Eq. F.1 in Appendix F, established in the case of one group and one-dimensional geometry. The resolution of this equation

Table 13.5 Input data for one group PWR core: homogeneous cylinder with a critical fissile radius of 168 cm and a reflector of 40 cm

$\Sigma_{a,1G}$ (1 Group)	D_{1G} (1 Group)	$(\nu\Sigma_f)_{1G}$ (1 Group)
$= \Sigma_{r2}$ (Th)	$= (L_2^2 + \tau_{1\to2}) \times \Sigma_{a,1G}$	$= k_{\infty,2G} \times \Sigma_{a,1G}$
0.179 54 cm	10.054 cm	0.181 41 cm

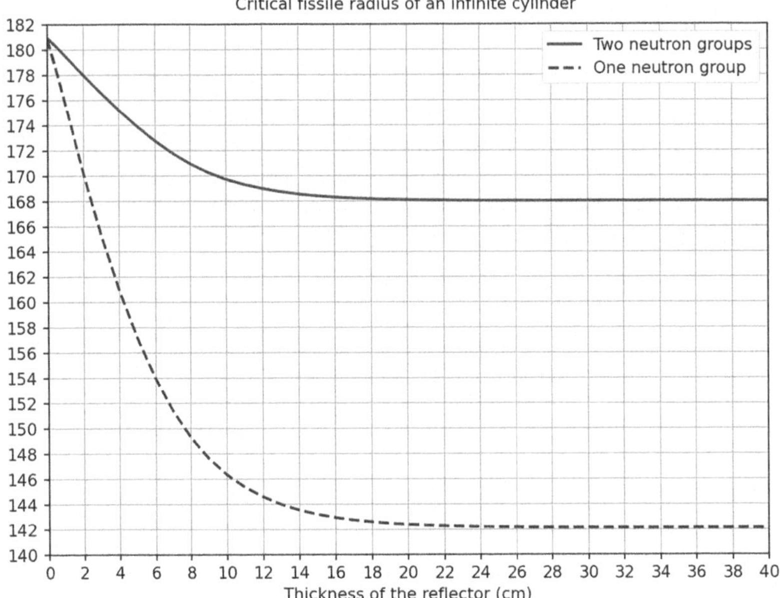

Fig. 13.7 Critical radius of an homogeneous cylinder surrounded by a water reflector, with thickness ranging from 0 to 40 cm. One and two neutron groups

enables to find the eigen vector associated to the maximal eigen value that defines k_{eff}. The eigen vector is called fundamental mode. Since we do not consider an homogeneous reactor with Dirichlet boundary conditions, this eigen vector differs from the fundamental mode of the Laplacian operator.

From the numerical equation, the set of the eigen functions could be established, and is a basis. This set is not infinite, since the size of operators is determined by the geometrical discretization. In the case of 1D problem in an infinite cylinder, thermal and fast flux, seen as functions, can also be decomposed onto a basis of Bessel functions of the first kind and order 0: $J_0(\frac{\mu_{m,0}r}{R})$ (see Appendix E) in the case of Dirichlet boundary conditions.

13.4.2 Regions with Various Enrichements

The fissile infinite cylinder is divided into three zones: a central disk of radius $R_{\mathrm{int,1}}$, a ring between $R_{\mathrm{int,1}}$ and $R_{\mathrm{int,2}}$, and an external ring between $R_{\mathrm{int,2}}$ and R_f. By varying the enrichment in each zone, il is possible to flatten the radial shape of the flux. A numerical calculation with 1D solver is performed with input data from Tables 13.3 and 13.4, except for $\nu_2\Sigma_{f2}$ which is set according the Table 13.6. The flattening of the thermal flux shape is shown on Fig. 13.8.

Table 13.6 $\nu_2\Sigma_{f2}$ for three zones

$r < R_{\text{int},1}$	$R_{\text{int},1} < r < R_{\text{int},2}$	$R_{\text{int},2} < r < R_f$
$\nu_2\Sigma_{f2} = 0.23199\,\text{cm}$	$\nu_2\Sigma_{f2} = 0.25203\,\text{cm}$	$\nu_2\Sigma_{f2} = 0.19633\,\text{cm}$

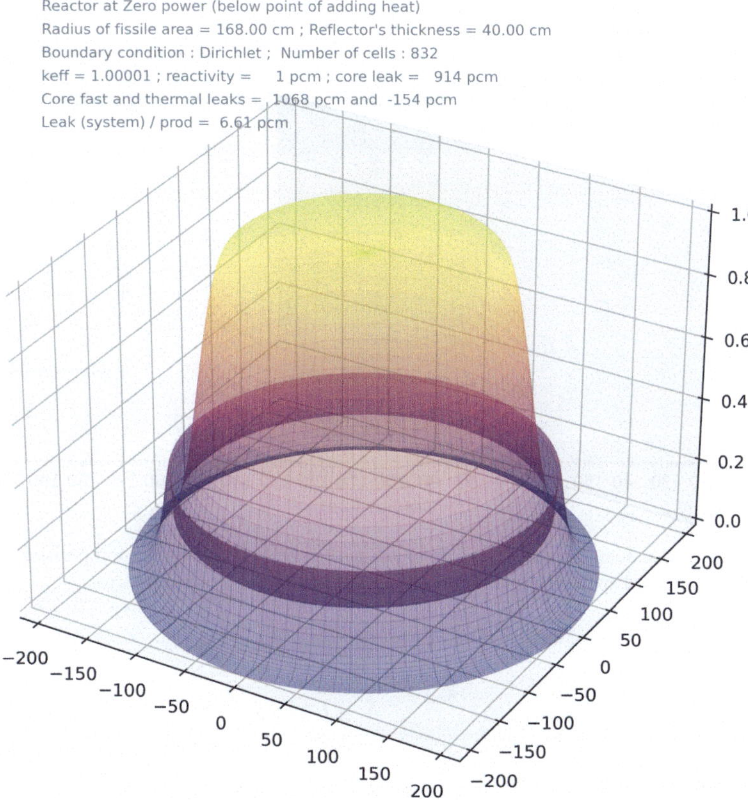

Fig. 13.8 Thermal flux shape flattened with zones of different enrichments. Two neutron groups calculation

13.4.3 *Reactivity Difference: Flattened and Not Flattened*

It's interesting to observe that the reactor with inputs from Table 13.6 is critical, and the average enrichment is lower than the average enrichment of the critical homogeneous cylinder with the same reflector. The homogeneous cylinder, with the average enrichment of table , is subcritical: $\rho = -2318\,\text{pcm}$. For both configurations, flattened shape and reflected bessel function, the leaks are not very different: respectively $914\,\text{pcm}$ and $1048\,\text{pcm}$

Now let's reason with one neutron group, and name A, F and L the absorption, fission and leak rate, already defined in Sect. 10.1.6.

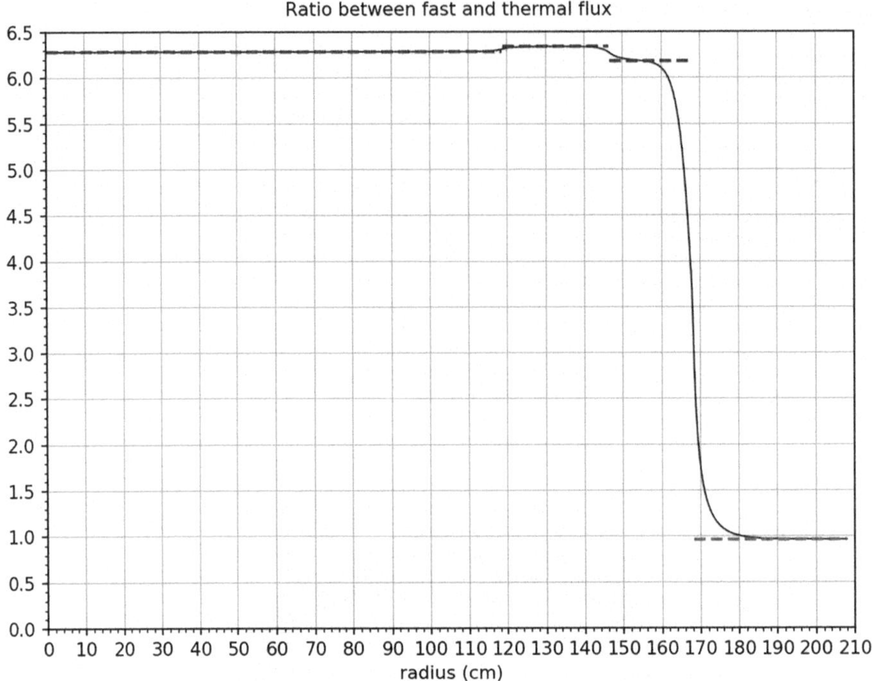

Fig. 13.9 Critical reactor flattened with reflector and three fissile regions: comparison between numerically calculated spectrum index $\gamma = \frac{\Phi_1}{\Phi_2}$ and the positive γ coupling factors

$$k_{\text{eff}} = \frac{F}{A + L}$$

$$\frac{1}{k_{\text{eff}}} = \frac{L}{F} + \frac{A}{F}$$

Two configurations of the core, index 1 and index 2 are compared, assuming that the leak rate are almost the same:

$$\frac{k_{\text{eff},2} - k_{\text{eff},1}}{k_{\text{eff},2} \times k_{\text{eff},1}} = \frac{A_1}{F_1} - \frac{A_2}{F_2}$$

$\dfrac{k_{\text{eff},2} - k_{\text{eff},1}}{k_{\text{eff},2} \times k_{\text{eff},1}}$ is the reactivity difference between the two states (Marguet 2018, p. 841).

It can be concluded that the flattening of the flux enables to enhance the ratio between the fission reaction rate and the absorption reaction rate, and consequently the reactivity of the core, even if the leaks are only slightly different.

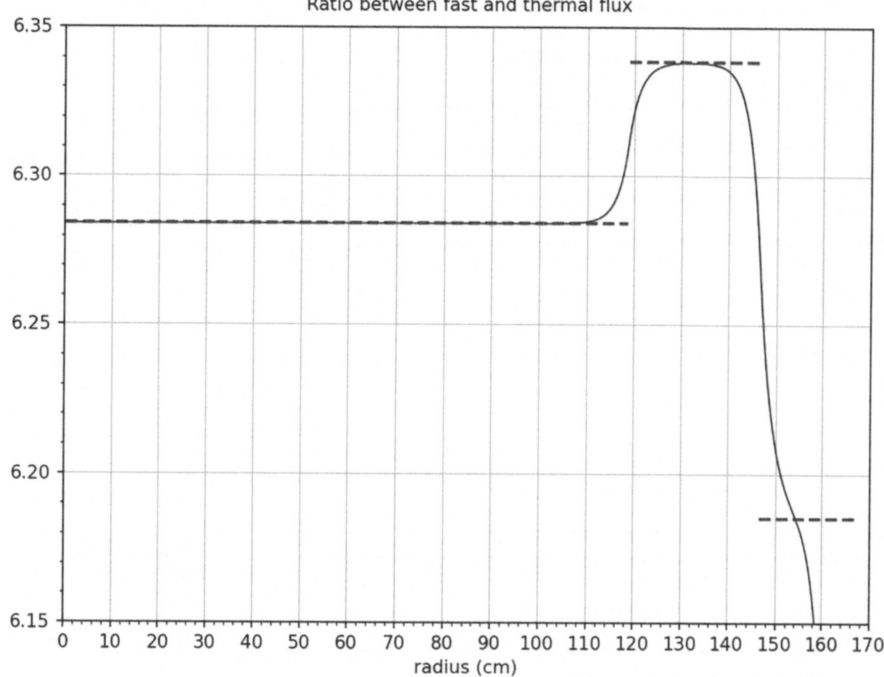

Fig. 13.10 Comparison between numerically calculated spectrum index $\gamma = \frac{\Phi_1}{\Phi_2}$ and the positive γ coupling factors: focus on fissile region

Table 13.7 Calculation of the coupling coefficients between fast and thermal flux and comparison with the curve Φ_1/Φ_2

Zones	$r < R_{\text{int},1}$	$R_{\text{int},1} \leq r < R_{\text{int},2}$	$R_{\text{int},2} \leq r < R_f$	$R_f \leq r < R_t$
$k_{\infty,\,\text{zone}}$	1.0001	1.0814	0.85544	0
Roots	$\mu_c^2 = 1.8307 \times 10^{-6}\ \text{cm}^2$	$\mu_c^2 = 1.4426 \times \text{cm}^2$	$-\mu_c^2 = -2.6182 \times \text{cm}^2$	$-\mu_w^2 = -2.6182\ \text{cm}^2$
Coupling	$\gamma_{[\mu_c^2]} = 6.2841$	$\gamma_{[\mu_c^2]} = 6.3384$	$\gamma_{[-\mu_c^2]} = 6.1853$	$\gamma_w = 0.96674$

13.5 Ratio Between Fast and Thermal Flux, in the Case of Zones with Various Enrichements

As an example, we go ahead studying the flattened critical reactor with a reflector, already presented in Sect. 13.4. The Figs. 13.9 and 13.10 enable to compare $\frac{\Phi_1}{\Phi_2}$, calculated with our 1D solver, and γ coupling coefficients.

Inside a fissile zone having $k_{\infty,\,\text{zone}} > 1$, the fast and thermal flux tend to a shape $J_0(\mu_c)$ and the ratio $\frac{\Phi_1}{\Phi_2}$ tends to $\gamma_{[\mu_c^2]}$, that can be calculated with Eq. 13.25. In this case, $\mu_c^2 D_{1,c}$ adds up the the removal of neutrons from the fast group, and represents the exportation of fast neutrons to adjacent regions.

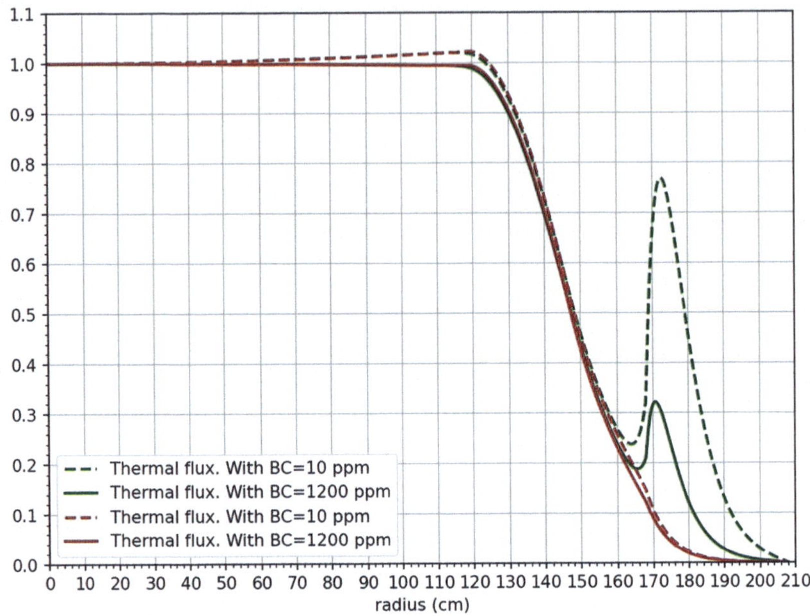

Fig. 13.11 Comparison between reflectors with $BC = 10\,\text{ppm}$ and $BC = 1200\,\text{ppm}$

Inside a fissile zone having $k_{\infty,\,\text{zone}} < 1$, the ratio $\frac{\Phi_1}{\Phi_2}$ tends to $\gamma_{[-\mu_c^2]}$, that can be calculated with Eq. 13.25 and $-\mu_c^2 D_{1,c}$ instead of $\mu_c^2 D_{1,c}$. In this case, $-\mu_c^2 D_{1,c}$ represents the importation of fast neutrons from adjacent regions.

In the reflector region, the ratio $\frac{\Phi_1}{\Phi_2}$ tends to γ_w, that can be calculated with Eq. 13.26 (Table 13.7).

13.6   Comparison Between Different Reflectors

In order to illustrate the influence of the reflector's properties on the flux shape, we take as an example the flattened critical reactor with a reflector, already presented in Sect. 13.4 (Fig. 13.11).

Another reflector is taken into account: water at 300°, 155 bar, and 10 ppm. The properties of the reflector are given in Table 13.8 of the fissile infinite cylinder are supposed to be unchanged.

Table 13.8 Input data for two groups: water reflector 300°, 155 bar, 10 ppm

$\tau_{1\to 2}$	$\Sigma_{r1,w}$ (Group 1)	$D_{1,w}$ (Group 1)	$\Sigma_{s12,w}$	L_2^2	$\Sigma_{r2,w}$ (Group 2)	$D_{2,w}$ (Group 2)
$42.932\,\text{cm}^2$	0.03 372 cm	1.4479 cm	0.033 363 cm	$20.908\,\text{cm}^2$	0.011678 cm	0.244 17 cm

13.7 Influence of the Thickness of Reflector on Flux Distribution

The thickness of the water reflector, in the case of the reactor studied in the Sect. 13.4 13.4, is reduced from 40 cm down to 0 cm by steps of –4 cm. The resulting calculations with the 1D solver are presented on Fig. 13.12. This figure shows that even if the reflector participates to the flattening, the flattenig is mainly due to the zoning of k_∞ of the different zones of the core.

At the interface with the reflector, the flux gradient is positive with a reflector, even for a width of 4 cm. This means that the net current of neutrons is negative, directed towards the fissile region. The layer of 10 cm at the interface contributes almost entirely to the back reflection towards the core of slowed down neutrons in the reflector. The additional water, above 10 cm has little effect.

13.8 Comparison Between Albedo and Reflector

We take the example of a flatenned reactor, with a water reflector (300°, 155 bar, 1200 ppm). Two calculations are performed with our 1D solver, and two neutron groups:

- A wall thickness of 40 cm, and vacuum boundary condition. This wall is almost equivalent to an infinite wall.
- Albedo boundary conditions, the coefficient α_f is calculated with the water reflector properties and an infinite thickness, and $\alpha_{th} = 1$.

  α_f enables to calculate:

$$A_f = \frac{(1 - \alpha_f)}{4\,(1 + \alpha_f)} \text{ and } B_f = \frac{D_{w,f}}{2}$$

Values are summarized in the Table 13.9.

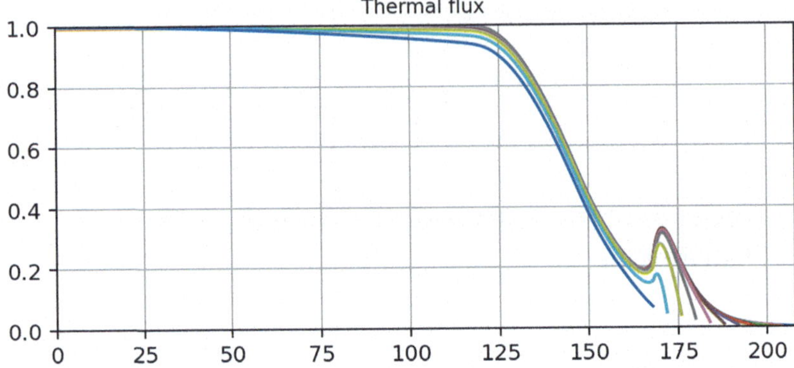

Fig. 13.12 Thermal flux, calculated with two groups, as a function of the radius and for different reflector's thicknesses (0–40 cm)

Table 13.9 Calculation of albedo: water reflector 300°, 155 bar, 1200 ppm

α_f	α_{th}	A_f	A_{th}	B_f	B_{th}
0.37468	1.0	0.11372	0.0	0.987 65	0.538 63

Figure 13.13 highlights that our proposed albedo coefficients calculation does not enable to calculate accurately the thermal flux in the vincinity of the boundary. The thermal flux calculated with a reflector has a positive slope in the vincinity of the boundary, which reflects the fact there are more thermal neutrons reflected back to the fissile region than thermal neutrons escaping the fissile region. And $\alpha_{th} = 1$ leads the a null slope at the boundary. However, fast fluxes, calculated with reflector and albedo, are almost the same.

13.9   Operating Point of the Reflector

The concept of core/reflector operating point point is developed in Marguet (2018, pp. 967–969).

The functions $\Phi_{1,c}(r)$, $\Phi_{1,w}(r)$, $\Phi_{2,c}(r)$ and $\Phi_{2,w}(r)$, established for each region (fissile region and water reflector) and each group in Sect. 13.3.3 enable to calculate the analytic functions for the diffusion currents $J_{1,c}(r)$, $J_{1,w}(r)$, $J_{2,c}(r)$ and $J_{2,w}(r)$.

$\Phi_1(r)$ and $\Phi_2(r)$ are defined by:

$$\begin{cases} \Phi_1(r) = \Phi_{1,c}(r) & \text{for } x \leq R_f \\[2mm] \Phi_1(r) = \Phi_{1,w}(r) & \text{for } x > R_f \end{cases}$$

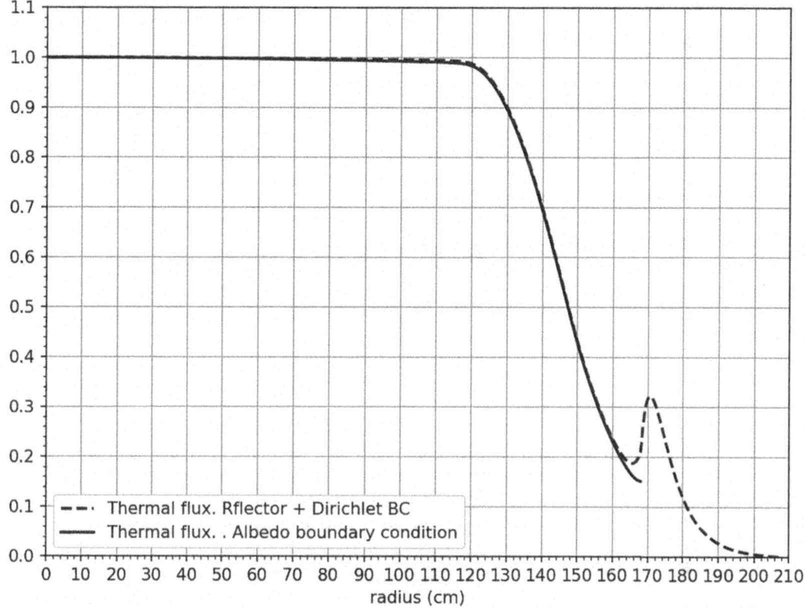

Fig. 13.13 Comparison between thermal flux calculated with reflector and albedo

And:

$$\begin{cases} \Phi_2(r) = \Phi_{2,c}(r) & \text{for } x \le R_f \\[2mm] \Phi_2(r) = \Phi_{2,w}(r) & \text{for } x > R_f \end{cases}$$

We remind the gradients of the Bessel functions depending on r in a cylindrical geometry:

$$\frac{d}{dr} J_0(r) = -J_1(r) \quad \frac{d}{dr} I_0(r) = I_1(r) \quad \frac{d}{dr} K_0(r) = -K_1(r)$$

The functions $\frac{J_{1,c}(r)}{\Phi_1(r)}$ and $\frac{J_{1,w}(r)}{\Phi_1(r)}$ are drawn as a function of the ratio $\frac{\Phi_2(r)}{\Phi_1(r)}$, on Fig. 13.14.

Similarly, the functions $\frac{J_{2,c}(r)}{\Phi_1(r)}$ and $\frac{J_{2,r}(r)}{\Phi_1(r)}$ are drawn as a function of the ratio $\frac{\Phi_2(r)}{\Phi_1(r)}$, on Fig. 13.15.

For the whole system, $\frac{\Phi_2(r)}{\Phi_1(r)}$ is bounded by $\gamma_{[\mu_c^2]}$ which is the value in the fissile region, away from the boundary with the reflector, and by γ_w which is the value within the reflector, away from the boundary with the fissile region.

Since there is one single homogeneous fissile region, $\frac{\Phi_2(r)}{\Phi_1(r)}$ is a decreasing function of r.

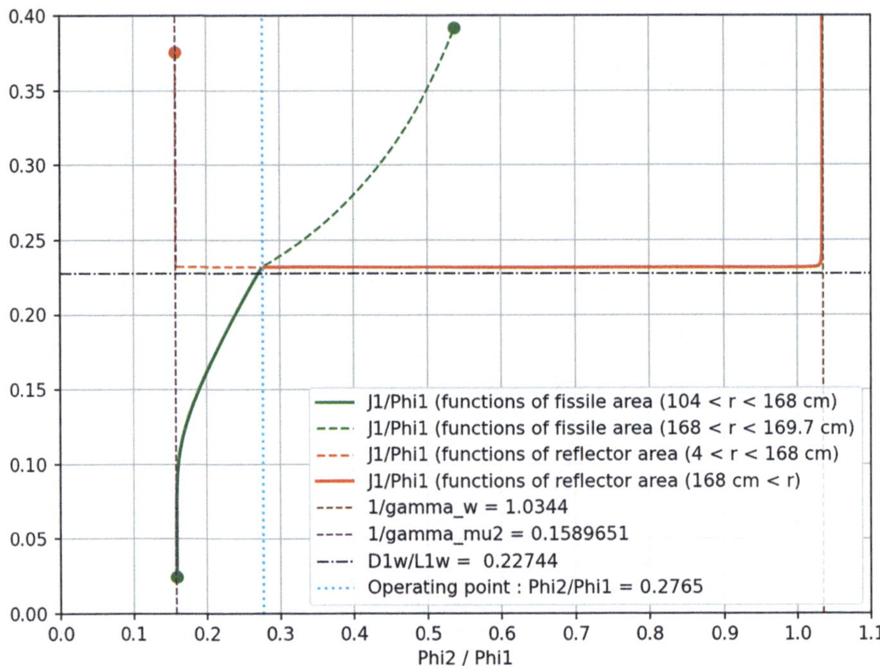

Fig. 13.14 Operating point of the reflector: fast group

Within a plane water wall, away from the interface with the core, the fast flux is a decreasing exponential $\Phi_1(x) = \Phi_1(0)\, e^{\frac{-x}{L_1}}$ and thus $\frac{J_{1,w}(x)}{\Phi_1(x)}$ is a constant: $\frac{D_{1,w}}{L_{1,w}}$.

In our considered cylindric reactor, due the low curvature ($R_f \gg D_{1,w}$), the situation in the cylindric water reflector is not far different from the case of the water wall; on Fig. 13.14 we can see that $\frac{J_{1,w}}{\Phi_1}$ is almost equal to $\frac{D_{1,w}}{L_{1,w}}$.

On Fig. 13.14, we can see that in the fissile region, on the left of the operating point, $\frac{J_{2,w}}{\Phi_1}$ is increasing, which means that $\Phi_1(r)$ is dropping sharper than an exponential function.

We can see that $\frac{J_{1,w}(r)}{\Phi_1(r)}$ and $\frac{J_{1,c}(r)}{\Phi_1(r)}$ are both always positive. The functions $\Phi_{1,c}(r)$ and $\Phi_{1,c}(r)$ have a negative gradient and the associated currents are directed towards the output of the system.

At the interface between the fissile region and the reflector, the flux and the current are continuous functions. Consequently, $\frac{J_{1,c}(R_f)}{\Phi_1(R_f)} = \frac{J_{1,w}(R_f)}{\Phi_1(R_f)}$. The intersection between the functions $\frac{J_{1,c}}{\Phi_1}$ and $\frac{J_{1,w}}{\Phi_1}$ both calculated for the whole system defines the value of $\frac{\Phi_2(R_f)}{\Phi_1(R_f)}$, at the interface: 0.2765.

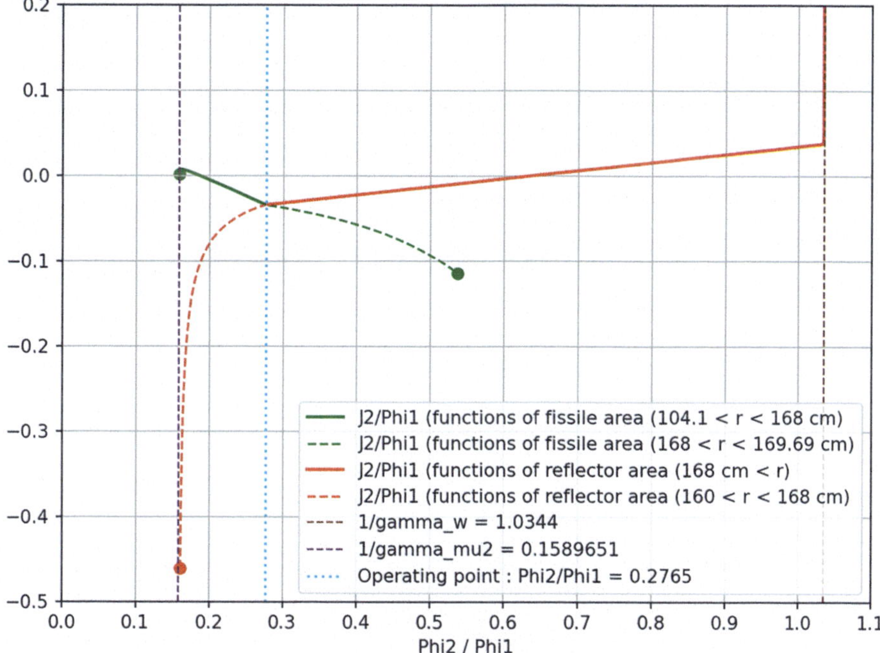

Fig. 13.15 Operating point of the reflector: thermal group

References

J. Bussac, P. Reuss, *Traité de neutronique: physique et calcul des réacteurs nucléaires avec application aux réacteurs à eau pressurisée et aux réacteurs à neutrons rapides*, vol. 25 (Hermann, 1985)

J.R. Lamarsh, *Introduction to Nuclear Engineering* (Prentice-Hall, 2001)

S. Marguet, *The Physics of Nuclear Reactors* (Springer International Publishing, 2018). ISBN 9783319595603. https://books.google.fr/books?id=9DlODwAAQBAJ

R.J.J. Stammler, M.J. Abbate, *Methods of Steady-State Reactor Physics in Nuclear Design* (1983)

Chapter 14
From Diffusion Theory to Point Kinetics

Abstract Reactor kinetics has the purpose to analyze the time-dependent behavior of the neutron population in a nuclear reactor in response to operational transients, abnormal conditions and accidents. This chapter presents the one group point kinetics theory. Despite the development of Multi-energy-group multi-region kinetics calculations, the topic of point kinetics is still relevant from a didactic point of view. What's more, point kinetics is still used in quasi static methods, and inverse point kinetics is the basis of reactivity meters and is used in the treatment of Dynamic Rod Weight Measurement tests. This chapter presents the merging of delayed neutron percursors into either 6 or 8 groups, and emphasises the assumptions that allow the kinetic equations to be written.

14.1 Delayed Neutrons

The delayed emission of neutrons at times ranging from a fraction of a second up to minutes provides a long time constant that slows the dynamic time response of a nuclear reactor, making it controllable by mechanical systems such as the control rods containing neutron absorbing materials. For this reason, excessively large uncertainties in the delayed neutron data used in reactor calculations (for determining β_{eff}) lead to costly conservatism in the design and operation of reactor control systems.

Dillmann et al. (2011, p. 8)

14.1.1 Emission of Delayed Neutrons

Fission fragments are highly unstable because of excess neutrons, and the usual mode of decay for such nuclei is β^- emission. Occasionaly, the β^- emission leads to the immediate emission of a neutron, in the case the excitation energy consecutive to the β^- decay is higher than the neutron separation energy.

H. Grard, *Diffusion of Neutrons in Nuclear Reactors*,
https://doi.org/10.1007/978-3-032-05088-5_14

The neutron emission is a result of β^- decay producing a daughter which has sufficient energy to throw off a neutron. The probability of a nuclide emitting a neutron as a result of a β^- decay is referred to as the P_n.

Bank (2000, p. 186)

Thus, the delayed neutron emitters exist on the extremely neutron rich side of the independent fission yield distribution.

Let's consider the case of the precursor ^{137}I.

A neutron may be emitted according to the following reactions:

$$^{137}_{53}\text{I} \xrightarrow{\beta^-} \,^{137}_{54}\text{Xe} \xrightarrow{n} \,^{136}_{54}\text{Xe} + \,^{1}_{0}\text{n}$$

The β^- decay of ^{137}I has a heat of reaction $Q_{\beta^-}(^{137}I) = 5880\,\text{keV}$.

The energy of separation of the last neutron of ^{137}Xe is $S_n(^{137}Xe) = 4025.56\,\text{keV}$.

$Q_{\beta^-}(^{137}I) > S_n(^{137}Xe)$, thus the emission of a neutron following the β^- is possible and may occur.

The heat of reaction of the β^- is split between the excitation energy of ^{137}Xe (according to discrete energy levels whose location and properties are governed by the rules of quantum mechanics), kinetic energy of the electron and energy of the anineutrino (variable part).

All the β^- decays leading to a sufficiently excited ^{137}Xe, with a level at least equal to $S_n(^{137}Xe)$ may lead to a neutron emission. In the case of neutron emission following a β^- decay, we can say than ^{137}I) undergoes a β^-, n decay.

β^-, n and β^-, γ are competing, we can define the probability:

$$P_n = \frac{\lambda_{\beta^-,n}}{\lambda_{\text{total}}}$$

In the case of ^{137}I: $P_n(^{137}I) = 7.14\%$.

The energy levels above the energy of separation are summarized in the table: With A: β^-, n and B: β^-, γ.

The Table 14.1 shows that the minimum level of excitation energy necessary for neutron emission is $4103.3\,\text{keV}$. In this case, the kinetic energy of the neutron is $4103.3 - 4025.56 = 77.74\,\text{keV}$. The maximal value of kinetic energy is $5408 - 4025.56 = 1382.44\,\text{keV}$. The energy spectrum of neutrons emitted from the precursor ^{137}I range from $77.74\,\text{keV}$ to $1382.44\,\text{keV}$. The example of ^{137}I illustrates that the spectrum of delayed neutrons, combining different precursors, in a lower range than the spectrum of prompt neutrons.

Table 14.1 Discrete energy levels of Xe-137 from https://www.nndc.bnl.gov/nudat3/getdataset.jsp?nucleus=137XE&unc=nds

Energy levels (keV)	XREF	Energy of γ (keV)
4025.73	B, E	396.03
		1788.18
		2491.29
		2807.76
4028.92	B	2092.80
		2155.64
		4028.92
4038.96	B	4038.96
4064.6	B	4064.6
4083.87	B	4083.87
4103.3	A, B	4103.3
4105.0	B	4105.0
4129.99	B	4129.99
4140.98	B	4140.98
4153	A	
4160.94	B	2943.1
		4160.94
4173.11	B	4173.11
4180.8	A	
4189.0	B	4189.0

Energy levels (keV)	XREF	Energy of γ (keV)
4199.1	A	4199.1
4211.6	B	4211.6
4260.4	B	4260.4
4270.3	B	4270.3
4276.53	B	2741.5
		4276.53
4282.6	A	
4288.1	B	4288.1
(..)	(..)	(..)
5170.2	B	5170.2
5179.7	A	
5208.9	A	
5230.3	A	
53 555	A	
53 795	A	
54 085	A	

14.1.2 *The Keepin Six-Group Model*

In practice, reactor kinetics codes consider a small set of 'lumped fission products' with a set of a representative decay constants and yields. This approach was pioneered by Keepin (Keepin et al. 1957), who found that a set of six 'lumped fission products' gave a good approximation to measurements.

Bank (2000, p. 189)

Keepin made measurements on the delayed neutrons from fast fission of six nuclides: ^{235}U, ^{233}U, ^{238}U, ^{239}Pu, ^{240}Pu and Th232 and from thermal fission of three nuclides: ^{235}U, ^{233}U and ^{239}Pu. The irradiation time of the fissile material was both "instantaneous" and "infinite" which means (Keepin et al. 1957):

- Short compared to the shortest delayed neutron period (the pulse was too short for any precursor to decay significantly during the irradiation.
- Long compared to the longest delayed neutron period (all precursors have reached equilibrium before the end of the irradiation).

Table 14.2 Delayed neutron half-lives, decay constants and yield for the thermal fission of U-235 (Keepin 1965, p. 90)

Group (index, i)	Half-life $T_{1/2}$, s	Decay constant λ, s^{-1}	Relative abundance $\alpha_i \equiv \frac{\beta_i}{\beta}$	Absolute group yield, %
1	55.72(128)	0.0124(3)	0.033(3)	0.052(5)
2	22.72(71)	0.0305(10)	0.219(9)	0.346(18)
3	6.22(23)	0.111(4)	0.196(22)	0.310(36)
4	2.30(9)	0.301(11)	0.395(11)	0.624(26)
5	0.610(83)	1.14(15)	0.115(9)	0.182(15)
6	0.230(25)	3.01(29)	0.042(8)	0.066(8)

Table 14.3 Delayed neutron half-lives, decay constants and yield for the thermal fission of Pu-239 (Keepin 1965, p. 90)

Group (index, i)	Half-life $T_{1/2}$, s	Decay constant λ, s^{-1}	Relative abundance $\alpha_i \equiv \frac{\beta_i}{\beta}$	Absolute group yield, %
1	54.28(234)	0.0128(5)	0.035(9)	0.021(6)
2	23.04(167)	0.0301(22)	0.298(35)	0.182(23)
3	5.60(40)	0.124(9)	0.211(48)	0.129(30)
4	2.13(24)	0.325(36)	0.326(33)	0.199(22)
5	0.618(213)	1.12(39)	0.086(29)	0.052(18)
6	0.257(45)	2.69(48)	0.044(16)	0.027(10)

After irradiation, the samples of fissile material were transfered by a pneumatic system for the counting of the delayed neutrons activity.

> It may be assumed that the decay of delayed neutron activity with time can be represented by a linear superposition of exponential decay periods. The general exponential decay problem was linearized (by Taylor's series expansion) and coded for iterative least-squares analysis (..). Six exponential periods were determined to be necessary and sufficient for optimum least-squares fit to the data.
>
> Keepin et al. (1957, p. 1046)

In the case of the thermal fission of ^{235}U (99.9%), Keepin established the following values, given on Table 14.2. The values established by Keepin In the case of the thermal fission of ^{239}Pu (99.8%) are significatively different: The values established by Keepin in the case of the thermal fission of ^{238}U (99.98%) are (Tables 14.3 and 14.4).

The Keepin's values show that $52 + 346 + 310 + 624 + 182 + 66 = 1580$ delayed neutrons are produced for $100\,000$ thermal fissions of ^{235}U. The total fraction of delayed neutrons can be calculated:

Table 14.4 Delayed neutron half-lives, decay constants and yield for the fast fission of U-238 (Keepin 1965, p. 86)

Group (index, i)	Half-life $T_{1/2}$, s	Decay constant λ, s^{-1}	Relative abundance $\alpha_i \equiv \frac{\beta_i}{\beta}$	Absolute group yield, %
1	52.38(129)	0.0132(3)	0.013(1)	0.054(5)
2	21.58(39)	0.0321(6)	0.137(2)	0.564(25)
3	5.00(19)	0.139(5)	0.162(20)	0.667(87)
4	1.93(7)	0.358(14)	0.388(12)	1.599(81)
5	0.490(23)	1.410(67)	0.225(13)	0.927(60)
6	0.172(9)	4.020(214)	0.075(5)	0.309(24)

$$\beta = \frac{1580}{100000 \times \overline{\nu}} = 650 \, \text{pcm}$$

where $\overline{\nu} = 2.4355$ is the average total number of neutrons produced by the fission of ^{235}U (see Sect. 3.3.2).

In the case of thermal fission of ^{239}Pu, $\overline{\nu} = 2.8836$.

$$\beta = \frac{610}{100000 \times \overline{\nu}} = 211.5 \, \text{pcm}$$

In the case of fast fission of ^{238}U, $\overline{\nu} = 2.819$.

$$\beta = \frac{4120}{100000 \times \overline{\nu}} = 1461.5 \, \text{pcm}$$

The relative abundance given in Table 14.2 enables ta calculate the delayed neutron fraction of each group:

$$\beta_1 = \beta \times \alpha_1 = 21.45 \, \text{pcm}; \beta_2 = 142.35 \, \text{pcm}; \beta_3 = 127.4 \, \text{pcm};$$

$$\beta_4 = 142.35 \, \text{pcm}; \beta_5 = 74.75 \, \text{pcm}; \beta_6 = 27.3 \, \text{pcm}$$

The mean number of neutrons emitted by a fission event, $\overline{\nu}$, is the sum of a number of delayed neutrons and prompt neutrons (overline symbol is removed to simplify the writing):

$$\nu = \nu_p + \nu_d$$

ν_d is relative to one fission event, and can be deduced directly from the sum of absolute group yields: $\nu_d = 0.0158$.

14.1.3   Relationship Between the Groups and the Nuclei

The first group is often referred to ^{87}Br because this precursor has a probability of decay $\lambda = 0.01245547\,\mathrm{s}^{-1}$ that is almost equal to $\lambda_1 = \frac{\ln 2}{T_{1/2}} = 0.0124398\,\mathrm{s}^{-1}$. However, the fission yield of this nuclide is $\gamma(^{87}Br) = 0.0127$, and $P_n(^{87}Br) = 2.6\%$. Thus 100 000 fissions produce 1270 nuclei of ^{87}Br, that produce $1270 \times 0.026 = 33.01$ delayed neutrons. The identification of the first group as the precursor ^{87}Br is not relevant since, for 100 000 fissions, the first group produces 52 delayed neutrons and ^{87}Br produces 33.01 delayed neutrons.

The fission products concentrations and the precursors concentrations must not be confused. All the precursors C_i are artifacts, and due to their mathematical construction, they have a branching ratio of 100% for β^-, n. The branching ratio of ^{87}Br for β^-, n is 2.6%. Consequently, at equilibrium, the concentration of ^{87}Br nuclei is $\frac{100}{2.6}$ times greater than the concentration of C_1 percursors.

The second group can be referred to isotopes ^{88}Br $(\lambda_{88}\mathrm{Br} = 0.042420\mathrm{s}^{-1})$ and ^{137}I$(\lambda_{137}\mathrm{I} = 0.028292\mathrm{s}^{-1})$.

14.1.4   One Limitation of the Six-Group Model

The presence of more than one fissile isotope (for example thermal fission of ^{235}U, ^{239}Pu and fast fission of ^{238}U) can be described approximatively by using a weighted-average set of, not only delayed-neutron yields, but also half-lives. Since the different types of fission to be taken into account are characterized by different sets of precursor half-lives, the approach is approximate and cannot be rigorous.

The buildup of fissile isotopes (^{239}Pu, ^{241}Pu) during the operating cycle of the reactor causes a shift in effective yields, and the above approximation has to be updated.

14.1.5   Eight-Group Model

A new eight-group reference structure for group constants has been defined. This work was agreed to be an important task by the Advisory Committee of the Colloquy on Delayed Neutron Data meeting held at Obninsk (Russia) on 9–10 April 1997. This uses the same set of eight-group half-lives for all fissioning systems, with the half-lives adopted for the three longest-lived groups corresponding to the three dominant long-lived precursors: ^{87}Br, ^{137}I, ^{88}Br.

Rudstam et al. (2002, p. 8)

The new delayed neutrons group data structure provides two benefits (Rudstam et al. 2002, p. 41):

- There is a single set of precursor half-lives (for all fissile isotopes and incident neutron energies). This is interesting in a the case of a PWR 1300 MWe at the beginning of an operating cycle, in which for example the thermal fission of ^{235}U represent 67% of the total number of fissions, produced, the thermal fissions of ^{239}Pu represent 26% and the fast fissions of ^{238}U represent 7%. This last value is close and consistent with the fast fission factor given as an example in Sect. 13.1.6. Since $\lambda_i(^{235}\mathrm{U}) \neq \lambda_i(^{239}Pu)$, the determination of parameters for the whole core (six λ_i and six β_i) is not immediate, because it is not possible to just add up the β_i of each nuclide contributing to fissions.
- The description of the delayed neutrons emission from the longest-lived precursors is more consistent and avoids distortions in the reactivity measurement analysis.

14.2 Effective Delayed Neutron Fraction

It is necessary to distinguish between the fractions delayed neutrons over fission neutrons in fission events, in reactor steady states and in transient processes.

The fraction of delayed neutrons over fission neutrons in fission events has been introduced in Sect. 14.1.2. However, delayed neutrons do not have the same properties and the same effect on the chain reaction as prompt neutrons released directly from fission. This is why some corrected fractions β_{eff} and $\beta_{i,\mathrm{eff}}$ should be introduced in Eqs. 14.13 and 14.15. In simple terms, we can say that if delayed neutrons have a higher effectiveness in generating fissions, then $\beta_{\mathrm{eff}} > \beta$.

The effective delayed neutrons fraction is one of the key concepts in the physics of nuclear reactors. The corresponding parameter β_{eff} largely determines the nuclear reactor dynamics as a measure of reactivity and an expression of prompt criticality.

The Sect. 14.7.6 highlights that during transients ($\rho \neq 0$) and resulting (14.15), the part of delayed neutrons among the total production of neutrons is not anymore β.

14.2.1 The Importance Factor

Since delayed neutrons are emitted at lower energies than prompt neutrons, with an initial energy in the range 0.1–1 MeV and an average energy of 0.4 MeV, they traverse a smaller energy range to become thermal neutrons, and are on the average subjected to less epithermal leakage and capture Hetrick (1971, p. 8). On the other hand, delayed neutrons are less likely to cause fast fission on ^{238}U, because their average energy is less than the minimum required for fast fission to occur, since the process of fast fission occurs at energies higher than 1 MeV, as specified in (12.5). These two effects (lower fast fission factor and higher fast non-leakage probability for delayed neutrons) tend to counteract each other.

However, in two energy groups core calculation, delayed neutrons and prompt neutrons are included in the same fast group and thus non differencied. In order to take into account the fact that they do not have the same effect on the chain reaction, a corrective factor γ, called importance factor, is calculated. If delayed neutrons have a higher importance, then $\gamma > 1$. The effective delayed neutron fraction enables to differentiate delayed neutrons from prompt neutrons. A further refinement may take into account that each delayed-neutron group has a different neutron emission spectrum, which results in a different correction factor for each β_i: in this case, we refer to family-dependent or group-dependent effective delayed fission neutron fractions.

The effective delayed neutron fraction is defined as the product of the average delayed neutron fraction and the importance factor γ:

$$\beta_{\text{eff}} = \gamma \beta \quad \text{and} \quad \beta_{i,\text{eff}} = \gamma_i \beta_i \quad \text{if each group has its own corrective factor}$$

where γ is the importance factor.

14.2.2   Calculation of the Importance Factor

For evaluating effective delayed neutron fractions, various approaches exist (Zwermann et al. 2021, p. 2).

A method usually applied is based on an adjoint and spectrum weighting of the delayed neutron production rate.

A method, called prompt k ratio method requires two forward transport calculations, one performed with unmodified reaction data, and the other with a set of microscopic reactions in which delayed neutrons emission is suppressed. With this method, β_{eff} is given by:

$$\beta_{\text{eff}} = 1 - \frac{k_p}{k}$$

where k is the effective multiplication factor of the system all fission neutrons, and k_p is the prompt multiplication factor, considering only prompt fission neutrons.

The Monte Carlo approach (Meulekamp and Van Der Marck 2006, p. 144) introduces neutrons with properties (r, Ω, E) in a critical system at zero power. The energy of emitted neutrons E is different for prompt and delayed neutrons. Each introduced neutron will produce other neutrons by inducing fissions. These neutrons will in turn cause fission and thereby lead to further neutrons, etc. During the simulation of the Monte Carlo history of an initial neutron and all the secondary particles that are produced by it, it is possible to count the total number of fissions along that entire fission chain, which is proportional to the iterated fission probability. Using the number of fissions counted for the initial introduction of prompt and delayed neutrons, it is possible to calculate the corrective factor.

14.2.3 Effective Value of β and Reactors

The effective value of β is different for different reactor types, even if they have the same nuclear material, because of the different epithermal leakage and capture probabilities for prompt and delayed neutrons, while the additional refinement of slightly different effectiveness for each delayed neutron group is a much smaller effect that is generally ignored (Hetrick 1971, p. 13).

In the case of a small reactor with highly enriched fuel, the enhancement of the non leakage probability predominates and $\gamma > 1$. For example, in the case of the high flux reactor operated by ILL, which has single fuel assembly with ^{235}U enriched at 93%, $\beta_{\text{eff}} = 708$ pcm. In this reactor, the part of fast fission in negligeable, the importance factor γ can be estimated to $\gamma = \frac{708}{650} = 1.089$.

In a PWR, γ is slightly lower than 1, for example $\gamma = 0.98$. A typical value of the effective delayed neutron fraction is $\beta_{\text{eff}} = 589$ pcm.

The Cabri research reactor has a small core with a square section of 60 cm, with PWR like fuel assemblies enriched at 6% and a negligible burn-up. The importance factor is $\gamma > 1$ and $\beta_{\text{eff}} = 720$ pcm

Please note that in the remainder of this chapter, β_{eff} will be written as β.

14.3 Preliminary Considerations Before Establishing the Equations of Point Kinetics

14.3.1 Separation of Time, and Space Energy

The assumption that time and space—energy variables can be separated, as shown in Eq. 14.1 is a necessary condition for establishing the point kinetics equations. This assumption means that the flux shape remains almost constant as a function of time. It is relevant at the condition there aren't any spatial disruptions of the power distribution. The distortion of the spatial shape due to insertion or withdrawal of absorbing rods can be quite significant, specially in the case of large reactors.

For monokinetic neutrons, we can write the variable separation as follows:

$$\Phi(\boldsymbol{r}, t) = n(t)\, g(\boldsymbol{r}) \tag{14.1}$$

where $n(t)$ has the dimension of neutron density (n/cm^3) and $g(\boldsymbol{r})$ has the dimension of a motion (cm/s).

14.3.2   Neutron Lifetime and Generation Time

We remind that neutron lifetime θ is the mean time from the birth of a prompt fission neutronto the loss (absorption or leakage) of the neutron in the reactor. The expression for θ in the one speed neutron approximation is:

$$\theta = \frac{1}{v\,(\Sigma_a + DB_g^2)} \tag{14.2}$$

where v is the speed of neutrons, Σ_a is the macroscopic absorption cross section, and D and B_g^2 are the neutron diffusion coefficient and geometrical buckling, respectively. DB_g^2 is added to the absorption Σ_a and represents the loss of neutrons due to leakageout of the core. It depends on the geometry of the core (shape and dimensions).

For an infinite reactor, then the lifetime of neutrons becomes $l_\infty = \dfrac{1}{v\,\Sigma_a} = \dfrac{\lambda_a}{v}$.

Where λ_a is the mean free path in the infinite absorbing and scattering medium. Lifetime is obviously equal to the path divided by the speed.

The diffusion length L_D can be introduced in Eq. 14.2, which becomes:

$$\theta = \frac{1}{v\Sigma_a(1 + L_D^2 B_g^2)} \tag{14.3}$$

If the reactor is critical, then $B_g^2 = B_m^2$

At criticality, the value of lifetime is called **generation time** l:

$$l = \frac{1}{v\Sigma_a(1 + L_D^2 B_m^2)} \tag{14.4}$$

where $B_m^2 = \dfrac{k_\infty - 1}{L_D^2}$

At criticality: $\nu\,\Sigma_f = \Sigma_a(1 + L_D^2 B_g^2)$. Thus we can write that:

$$
\begin{aligned}
l &= \frac{1}{v\nu\Sigma_f} \\
&= \frac{1 \times \Sigma_a(1 + L_D^2 B_g^2)}{v\Sigma_a(1 + L_D^2 B_g^2) \times \nu\Sigma_f} \\
&= \frac{\theta}{k_{\text{eff}}}
\end{aligned}
$$

The generation time l, which is the value of lifetime at criticality, can be defined by $\frac{\theta}{k_{\text{eff}}}$.

A the condition that reactor is close to criticality, generation time can be used instead of lifetime.

We can estimate the generation time of neutrons in a PWR, with the values established for an aquivalent one group model, in the Sect. 10.5.2, Table 10.4. This homogeneous infinite cylinder is critical in Dirichlet conditions, without reflector, and with a fissile radius equal to 167.9 cm.

At 579.15 K, the mean velocity of the neutrons is $\overline{v} = 3486.7\,\mathrm{m} \cdot \mathrm{s}^{-1}$.

$$l = \frac{1}{3486.7 \times 100 \times \left(0.17954370 + 10.05445 \times \left(\frac{2.4048}{167.9}\right)^2\right)}$$
$$= 1.5974 \times 10^{-5}\,\mathrm{s}$$

Or also:

$$l = \frac{1}{3486.7 \times 100 \times 0.18161}$$
$$= 1.5974 \times 10^{-5}\,\mathrm{s}$$

14.4   Kinetics and Diffusion

14.4.1   Equation Without Delayed Neutrons nor independent source

For monokinetic neutrons, we can write that:

$$\frac{1}{v}\frac{\partial}{\partial t}\Phi(\boldsymbol{r},t) = D\Delta\Phi(\boldsymbol{r},t) - \Sigma_a\Phi(\boldsymbol{r},t) + \nu\Sigma_f\Phi(\boldsymbol{r},t)$$

Since $k_\infty = \dfrac{\nu\Sigma_f}{\Sigma_a}$, we can write:

$$\frac{1}{v}\frac{\partial}{\partial t}\Phi(\boldsymbol{r},t) = D\Delta\Phi(\boldsymbol{r},t) + (k_\infty - 1)\Sigma_a\Phi(\boldsymbol{r},t) \tag{14.5}$$

We assume that time and space - energy variables can be separated:

$$\Phi(\boldsymbol{r},t) = g(\boldsymbol{r})n(t)$$

The dimensions of $n(t)$ and $g(\boldsymbol{r})$ are cm^{-3} and $\mathrm{cm} \cdot \mathrm{s}^{-1}$ respectively. The Eq. 14.5 can now be written as:

$$\frac{1}{n(t)}\frac{1}{v}\frac{dn(t)}{dt} = D\frac{1}{g(\boldsymbol{r})}\Delta g(\boldsymbol{r}) + (k_\infty - 1)\Sigma_a \tag{14.6}$$

The reactor is assumed to be critical, or close to be critical. Thus the shape of the flux $g(r)$ is the fundamental mode and we can write that:

$$\Delta g(r) + B_g^2 g(r) = 0$$

The space-independent kinetic equations govern transient response of the system "as a whole", assuming the fundamental mode prevails throughout the course of the transient (i.e., assuming that any change in *shape* of the neutron flux, as distinct from its amplitude, is a higher order effect).

Keepin (1965, p. 166)

Then Eq. 14.6 becomes:

$$\frac{1}{n(t)}\frac{1}{v}\frac{dn(t)}{dt} = -DB_g^2 + (k_\infty - 1)\Sigma_a$$

$$= \Sigma_a \left(-\frac{D}{\Sigma_a}B_g^2 + k_\infty - 1\right)$$

$$= \Sigma_a \left(k_\infty - 1 - L_D^2 B_g^2\right)$$

$$= \Sigma_a \left(1 + L_D^2 B_g^2\right)\frac{k_\infty - 1 - L_D^2 B_g^2}{1 + L_D^2 B_g^2}$$

$$= \Sigma_a \left(1 + L_D^2 B_g^2\right)\left(\frac{k_\infty}{1 + L_D^2 B_g^2} - 1\right)$$

$$\frac{1}{n(t)}\frac{dn(t)}{dt} = \underbrace{v\Sigma_a(1 + L_D^2 B_g^2)}_{\frac{1}{\theta}}\underbrace{\left(\frac{k_\infty}{1 + L_D^2 B_g^2} - 1\right)}_{k_{\text{eff}}}$$

$$\frac{1}{n(t)}\frac{dn(t)}{dt} = \frac{k_{\text{eff}} - 1}{\theta} \tag{14.7}$$

Equation 14.7 is a simplified, established without delayed neutrons, space-independent kinetics equation, also called point-reactor. The use of space-independent kinetics equations implicitly assumes that the spatial distribution of flux is unchanged, i.e., that essentially the fundamental-mode flux distribution predominates throughout a given experiment (Keepin 1965, p. 243).

14.4.2 *Equation with Delayed Neutrons and an Independent Source*

For monokinetic neutrons, we can write that:

$$\frac{1}{v}\frac{\partial}{\partial t}\Phi(r,t) = D\Delta\Phi(r,t) - \Sigma_a\Phi(r,t) + S(r,t) \tag{14.8}$$

where the source S has three components:

- Prompt neutrons of fission
- Delayed neutrons $\beta = \sum_i \beta_i$, where $i = 6$ or $i = 8$

- Independent source of neutrons: independent means that neutrons are not produced by neutron induced fissions (see Sect. 14.8.1)

β is the total delayed neutrons fraction. The source S is equal to:

$$S(r,t) = \underbrace{k_\infty(1-\beta)\Sigma_a\Phi(r,t)}_{\substack{\text{prompt neutrons} \\ \text{of fission}}} + \sum_i \lambda_i \cdot \underbrace{G_i(r,t)}_{\substack{\text{density of} \\ \text{precursors}}} + S_{\text{indep}(r,t)}$$

$\lambda_i \cdot G_i(r,t)$ is a number of neutrons /cm^3/s.

The reactor is assumed to be critical, or close to be critical. Thus the shape of the flux $g(r)$ is the fundamental mode and we can write that:

$$\Delta g(r) + B_g^2 g(r) = 0$$

Thus Eq. 14.8 becomes:

$$\frac{1}{v}\frac{\partial}{\partial t}\Phi(r,t) = \left[(k_\infty(1-\beta)-1)\Sigma_a - DB_g^2\right]\Phi(r,t) + \sum_i \lambda_i \cdot G_i(r,t) + S_{\text{indep}(r,t)}$$

$$\tag{14.9}$$

We remind that $\Sigma_a L_D^2 = D$, then we can write:

$$(k_\infty(1-\beta)-1)\Sigma_a - DB_g^2 = (k_\infty(1-\beta)-1)\Sigma_a - \Sigma_a L_D^2 B_g^2$$

$$= \Sigma_a\left[k_\infty(1-\beta) - (1 + L_D^2 B_g^2)\right]$$

$$= \Sigma_a(1 + L_D^2 B_g^2)\left(\frac{k_\infty(1-\beta)}{1 + L_D^2 B_g^2} - 1\right)$$

$$= \Sigma_a(1 + L_D^2 B_g^2)(k_{\text{eff}}(1-\beta) - 1)$$

An thus, Eq. 14.9 can be written as:

$$\frac{1}{v}\frac{\partial}{\partial t}\Phi(\boldsymbol{r},t) = \Sigma_a \Phi(\boldsymbol{r},t)(1 + L_D^2 B_g^2)(k_{\text{eff}}(1-\beta)-1) + \sum_{i=1}^{6}\lambda_i \cdot G_i(\boldsymbol{r},t) + S_{\text{indep}}(\boldsymbol{r},t) \quad (14.10)$$

From Eq. 14.3, we can introduce θ and then Eq. 14.10 becomes:

$$\frac{\partial}{\partial t}\Phi(\boldsymbol{r},t) = \frac{k_{\text{eff}}(1-\beta)-1}{\theta}\Phi(\boldsymbol{r},t) + \sum_{i=1}^{6} v\lambda_i \cdot G_i(\boldsymbol{r},t) + v\,S_{\text{indep}}(\boldsymbol{r},t) \quad (14.11)$$

We assume that time and space - energy variables can be separated:

$$\Phi(\boldsymbol{r},t) = g(\boldsymbol{r})n(t)$$

We assume that the spatial distribution of precursors has the same shape, defined by $g(\boldsymbol{r})$, than the flux. Let's introduce a new function $C_i(t)$ which has the dimension of a density /cm^3, and then we can write:

$$vG_i(\boldsymbol{r},t) = g(\boldsymbol{r})C_i(t)$$

In the same manner, we assume that the independent source has the shape of the fundamental mode and a constant magnitude. This assumption enables to write:

$$v\;\underbrace{S_{\text{indep}}(\boldsymbol{r},t)}_{\text{n/cm}^3/\text{s}} = \underbrace{g(\boldsymbol{r})}_{\text{cm/s}}\cdot\;\underbrace{S_0}_{\text{n/cm}^3/\text{s}}$$

Then Eq. 14.11 becomes:

$$g(\boldsymbol{r})\frac{dn(t)}{dt} = g(\boldsymbol{r})\frac{k_{\text{eff}}(1-\beta)-1}{\theta}n(t) + \sum_{i}\lambda_i \cdot g(\boldsymbol{r})C_i(t) + g(\boldsymbol{r})S_0 \quad (14.12)$$

From Eq. 14.12, a factorisation and then an elimination of $g(\boldsymbol{r})$ can be performed. In a similar way, an equation can be established for precursors.

This enables us to establish a system of coupled differential equations:

$$\begin{cases} \dfrac{dn(t)}{dt} = \dfrac{k_{\text{eff}}(1-\beta)-1}{\theta}n(t) + \displaystyle\sum_{i}\lambda_i \cdot C_i(t) + S_0 \\[2ex] \dfrac{dC_i(t)}{dt} = \beta_i\dfrac{k_{\text{eff}}}{\theta}n(t) - \lambda_i \cdot C_i(t) \end{cases} \quad (14.13)$$

The system of Eq. 14.13 is coined point kinetics, because from a formal point of view, it is the system that could be established if the reactor was considered as a single point of the space, characterized by with k_{eff}, β_i, λ_i and a local quantity of neutrons $n(t)$.

The system of Eq. 14.13 can be written on a different form, by showing the reactivity ρ instead of k_{eff}, and a new variable $l = \frac{\theta}{k_{\text{eff}}}$ calculated from the lifetime of neutrons θ:

$$\begin{cases} \dfrac{dn(t)}{dt} = \dfrac{k_{\text{eff}} - 1 - k_{\text{eff}}\beta}{k_{\text{eff}} \times \frac{\theta}{k_{\text{eff}}}} n(t) + \sum_i \lambda_i \cdot C_i(t) + S_0 \\[2ex] \dfrac{dC_i(t)}{dt} = \dfrac{\beta_i}{\frac{\theta}{k_{\text{eff}}}} n(t) - \lambda_i \cdot C_i(t) \end{cases} \tag{14.14}$$

$$\begin{cases} \dfrac{dn(t)}{dt} = \dfrac{\rho - \beta}{l} n(t) + \sum_i \lambda_i \cdot C_i(t) + S_0 \\[2ex] \dfrac{dC_i(t)}{dt} = \dfrac{\beta_i}{l} n(t) - \lambda_i \cdot C_i(t) \end{cases} \tag{14.15}$$

We remind that the variable $l = \frac{\theta}{k_{\text{eff}}}$ defines the generation time, which is the life time of neutrons at criticality. In Eqs. 14.13 and 14.15, the effective delayed neutrons fractions should be used instead of β and β_i, as explained in Sect. 14.2.

Note that $n(t)$ may have any units, as long as the units of n and c are the same. Since the neutron density is assumed to have a fixed spatial distribution, n may be regarded as any integral or volume-averaged property that is proportional to the instantaneous neutron density at some point in the reactor, such as the total number of neutrons, fission rate, power, or average power density.

Hetrick (1971, p. 7)

14.4.3 Simplified Neutron Cycle Illustrating Point Kinetics

It is interesting to represent the kinetics coupled Eq. 14.15 in the form of a neutrons cycle diagram, as show on Fig. 14.1. During a steady state with $\rho = 0$, $n(t)$ is constant and the balance of neutrons can be described with rates of change per unit time:

- Loss of neutrons $\dfrac{-n(t)}{l}$

- Prompt neutrons production $n(t)(1 - \beta)\dfrac{1}{l}$

- Precursors production $\beta\dfrac{n(t)}{l}$

- Delayed neutrons production $\sum_i \lambda_i C_i = \beta\dfrac{n(t)}{l}$

We can see that the loss of neutrons is balanced by the sum of prompt neutrons production and delayed neutrons production. If ρ becomes different from 0, then this balance is broken.

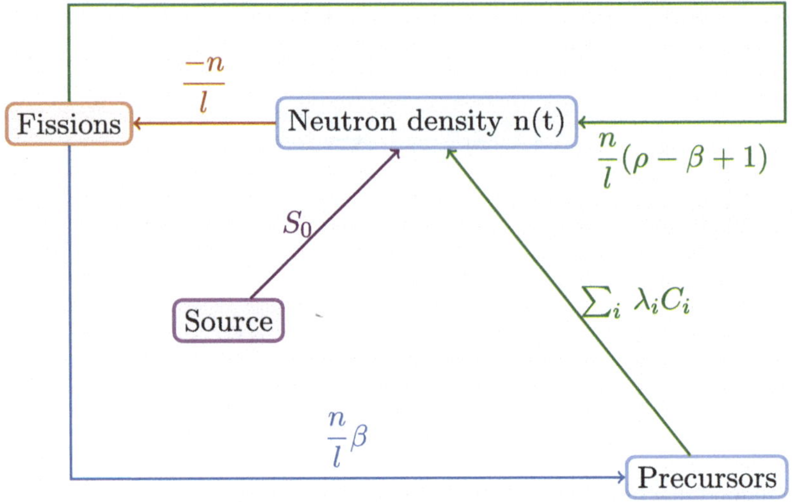

Fig. 14.1 Prompt and delayed neutrons cycle, indicating rates of change per unit time

14.5   Case of a Constant Reactivity Without Source. The Nordheim Function

14.5.1   *Steady State at Criticality*

At criticality, we can write that:

$$\sum_i \frac{dC_i(t)}{dt} + \frac{dn(t)}{dt} = 0 \tag{14.16}$$

The Eq. 14.16 shows that if the precursors are in excess by comparison with the neutrons, then the precursor population decreases and the neutron population increases until an equilibrium, or asymptoteic solution is achieved. This equilibrium is characterized by a relationship between neutrons and precursors:

$$C_{i,0} = \frac{\beta_i}{l \times \lambda_i} n_0 \tag{14.17}$$

At criticality and at equilibrium, the production of neutrons by decay of precursors does exactly compensate for the deficit of prompt neutrons between two generations, which is the fraction β of the neutron population. It is important to notice that a critical reactor is undercritical for prompt neutrons, which can be written $\rho_p = -\beta$.

14.5.2 *Evolution of the System from Criticality to a Constant Reactivity $\rho \neq 0$*

14.5.2.1 The Equation of Nordheim

The problem to solve is a system of differential equations with constant coefficients. The reader will find, in Appendix D, a demonstration of the general solution, and a method to establish semi analytical solutions.

We assess solutions in the form $n(t) = n_0 e^{\omega t}$ and $C_i(t) = C_{i,0} e^{\omega t}$, and we introduce these functions in the system of Eq. 14.15:

$$\begin{cases} n_0 \omega e^{\omega t} = \dfrac{\rho - \beta}{l} n_0 e^{\omega t} + \sum_i \lambda_i \cdot C_{i,0} e^{\omega t} \\[3mm] C_{i,0} \omega e^{\omega t} = \dfrac{\beta_i}{l} n_0 e^{\omega t} - \lambda_i \cdot C_{i,0} e^{\omega t} \end{cases} \tag{14.18}$$

The second equation of the system (14.18) gives:

$$C_{i,0} = \frac{\beta_i}{l(\omega + \lambda_i)} n_0$$

And then the first equation becomes:

$$n_0 \omega = \frac{\rho - \beta}{l} n_0 + n_0 \sum_i \frac{\lambda_i \beta_i}{l(\omega + \lambda_i)}$$

$$\rho = \omega l + \beta - \sum_i \frac{\lambda_i \beta_i}{\omega + \lambda_i}$$

$$= \omega l + \sum_i \frac{\beta_i(\omega + \lambda_i) - \lambda_i \beta_i}{\omega + \lambda_i}$$

$$= \omega l + \sum_i \frac{\omega \beta_i}{\omega + \lambda_i}$$

And then:

$$\rho = \omega \left[l + \sum_i \frac{\beta_i}{\omega + \lambda_i} \right] \tag{14.19}$$

Another form of the same equation is given in (14.20):

$$\rho = \beta + l\,\omega - \sum_i \frac{\beta_i \lambda_i}{\omega + \lambda_i} \tag{14.20}$$

The equation written in the two forms (14.19) and (14.20) is also called inhour equation. The name originated in the early days of reactor technology, one inhour being defined as the amount of positive reactivity which corresponds to an exponential power rise having a period (time constant) of one hour (Hetrick 1971, p. 20).

The reactor period, or e-folding time, is defined as the time required for the neutron density to change by a factor $e = 2.718$.

14.5.2.2   Graphic Representation of the Equation of Nordheim

The graphic representation of the function of Nordheim $\rho = f(\omega)$ with 6 groups (example of a PWR) is shown on Fig. 14.2 with a semi-log graph for ω axis and on Fig. 14.3 with a linear graph. The linear graph has the advantage to highlight the linear oblic asymptote, and the disadvantage that the vertical asymptotes are very tight and cannot be easily distinguished.

On both figures, the 6 vertical asymptote, located at the abscissas $-\lambda_i$, are drawn. The oblic asymptote intersects the ω axis at $\omega_6 = \frac{-\beta}{l}$ and the ρ axis at β. The slope of this oblic asymptote is l. The curve passes through the point ($\omega = 0, \rho = 0$).

The Nordheim function of the corresponding 1 group model (see Sect. 14.6) is drawn on each figure.

14.5.2.3   General Solution of the Equation of Nordheim

The graphic representation of the function of Nordheim shows that in the case of 6 groups, the are 7 values for ω whatever is the constant reactivity (and 9 values for ω in the case of 8 groups).

The general solution is a sum of exponentials:

$$\begin{cases} n(t) = \sum_{i=1}^{6 \text{ or } 8} a_i e^{\omega_i t} \\[2ex] C_i(t) = \sum_{i=1}^{6 \text{ or } 8} b_i e^{\omega_i t} \end{cases} \tag{14.21}$$

The coefficients a_i and b_i are determined by the initial conditions, that is the initial values of $n(t)$ and $C_i(t)$.

If $\rho \geq 0$ then $\omega_0 \geq 0$ and $\omega_{i \neq 0} < 0$.

If $\rho < 0$ then $\omega_i < 0$. In this case, ω_0 is the less negative solution and is called the dominant ω value.

In all cases, after the 6 or 8 more negative exponential vanish, the evolution of $n(t)$ and $C_i(t)$ is determined by $e^{\omega_0 t}$.

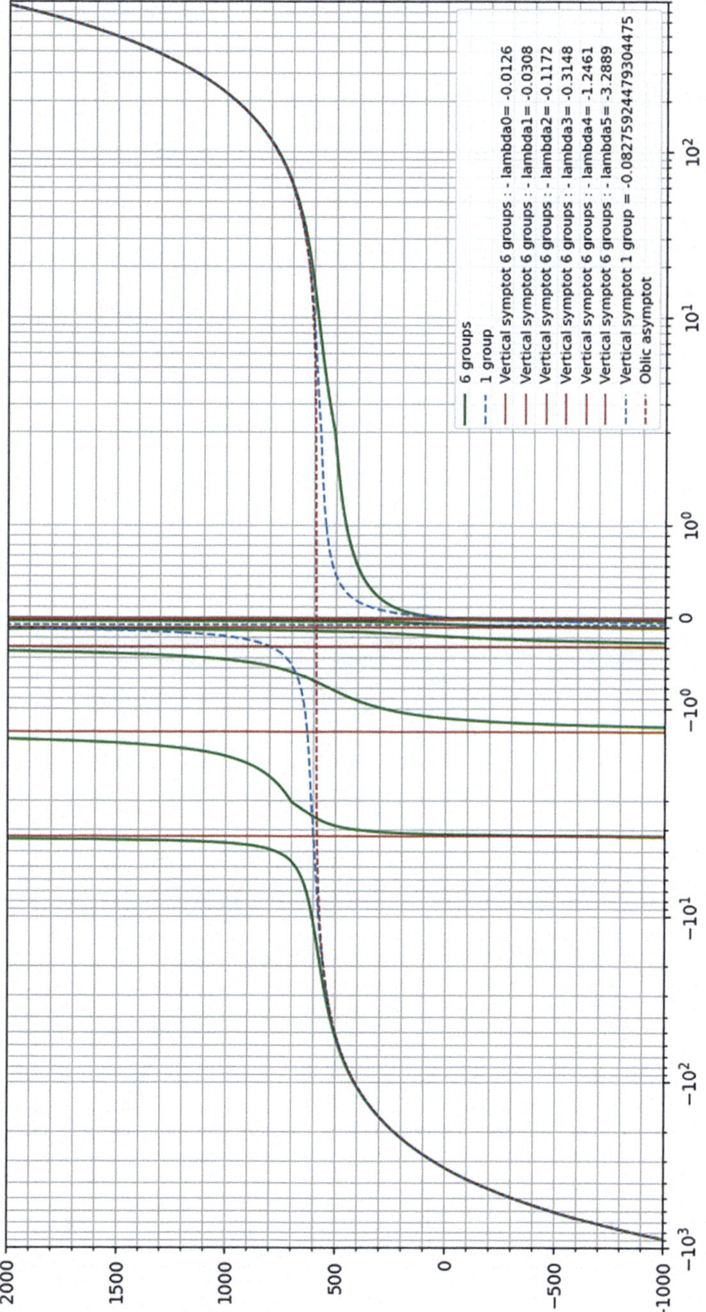

Fig. 14.2 Nordheim function for 6 neutrons groups (example of a PWR) and the equivalent 1 group. Semi log scale

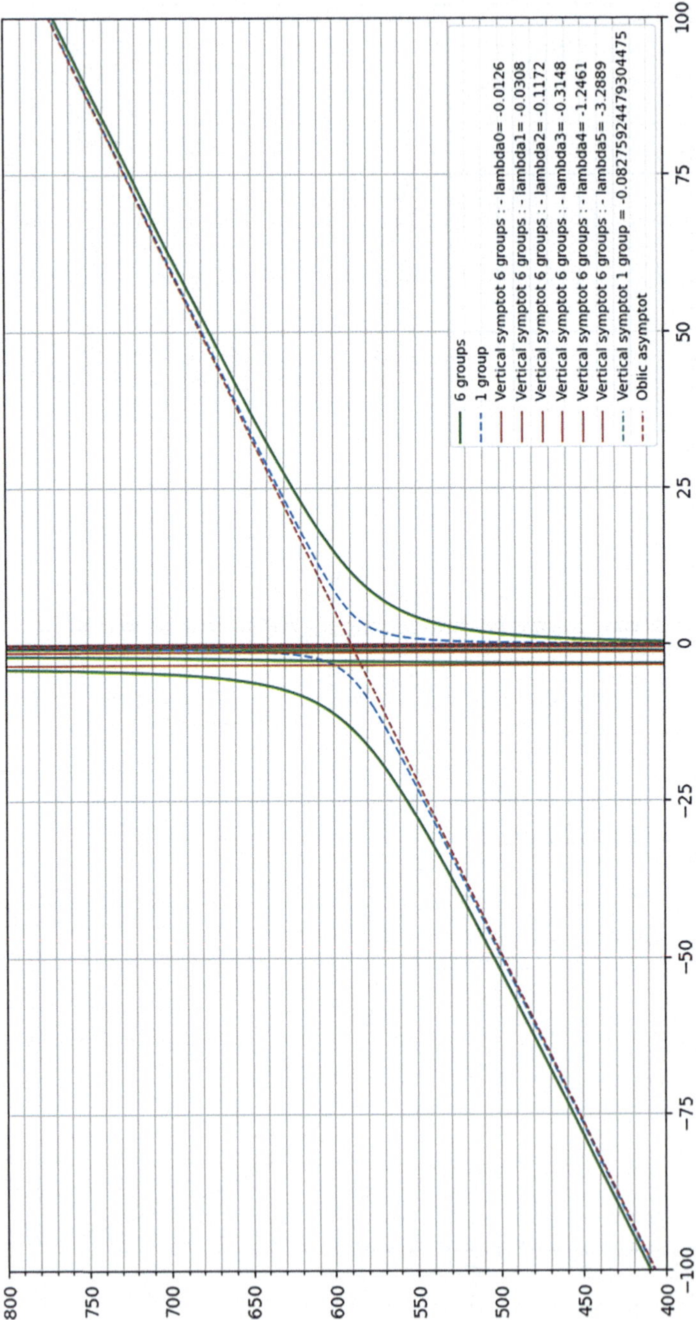

Fig. 14.3 Nordheim function for 6 neutrons group (example of a PWR) and the equivalent 1 group

14.6 Analytical Solution with One-Group Theory

1 group kinetics can be derived from 6 or 8 groups kinetics characterized by λ_i, β_i and $\sum_i \beta_i = \beta$. A probability of decay per unit time, λ, is defined by the equation:

$$\frac{\beta}{\lambda} = \sum_i \frac{\beta_i}{\lambda_i}$$

The equation of Nordheim becomes:

$$\rho = \omega \left[l + \frac{\beta}{\lambda + \omega} \right]$$

It can be put in the form of a second order equation:

$$l\omega^2 + \omega \left(\lambda l + \beta - \rho\right) - \lambda\rho = 0 \tag{14.22}$$

If we consider, as an example, the kinetics parameters of a PWR with 6 groups given in Table 14.5, we calculate $\lambda = 0.082\,759\,\text{s}^{-1}$.

Then we observe that, if the reactivity is $\rho < 0.9\beta$, then $(\beta - \rho) \gg \lambda l$. The involved values are $\beta = 588.9 \times 10^{-5}$ and $\lambda l = 1.5120 \times 10^{-6}$. This enables to simplify the Eq. 14.22:

$$l\omega^2 + (\beta - \rho)\,\omega - \lambda\rho = 0 \tag{14.23}$$

The two solutions of Eq. 14.23 are:

$$\omega_{0,1} = \frac{1}{2} \left[-\frac{\beta - \rho}{l} \pm \sqrt{\left(\frac{\beta - \rho}{l}\right)^2 + 4\,\frac{\rho\lambda}{l}} \right] \tag{14.24}$$

Table 14.5 Kinetics parameters used for a PWR (example). With $l = 1827\,\text{s}$

Group (index, i)	β_i pcm	$\lambda_i\,\text{s}^{-1}$
1	17.2	0.0126
2	123.2	0.0308
3	111.2	0.1172
4	229.5	0.3148
5	80.5	1.2461
6	27.3	3.2889

The Eq. 14.24 can also be simplified taking into account that, if $\rho < 300\,\text{pcm}$, then $\left(\dfrac{\beta - \rho}{l}\right)^2 \gg \left|4\,\dfrac{\rho\lambda}{l}\right|$, we can write again the two solutions write in a simpler form:

$$
\begin{cases}
\omega_1 = -\dfrac{\beta - \rho}{l} \\[2ex]
\omega_0 = \dfrac{\lambda\rho}{\beta - \rho}
\end{cases}
\tag{14.25}
$$

Without any simplification, directly from Eq. 14.22, we can write that ω_0 and ω_1 are characterized by the two relations:

$$
\begin{cases}
\omega_0 + \omega_1 = -\left(\lambda + \dfrac{\beta - \rho}{l}\right) \\[2ex]
\omega_0 \times \omega_1 = \dfrac{-\lambda\rho}{l}
\end{cases}
\tag{14.26}
$$

The graphic representation of this Nordheim function with 1 group does only have 1 vertical asymptote at the abscissa $-\lambda$. It has the same oblic asymptote than the 6 or 8 function from which it derives (see Figs. 14.2 and 14.3). With 1 group model, it is possible to establish analytical formulas for $n(t)$ and $C(t)$ as solutions of the system of two differential equations:

$$
\begin{cases}
n(t) = \dfrac{n(0)}{\omega_0 - \omega_1}\left[\left(\dfrac{\rho}{l} - \omega_1\right) e^{\omega_0 t} + \left(\omega_0 - \dfrac{\rho}{l}\right) e^{\omega_1 t}\right] \\[3ex]
C(t) = \dfrac{C(0)}{\omega_0 - \omega_1}\left[-\omega_1 e^{\omega_0 t} + \omega_0 e^{\omega_1 t}\right]
\end{cases}
\tag{14.27}
$$

14.7   Analysis of Some Transients Without Source

14.7.1   Positive Step of Reactivity, Below β

A positive step below β, in this case 100 pcm, is numerically calculated with kinetics parameters characteristics of a PWR.

The kinetics parameters used in this book for a PWR at the beginning of its operating cycle are given in the Table 14.5. Both β_i and λ_i are weighted-average values, taking into account the relative contributions of the different kinds of fission occurring in the core. Table 14.5 is consistent with the relative contributions given in Sect. 14.1.5. The Fig. 14.4 exhibits a sharp evolution of $n(t)$ triggered by the reactivity step, followed by a slow evolution. This sharp evolution is coined prompt jump and is similar for the 1 group and the 6 groups calculations. Thus the 1 group model enables to analyse the prompt jump of $n(t)$.

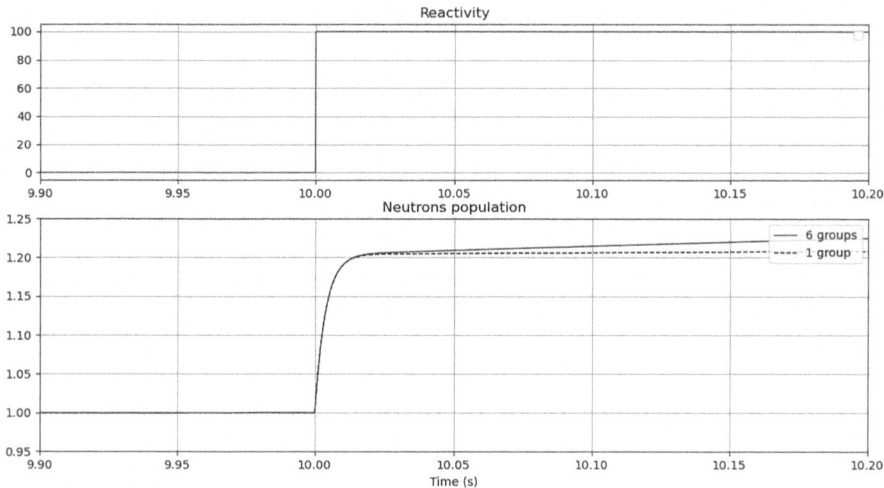

Fig. 14.4 Response to a reactivity step $\rho = 100$ pcm (example of a PWR, for 6 neutrons group) and the equivalent 1 group

$n(t)$ is the sum of two exponentials, according to the system of Eq. 14.27. The prompt jump is described by $\frac{n(0)}{\omega_0 - \omega_1} \left(\omega_0 - \frac{\rho}{l} \right) e^{\omega_1 t}$, and ω_1 is the negative solution of the 1 group equation.

It should be noted that $\frac{n(0)}{\omega_0 - \omega_1} \left(\omega_0 - \frac{\rho}{l} \right)$ is negative, thus the exponential ω_1 corresponding to the prompt jump ($\$\rho > 0$) is concave, and the positive exponential ω_0 is convex.

The relationships (14.26), applied to the values of Table 14.5 and a reactivity $\rho = 100$ pcm enable to calculate the two solutions of the Nordheim 1 group equation:

$$\omega_0 = 0.0169213 \, \text{s}^{-1} \quad \text{and} \quad \omega_1 = -267.697 \, \text{s}^{-1}$$

The Fig. 14.5 shows that the most negative ω, ω_1 for 1 group and ω_6 for 6 groups are very close. This is easy to understand if we see that the two Nordheim functions are both almost overlapped with the oblic asymptot, as shown on Fig. 14.6. Before the reactivity step, at $\rho = 0$ pcm, the most negative ω is $\frac{-\beta}{l} = -322.33 \, \text{s}^{-1}$.

The 1 group model illustrates that the prompt jump is governed by the most negative ω.

The two dominant ω_0, the one calculated with 1 group and the one calculated with 6 groups are significantly different, thus it is visible with a few minutes time frame that the two exponentials do not have the same strength. On the Fig. 14.7, the 1 group $n(t)$ looks like two straight lines, the almost vertical one for the prompt jump is associated to ω_1 and the long term evolution is associated to the exponential ω_0.

With the 6 groups calculation, there is a transition phase between the prompt jump and the long term domination exponential, this phase lasts as long as the 5 negative exponentials ω_1 to ω_5 (ranging from $-0.013647 \, \text{s}^{-1}$ to $-3.1307 \, \text{s}^{-1}$) do not vanish.

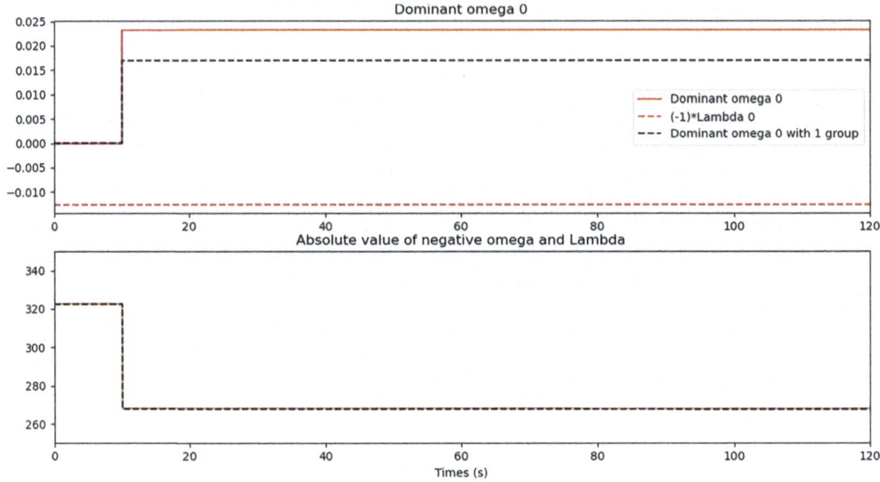

Fig. 14.5 Dominant ω_0 and most negative ω during a reactivity step of $\rho = 100$ pcm. Comparison between 1 and 6 groups calculations

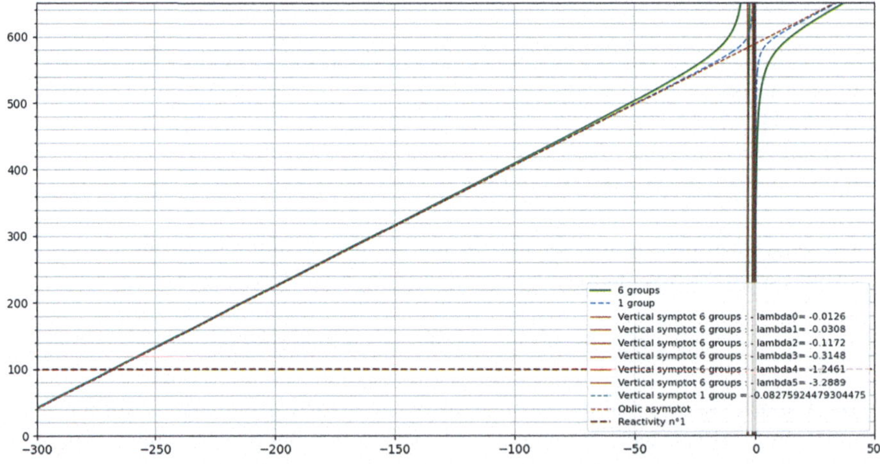

Fig. 14.6 Nordheim function, comparison between 1 and 6 groups, in the case of a reactivity step of $\rho = 100$ pcm

14.7.2   *Parametric Study with Various Positive Steps of Reactivity, Below* β

Th Fig. 14.8 shows the response to various reactivity steps, positive and negative one. In the case of a negative step, $n(t)$ undergoes a prompt drop.

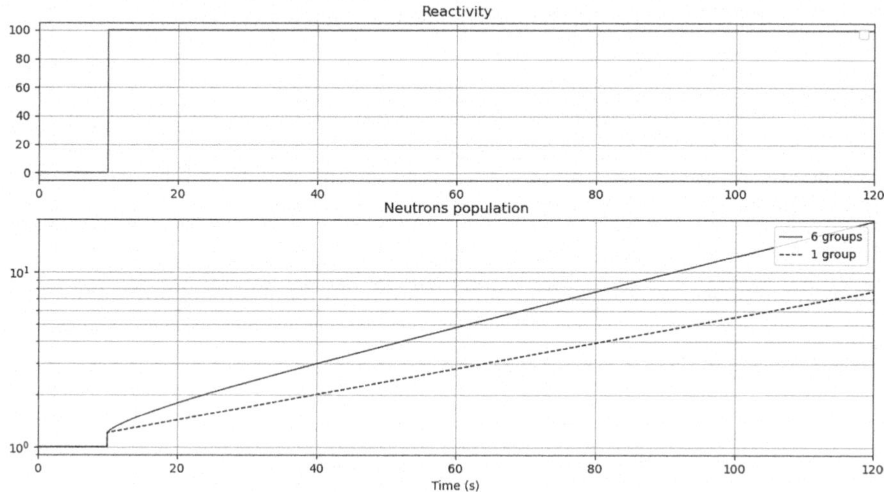

Fig. 14.7 Response to a reactivity step $\rho = 100\,$pcm, logarithmic scale (example of a PWR, for 6 neutrons group) and the equivalent 1 group

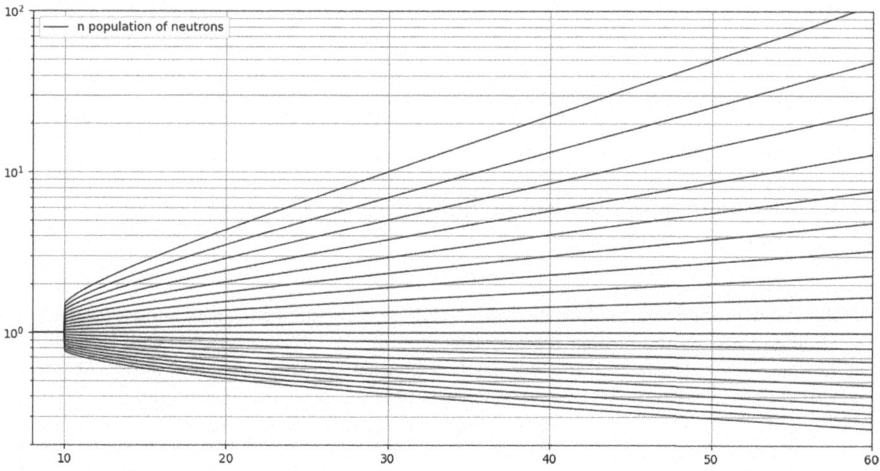

Fig. 14.8 Response to reactivity steps. Parametric study with 20 steps between $-180\,$pcm and $-200\,$pcm

14.7.3 *Understanding the Prompt Jump or the Prompt Drop*

14.7.3.1 Estimation of the Prompt Jump with the One-Group Model

We assume that the prompt jump occurs within a short time frame (a few hundredths of a second), between t_0 and t_{0+}. It can be written that $e^{\omega_1 t_{0+}} \approx 0$ because of the strongly negative value of ω_1, and $e^{\omega_0 t_{0+}} \approx 1$ because of the very slightly positive value of ω_0. Thus we can approximate the prompt jump with:

$$\frac{n(t_{0+})}{n(t_0)} = \frac{1}{\omega_0 - \omega_1} \left(\frac{\rho}{l} - \omega_1 \right)$$

We need to approximate ω_1 and $\omega_0 - \omega_1$. If we assume that the point of coordinates (ω_1, ρ) is on the oblic asymptote, it is possible to write that:

$$l \approx \frac{\rho}{\omega_1 + \frac{\beta}{l}} \quad \text{and then} \quad \omega_1 \approx \frac{\rho - \beta}{l}$$

If we assume that $|\omega_0| \ll |\omega_1|$, then we can write:

$$l \approx \frac{\beta - \rho}{\omega_0 - \omega_1} \quad \text{and then} \quad \frac{1}{\omega_0 - \omega_1} \approx \frac{l}{\beta - \rho}$$

Combining these approximations:

$$\frac{n(t_{0+})}{n(t_0)} \approx \frac{l}{\beta - \rho} \left(\frac{\rho}{l} - \frac{\rho - \beta}{l} \right)$$

$$\frac{n(t_{0+})}{n(t_0)} \approx \frac{\beta}{\beta - \rho}$$

14.7.3.2 Estimation of the Prompt Jump with the Constant Delayed Source Assumption

Within a few prompt-neutron lifetimes after the drop, the system adjusts to a lower neutron level determined by the prompt reproduction number, $k_{p,1} < k_{p,0}$, and remains constant at this "quasistatic level" until it is ultimately decreased by delayed-neutron decay. At this quasistatic level, the original precursor concentrations $C_{i,0}$ are still unchanged and $\frac{dn}{l} dt \approx 0$.

Keepin (1965, p. 243)

We consider a steady state at criticality until t_0, followed by a reactivity step at t_0.

The constant delayed source assumption consists in writing that during the short time duration of the prompt jump, between t_0 and t_{0+}, the production of delayed neutrons by unit of time remains unchanged.

We also assume that the lifetime of neutrons is unchanged and equal to the generation time l.

At t_0, the reactivity becomes $\rho > 0$ and the rate of production of prompt neutrons becomes $n(t_0)(\rho - \beta + 1)\frac{1}{l}$ as illustrated in Fig. 14.1. And the sum of prompt neutrons production and delayed neutrons production exceeds the loss of neutrons.

Thus $n(t)$ increases, until it reaches at t_{0+} a value n_1, or $n(t_{0+})$, for which the loss of neutrons is balanced again by the sum of prompt neutrons production and delayed neutrons production assumed constant:

- Loss of neutrons $-n_1\dfrac{1}{l}$

- Prompt neutrons production $n_1(1 - \beta + \rho)\dfrac{1}{l}$

- Precursors production $n_1\beta\dfrac{1}{l}$

- Delayed neutrons production $n(t_0)\beta\dfrac{1}{l}$

This enables to write the equation:

$$n_1\frac{1}{l} \approx n_1(1 - \beta + \rho)\frac{1}{l} + n(t_0)\beta\frac{1}{l}$$

$$n_1(\beta - \rho) \approx n(t_0)\beta$$

And then:

$$\frac{n(t_{0+})}{n(t_0)} \approx \frac{\beta}{\beta - \rho} \tag{14.28}$$

The pseudo equilibrium that is reached at t_{0+} is not sustainable because the constant delayed source assumption is only valid during a very short time interval. Because of the increase in the production of precursors by fissions due to $\rho > 0$, the delayed neutrons production will also increase. The Eq. 14.28 is applicable for positive or negative reactivity steps. In the case of a negative reactivity step, the transient is coined prompt drop.

14.7.4 Positive Step for Various Reactors

14.7.5 Small Reactivities

We perform reactivity steps between -10 pcm and 10 pcm, every 2 pcm, with 6 groups.

The Fig. 14.9 show that, for small reactivities around the origin point, the Nordheim curve can be assimilated to its tangent with a negligeable error, and the reactivity is approximately a linear function of ω_0. For small reactivities, we can write that $\omega \ll \lambda_i$, and the analytical formula of this approximation is (Fig. 14.10):

$$\rho \approx \omega\,(l + \sum_i \frac{\beta_i}{\lambda_i}) = \omega\bar{l}$$

It is easy to check that $\bar{l}$ is the average lifetime of neutrons:

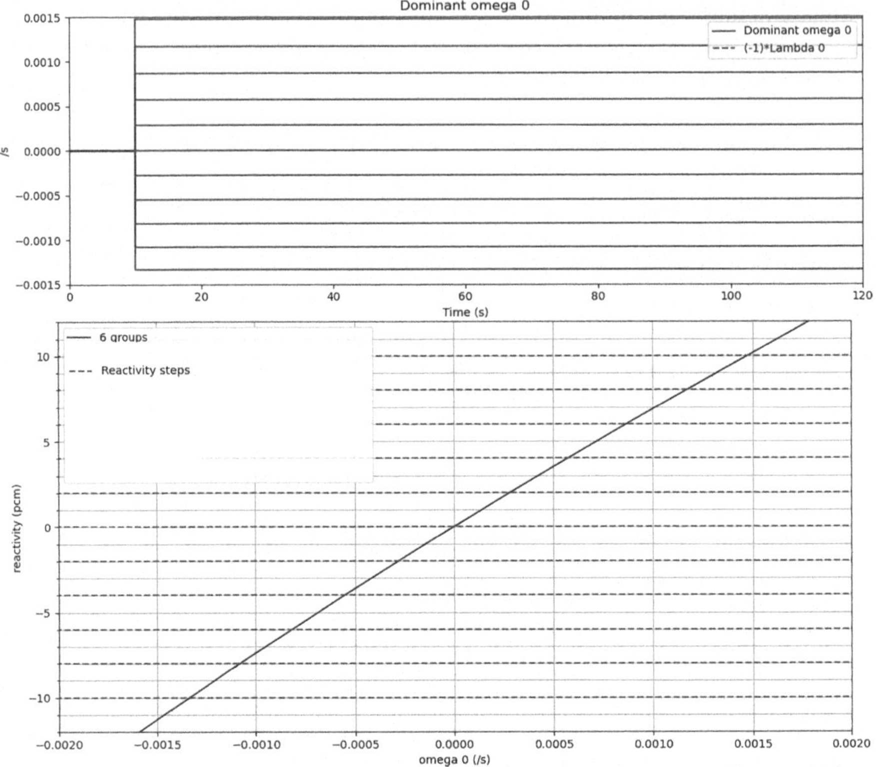

Fig. 14.9 Linear relationship between ω and ρ for small reactivities

$$\bar{l} = l\left(1 - \sum_i \beta_i\right) + \sum_i \left(l + \frac{1}{\lambda_i}\,\beta_i\right)$$

$$= l + \sum_i \frac{\beta_i}{\lambda_i}$$

14.7.6   Positive Reactivity Slots

We perform a reactivity slot of an amplitude 100 pcm and a duration of 60 s. In the Fig. 14.11, the asymptotic value of the population of precursors at t time is the value at equilibrium with the neutron population $n(t)$. In the Fig. 14.12, the proportion of delayed neutrons produced by each group, as a part of the total number of neutrons produced by the chain reaction, is calculated. Hence the proportion of

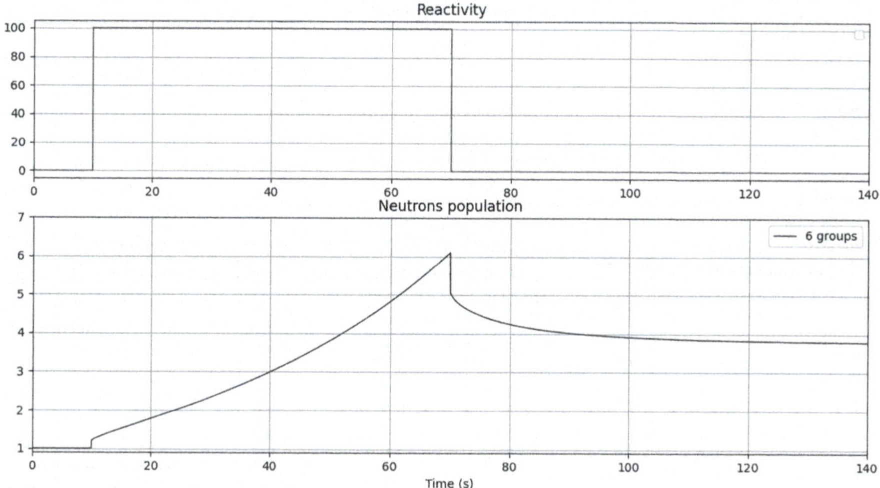

Fig. 14.10 Response to a reactivity slot with 6 groups

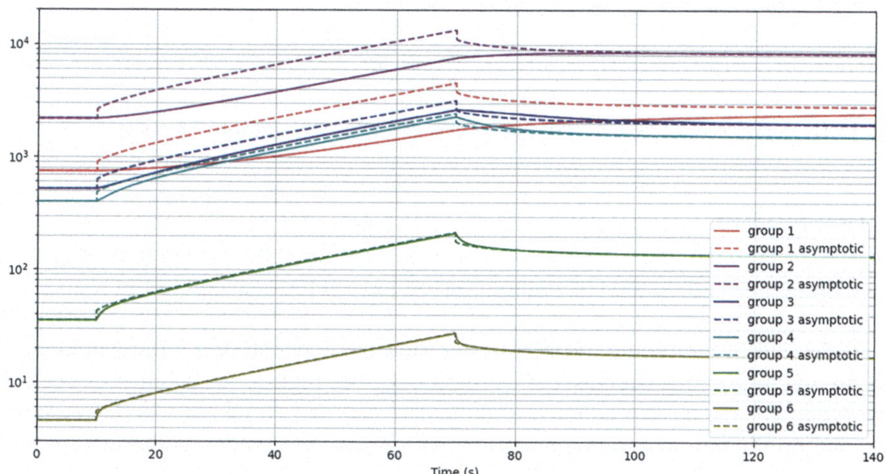

Fig. 14.11 Analysis of precursors in the case of a reactivity slot with 6 groups

delayed neutrons produced by the whole of precursors. The first plot of the figure
enables to compare the rate at which delayed neutrons are produced by each group.

   The Fig. 14.11 shows that, before the transient, the most abundant precursors are
groups 2 and 1. This can be easily understood with Eq. 14.17: group 2 is characterized
by a high relative abundance associated with a low probability of decay, and group
1 is characterized by the lowest probability of decay, however its relative abundance
is low.

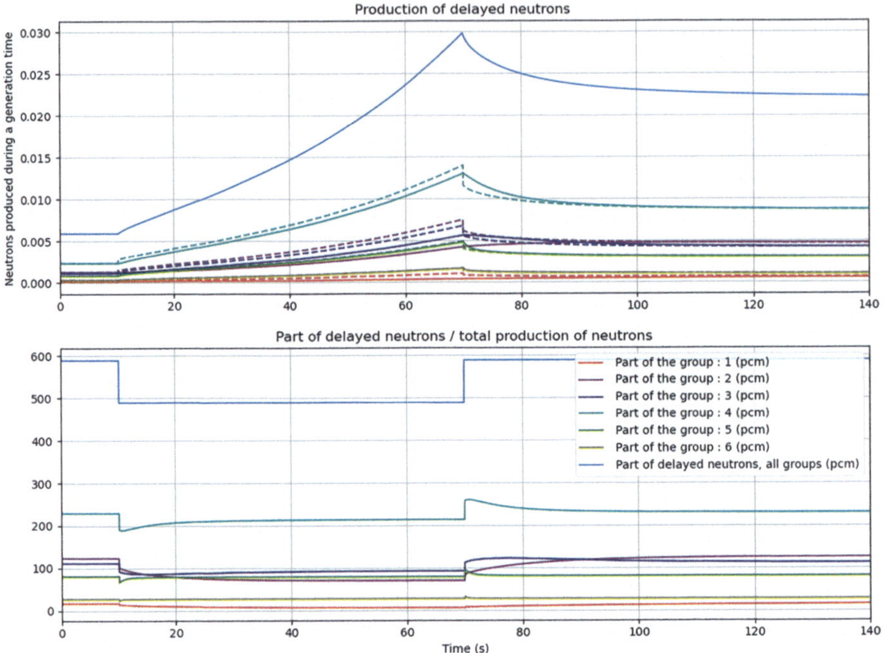

Fig. 14.12 Analysis of delayed neutrons in the case of a reactivity slot with 6 groups

The Fig. 14.12 shows that the most productive groups, listed in descending order, are 4, 2, 3, 5, 6 and 1. This is the descending order of β_i, and this is easy to understand since the neutron production by unit of time of the group i is, at equilibrium, $\lambda_i C_i = \frac{\beta}{l} n(t)$.

At the end of the prompt drop, after the slot, the concentration of precursors of the groups 4, 5 and 6 decrease, since they have the highest probability of decay. The group 4 is the most productive and the decrease of $C_4(t)$ leads to a decrease in the total production of delayed neutrons. As a result, $n(t)$ is dragged down by the decrease of $\sum_i \lambda_i C_i(t)$.

$n(t)$ and $C_i(t)$ decrease until a new balance is achieved, at criticality.

On the Fig. 14.12, we can see that the proportion of delayed neutron in the total population of neutrons is β when the reactor is critical, and below β when the reactor is overcritical, and in fact it is $\beta - \rho$.

14.7.7   *Analytical Approach for Calculation of the Delayed Neutrons Part, Without Source*

From the balance of neutrons established at t_{0^+} in Sect. 14.7.3.2, we can write that:

$$n_1 = \frac{\beta}{\beta - \rho} n(t_0) = \frac{\beta}{\beta - \rho} \frac{\lambda_i \, l}{\beta_i} C_i(t_0) \quad \forall i \in \{1, 2, 3..6 \text{ or } 8\}$$

$$\lambda_i C_i(t_0) = n_1 \frac{\beta - \rho}{\beta} \frac{\beta_i}{l}$$

Precursors do not undergo prompt jumps or drops, and given the small duration of the time interval of the prompt jump/drop $[t_0, t_{0+}]$, we can write that $C_i(t_0) \approx C_i(t_{0+})$. An then:

$$\sum_i \lambda_i C_i(t_{0+}) = n(t_{0+}) \frac{\beta - \rho}{\beta} \frac{\sum_i \beta_i}{l}$$

$$\sum_i \lambda_i C_i(t_{0+}) dt = n(t_{0+}) (\beta - \rho) \frac{dt}{l} \tag{14.29}$$

The Eq. 14.29 gives the production of delayed neutrons during dt, at time t_{0+}, so at the end of the prompt jump.

Then, we assume that, after the prompt jump, the delayed neutrons production by precursors and the neutrons population undergo the same evolution:

$$\sum_i \lambda_i C_i(t) dt = n(t) (\beta - \rho) \frac{dt}{l} \quad \text{with } t > t_{0+} \tag{14.30}$$

The Eq. 14.30 is equivalent to write that the evolution of $n(t)$ is very slow: $\frac{dn(t)}{dt} \approx 0$. We can now calculate the part of delayed neutrons among the total production of neutrons, i.e. the delayed neutron generation fraction:

$$\frac{\sum_i \lambda_i C_i(t) dt}{\sum_i \lambda_i C_i(t) dt + n(t)(1 - \beta + \rho)\frac{dt}{l}}$$
$$= \frac{\beta - \rho}{\beta - \rho + (1 - \beta + \rho)}$$
$$= \beta - \rho$$

14.7.8 Ramps of Reactivity

The Fig. 14.13 show that if we calculate $\frac{n(t_{\text{end of ramp}})}{n(t_0)}$, this ratio is equal to $\frac{\beta}{\beta - \rho}$ in the case of an instantaneous ramp, i.e. its duration has the same order of magnitude than the generation time. The drop of $n(t)$ increases with the duration of the ramp.

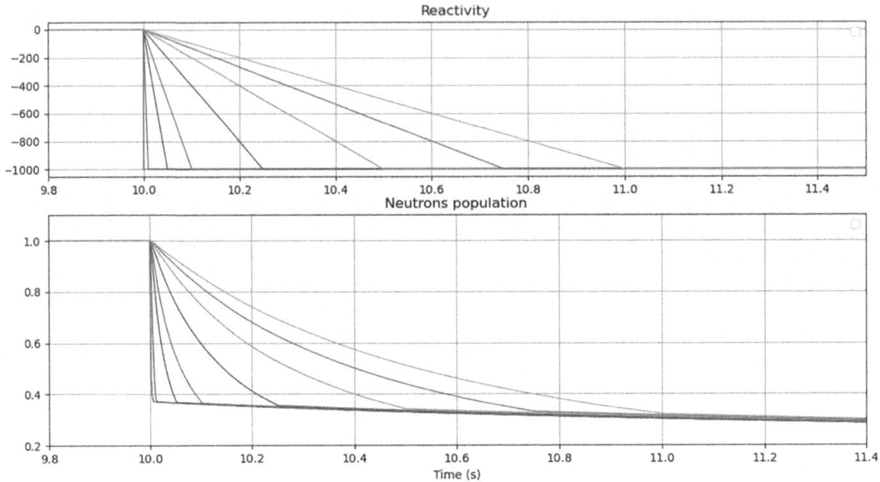

Fig. 14.13 Negative ramps of various durations calculated with 6 groups

14.7.9   Kinetics in the Case of Prompt Criticallity

14.7.9.1   Positive Step Above β

We calculate 10 different reactivity steps ranging from 200 to 1100 pcm, with 6 groups and $\beta = 588.9$ pcm. The Fig. 14.14 shows that for the steps 200, 300, 400 and 500 pcm, the part of delayed neutrons is accurately estimated with the formula $\beta - \rho$ (389, 289.2, 189.9 and 92.26 pcm). However, for higher reactivities, and specifically above β, the formula $(\beta - \rho)$ is not anymore relevant and the part of delayed neutrons tend to 0.

On the Fig. 14.15, we can see that the prompt jump vanishes when the reactivity gets close to β and greater. An interesting mental picture that enables to explain why

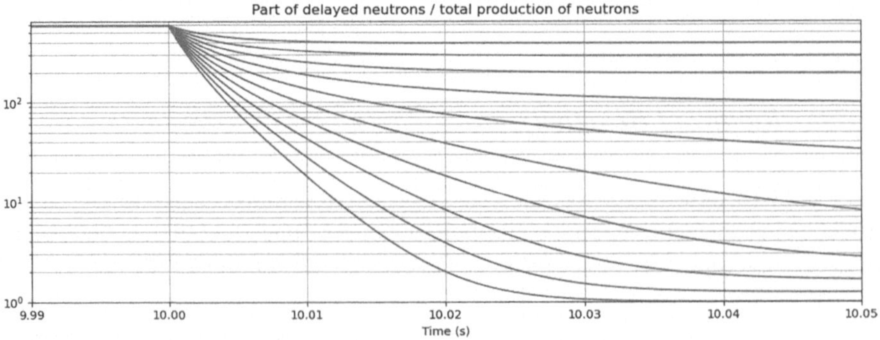

Fig. 14.14 Part of delayed neutrons during reactivity steps, below and above $\rho = \beta$

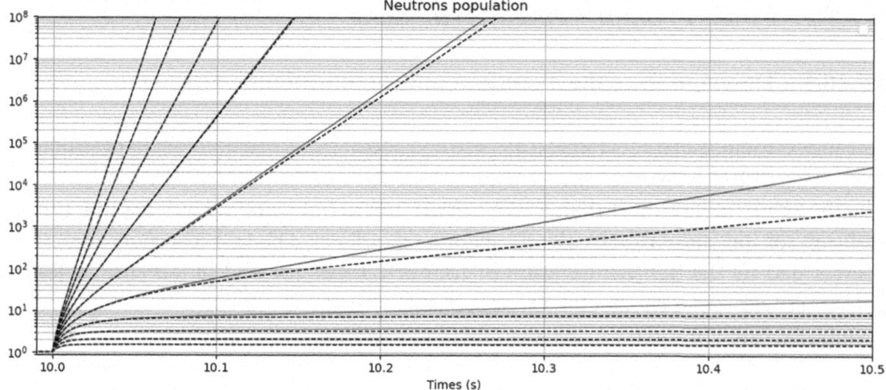

Fig. 14.15 $\rho(t)$ and $n(t)$ during 10 reactivity steps (200–1100 pcm). 1 group calculation in black dotted line

the prompt jump vanishes is to take into account the two values ω_0 and ω_1 of the 1 group model, presented in Sect. 14.6. Let's consider the case of ρ is significantly higher than β. From the relationship (14.26) between ω_0 and ω_1, we can write that:

$$\begin{cases} \omega_0 \to -\lambda & \text{vertical asymptot} \\[2mm] \omega_1 \to \dfrac{\rho - \beta}{l} & \text{oblic asymptot} \end{cases}$$

In this prompt criticality situation, unlike to what happens when $\rho \ll \beta$, the slow exponential, i.e. the one having the lowest ω in absolute value, is the negative exponential, $e^{-\lambda t}$. And the fast exponential, i.e. the one having the highest ω in absolute value, is the positive exponential $e^{\frac{\rho-\beta}{l}t}$. The dominant exponential is faster than the one supposedly giving the prompt-jump. This is why the prompt jump is not visible and with 1 group, we can write the following approximation:

$$n(t) = \frac{n(t_0)}{\frac{\rho-\beta}{l} + \lambda} \left[\left(\frac{\rho}{l} + \lambda \right) \underbrace{e^{\frac{\rho-\beta}{l}t}}_{\text{fast}} + \frac{-\beta}{l} \underbrace{e^{-\lambda t}}_{\text{slow}} \right]$$

$$n(t) \approx n(t_0)e^{\frac{\rho-\beta}{l}t}$$

The prompt critical kinetics with delayed neutrons is equivalent to kinetics without delayed neutrons, on condition that the reactivity is $\rho - \beta$ instead of ρ.

14.7.10   Cabri Prompt Critical Transient

14.7.10.1   Description of the Transient and Assumptions

A prompt critical transient (Pulse transient) has been performed on the experimental
Cabri. The initial power of the reactor was 0.0987 MW (almost 100 kW), which
gives the same linear power than a PWR operating in hot stand by at 0.5%PN, initial
condition for the study of RIA (Rod Ejection Accident). The fast reactivity injection
is performed by the depressurization of four ^{3}He tubes located in the core; it takes
approximatively 0.1 s to inject 3000 pcm in the core.

   Some features of the core are given below:

- Fuel rods within graphite reflector walls
- An Am-Be neutron source provides neutrons. This source consists in alpha emitters
  ^{241}Am intimately mixed with targets ^{9}Be, in order to produce (α, n) reactions. It
  emits neutrons of an average energy 4.2 MeV.
- Dimensions: 60, 60 and 80.15 cm (active height)
- 1487 fuel rods: UO_2 (^{235}U 6%)
- Mass of UO_2 in the core 766.87 kg
- Heat capacity $C_p = 80\,\text{J} \cdot \text{mol}^{-1} \cdot \text{K}^{-1}$

The fast and over critical reactivity injection leads to a rapid rise in $n(t)$, the heat
generated by the chain reaction does increase the fuel temperature, which results in
Doppler negative feedback effect, and a reactivity loss.

   We assume that the nuclear power produced in the Cabri core during the transient
far exceeds the power transferred by thermal exchange to the moderator, and thus that
the power transferred to the moderator can be neglected. This adiabatic assumption
enables us to calculate the energy deposit within the fuel rods during the power peak,
by integration of the power. If we make the assumption of an homogeneous axial and
radial power in the whole core, and then assimilate the fuel rods with a 1D string,
we can calculate the temperature reached in the fuel as a function of time during the
transient, and then estimate the Doppler feedback effect.

   The Doppler feedback effect can be calculated with the following formula:

$$\alpha_{\text{Dop}}(T\,°\text{C}) \approx \alpha_{\text{Dop}}(300\,°\text{C})\sqrt{\frac{300 + 273.15}{T + 273.15}} \tag{14.31}$$

With $\alpha_{\text{Dop}}(300°\text{C}) = -3.25 \text{ pcm}/°\text{C}$

14.7.10.2   Power Peak

The Fig. 14.16 shows how fast $n(t)$ builds up, due to prompt criticality. At $t =$
0.068 s, the reactivity is peaking and the corresponding dominant exponential is $\omega_0 =$

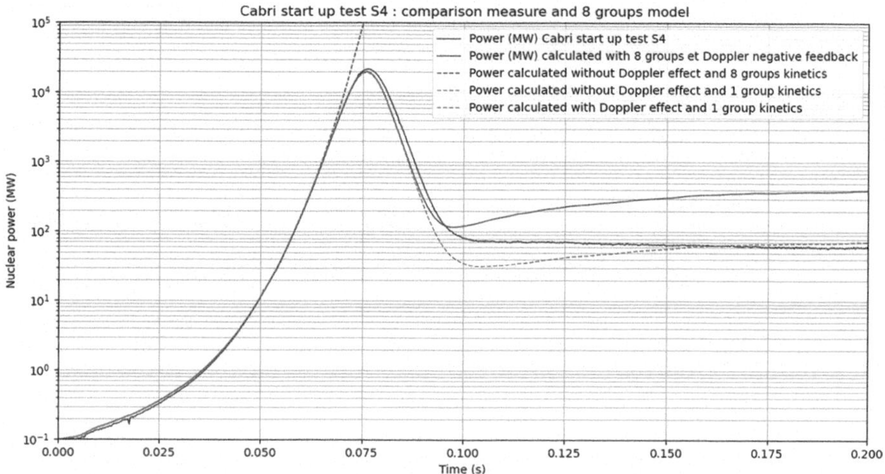

Fig. 14.16 Nuclear power calculated with and without Doppler negative feedback

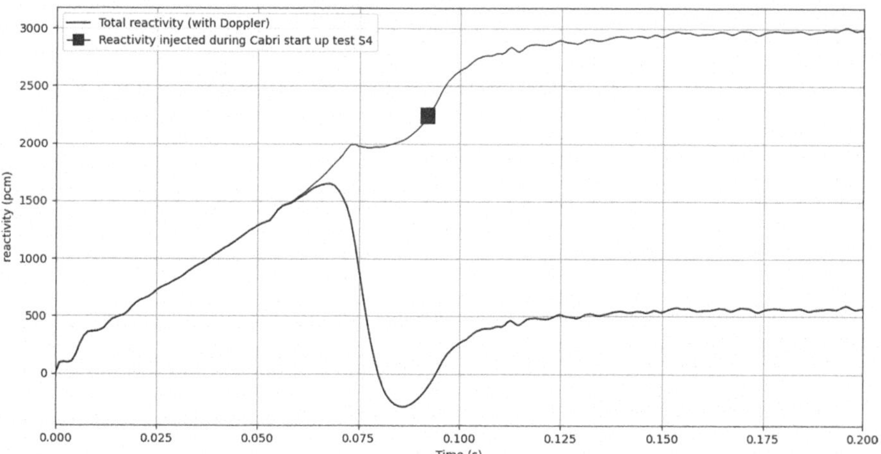

Fig. 14.17 Injected reactivity and total reactivity, taking into account Doppler negative feedback

363.8 s^{-1} (see Fig. 14.18) and thus the doubling time is $T_d = \frac{\ln 2}{\omega_0} = 1.9053 \times 10^{-3}$ s. The 8 groups kinetics, with a simplified Doppler feedback calculation, accurately reproduces the transient until $t = 0.1$ s (Figs. 14.17 and 14.19).

An interesting observation must be done: the nuclear power peaks at $t = 0.076$ s and the reactivity reaches zero at $t = 0.08$ s. During the peak, the contribution of delayed neutrons to the chain reaction is negligeable, and the part of delayed neutrons is of the order of magnitude of a pcm and prompt neutrons can be considered as the only ones to sustain the chain reaction. The nuclear power peaks when the diminishing ρ reaches β, that is prompt criticality, i.e. a chain reaction only sustained by prompt

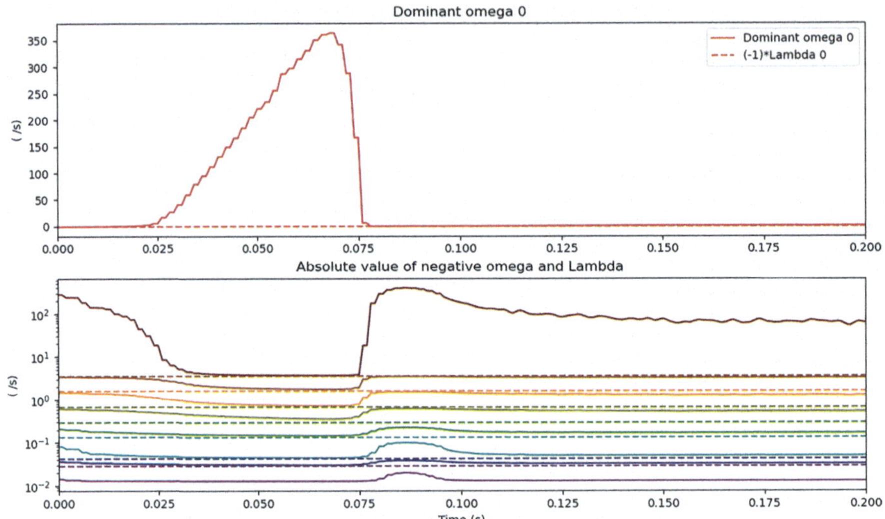

Fig. 14.18 ω solutions with 8 groups kinetics, calculated with $\rho(t)$

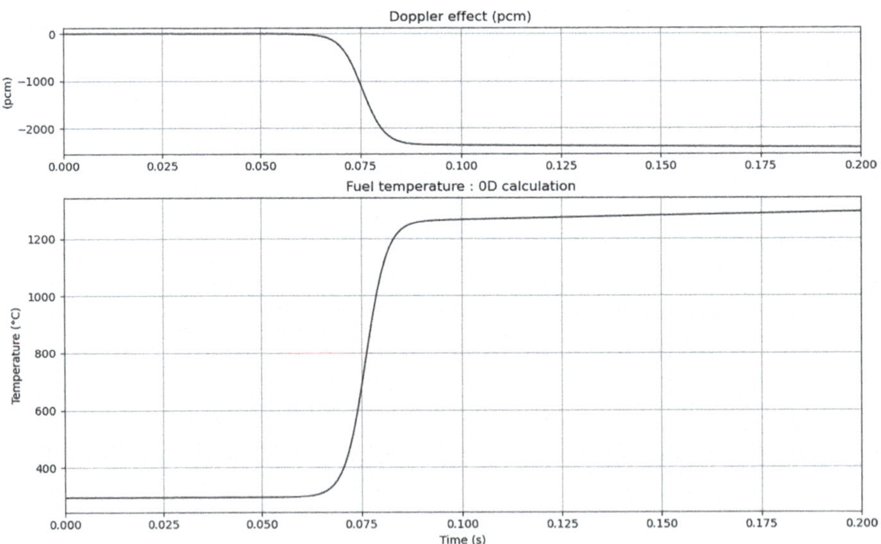

Fig. 14.19 Temperature rise in the fuel and associated Doppler antireactivity

neutrons is critical. After a prompt critical power excursion, the decrease in nuclear power does not need a negative reactivity, it only needs $\rho < \beta$.

Concerning the accuracy of the 1 group approximation, It can be seen in the Fig. 14.16 that the 1 group calculation of $n(t)$ is overlaid with the 8 group calculation when $\rho > \beta$.

14.8 Kinetics with Neutron Source

14.8.1 Source Neutrons

The term source neutrons, or independent source of neutrons (also called extraneous), refers to neutrons which are not dependent on the neutron density of the system, and arising from events other than neutron collisions, i.e., not from fission, (n, 2n), and similar neutron reactions (Bell and Glasstone 1970, p. 7). There are two categories of source neutrons:

- Nuclear fuel, especially when irradiated, produces inherent source neutrons by the following reactions:

  - Spontaneous fission of very heavy elements, especially transuranic elements. For example, $^{240}_{94}$Pu has a very high rate of spontaneous fission. For high burnup fuel, the neutrons are provided predominantly by the spontaneous fission of curium nuclei ($^{242}_{96}$Cm and $^{244}_{96}$Cm). Spontaneous fission is the major provider of inherent source neutrons.
  - (α, n) reaction with $^{18}_{8}$O. Oxygen element is present in the uranium oxide fuel. The neutron-producing reaction is: $^{18}O(\alpha,n)^{21}Ne$. Alpha particles are commonly emitted by all of the heavy radioactive nuclei occurring in fuel.
  - Neutrons can also be produced by (γ, n) reactions, and therefore they are usually referred to as photoneutrons. A high-energy photon (gamma-ray) can eject a neutron from a nucleus if its energy exceeds the neutron separation energy S_n. The lowest thresholds are those of ^{9}Be (1.666 MeV) and ^{2}H (2.226 MeV). Therefore photoneutrons are very important in nuclear reactors moderated by D_2O (CANDU reactors).

- External Source of Neutrons. Sometimes source neutrons must be artificially added to the system. An external source of neutrons contains a material that emits neutrons and cladding to provide a barrier between the reactor coolant and the material. External sources are usually loaded directly into the reactor core.

  - Spontaneous fission source. ^{252}Cf is one of the most important neutron sources. Its neutron energy spectrum is very similar to that of a reactor fission spectrum. The high neutron yield per unit mass enables to use small-sized neutron sources. ^{252}Cf is produced in high-flux reactors, and its effective half-life is 2.645 years.
  - Gamma-neutron source. A typical photo-neutron source consists of an inner aluminum-encapsulated gamma-emitting core surrounded by a neutron emitting target. The overall shape of the source might be cylindrical or spherical. In the case of an antimony-beryllium source, the core is antimony activated in a reactor. Emitted neutrons emitted by a Sb-Be source are very close to being monoenergetic (0.024 MeV).

Source neutrons are very important during reactor startup and shutdown operations when the reactor is subcritical because they allow monitoring the subcriticality, usually via source range excore neutron detectors. Neutron levels shall be high enough to be detected by nuclear instruments at all times, in order to produce a specified minimum count rate. During the subcritical startup operation of a PWR loaded with an initial core, i.e. fresh fuel assemblies, ^{252}Cf is used to supply source neutrons. In the case of reload cores, the inherent neutrons from spontaneous fissions of actinides such as ^{242}Cm and ^{244}Cm may, or may not, be sufficient to yield the minimum detector count rates, depending on the geometry of the core, on the burn-up of he fuel assemblies loaded on the periphery of the core facing detectors, and on the path neutrons have to travel to reach the detectors. If the inherent source neutrons are not sufficient, PWR cores are reloaded with Sb-Be secondary neutron sources.

14.8.2 Subcritical Multiplication and Source Neutrons

We remind the system of Eq. 14.13:

$$\begin{cases} \dfrac{dn(t)}{dt} = \dfrac{k_{\text{eff}}(1-\beta)-1}{\theta} n(t) + \sum_{i=1}^{6} \lambda_i \cdot C_i(t) + S_0 \\[2ex] \dfrac{dC_i(t)}{dt} = \beta_i \dfrac{k_{\text{eff}}}{\theta} n(t) - \lambda_i \cdot C_i(t) \end{cases}$$

These equation rely on the assumption of space and energy variables separation (the validity of which is commented in Sect. 14.3.1) and that the flux shape is the fundamental mode, which is obviously not the case during approach to criticality (startup after refuelling) if external sources of neutrons have been introduced in some of the reloaded fuel assemblies.

$n(t)$ can be considered as the total number of neutrons in the core and S_0 as the emissivity of the independent source (n/s).

The conditions for a stationary state are:

$$\begin{cases} 0 = \dfrac{k_{\text{eff}}-1}{\theta} n(t) - \dfrac{\sum_{i=1}^{6} \beta_i\, k_{\text{eff}}}{\theta} n(t) + \sum_{i=1}^{6} \lambda_i \cdot C_i(t) + S_0 \\[2ex] \beta_i \dfrac{k_{\text{eff}}}{\theta} n(t) = \lambda_i \cdot C_i(t) \end{cases}$$

$$\begin{cases} \dfrac{k_{\text{eff}}-1}{\theta} n(t) = -S_0 \\[2ex] \beta_i \dfrac{k_{\text{eff}}}{\theta} n(t) = \lambda_i \cdot C_i(t) \end{cases}$$

Then:

$$n = \frac{S_0 \theta}{1 - k_{\text{eff}}}$$

We remind that the generation time l is:

$$l = \frac{\theta}{k_{\text{eff}}}$$

This enables to write the stationary quantity n in another form:

$$n = \frac{S_0 \, \theta \, k_{\text{eff}}}{1 - k_{\text{eff}}} l = -\frac{S_0 \, l}{\rho} \tag{14.32}$$

An important conclusion is that a subcritical reactor, with an independent source of neutrons, tends towards a stationary state. With the adequate assumptions, the population of neutrons in the reactor tends to $-\frac{S_0 l}{\rho}$. This result can be understood very easily: the emission of the source compensate exactly for the lack of neutrons between two generations, due to subcriticality.

14.8.3 *Case of a Critical Reactor with Source*

The system of Eq. 14.15 can be written at criticality:

$$\begin{cases} \dfrac{dn(t)}{dt} = \dfrac{-\beta}{l} n(t) + \sum_i \lambda_i \cdot C_i(t) + S_0 \\[2mm] \dfrac{dC_i(t)}{dt} = \dfrac{\beta_i}{l} n(t) - \lambda_i \cdot C_i(t) \end{cases} \tag{14.33}$$

It can be demonstrated that $n(t)$ and $C_i(t)$ are the following linear functions:

$$\begin{cases} n(t) = \dfrac{l}{l} S t + n_0 \\[2mm] C_i(t) = \dfrac{\beta_i}{\lambda_i l} S t + C_{i,0} \end{cases} \tag{14.34}$$

An thus $n(t) + \sum_i C_i(t) = S t + (n_0 + \sum_i C_{i,0})$. The time integration of the source, $S t$, is directly reflected in $n(t) + \sum_i C_i(t)$.

14.8.4   Case of a Subcritical Reactor with Source

The stationnary state of a subcritical reactor with a source, given by the Eq. 14.32, is studied in the chapter Subcritical reactor. Each time the reactivity of a subcritical reactor is modified during an approach to criticality(by dilution of the boric acid concentration, or withdraw of rods), after a transient, the reactor tends to the a new stationnary state defined by the new value of reactivity.

14.8.5   Illustration of the Dynamics of the Subcritical Reactor Followed by the Divergence

The Fig. 14.20 shows the simulation of an approach to criticality performed on the high flux reactor of ILL. Some more informations about approach to criticality can be found in Sect. 15.3. The control rod is extracted (by descent, the fully inserted position is 0 mm and the fully extracted position is 1000 mm) through a tiered approach. Each time the control rod is withdrawn, the neutron flux rises to a new level. The jump up

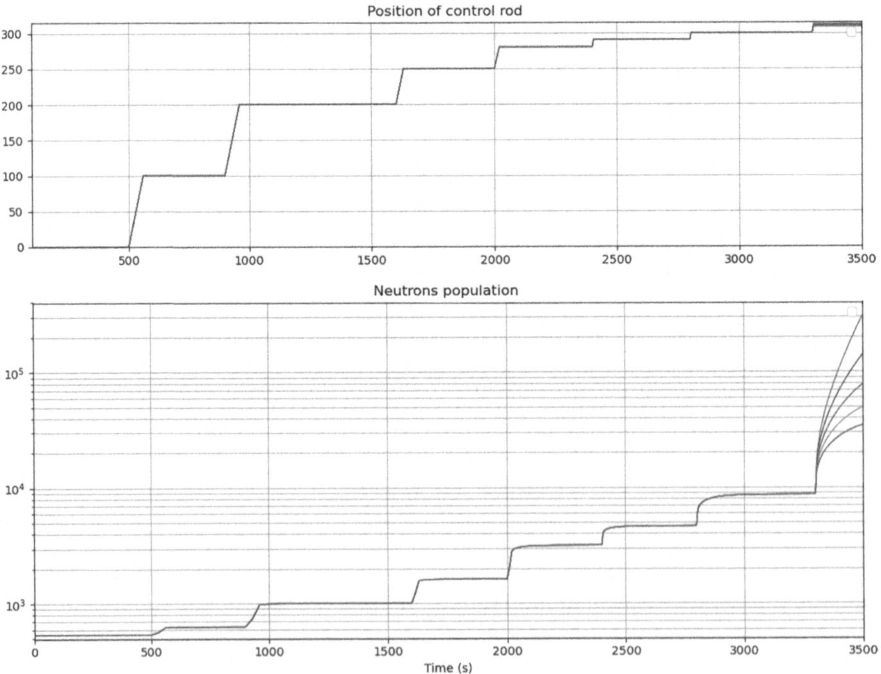

Fig. 14.20 HFR reactor. Simulation of subcritical approach

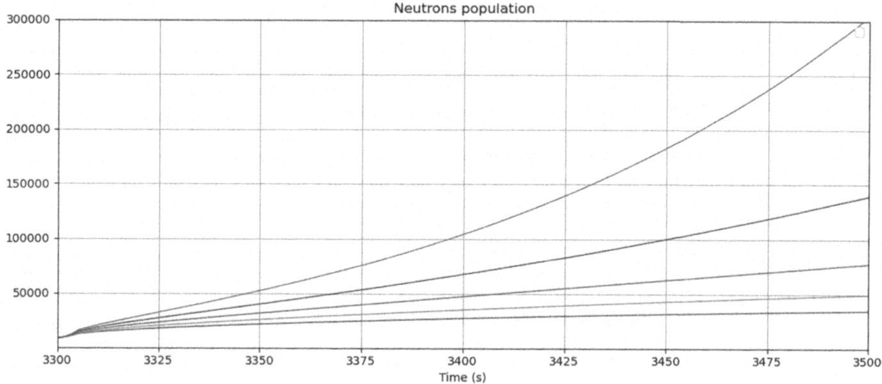

Fig. 14.21   HFR reactor of ILL. Simulation of a divergence following a subcritical approach. $n(t)$ is calculated for five final positions of the control rod 308.2, 309.2, 310.2, 311.2 and 312.2 mm

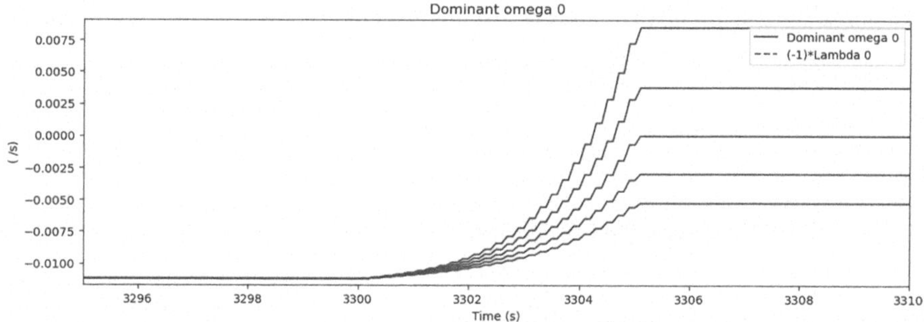

Fig. 14.22   HFR reactor of ILL. Dominant ω_0 for five final positions of the control rod 308.2, 309.2, 310.2, 311.2 and 312.2 mm

in flux is larger the closer the reactor gets to criticality, and it also takes progressively longer to reach a steady value.

The approach to criticality, along with the inverse count rate method, enables to estimate the critical position of the control rod, before reaching. Pratically, the reactor operator of ILL HFR inserts the control rod in the core after he estimates the critical position (that is 310.2 mm in this example, cycle 188 of the reactor). In our simulation, the control rod is finally extracted in order to trigger the divergence of the reactor.

The evolutions of $n(t)$ for five final positions of the control rod, 308.2, 309.2, 310.2, 311.2 and 312.2 mm, are compared to each other. We can observe on Fig. 14.21 the evolution of $n(t)$ when the control rod is extracted to a final position. In the case the final position is 310.2 mm, the reactor is critical and the evolution of $n(t)$ is linear, consistent with (14.34), and the dominant ω_0 is zero as shown on Fig. 14.22. In the cases where the control rod is extracted beyond its critical position, the dominant ω_0 is positive and $n(t)$ is an increasing exponential with a convex shape.

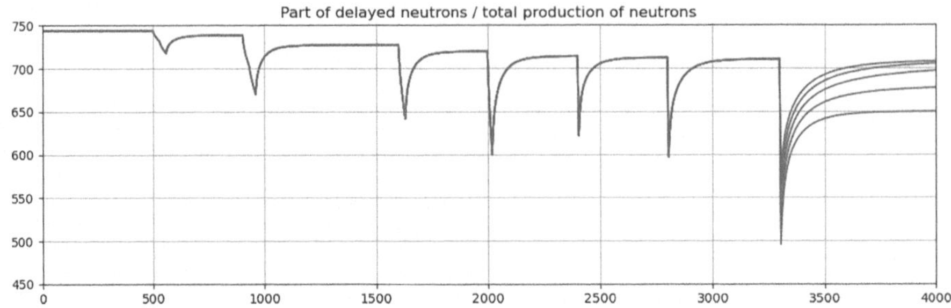

Fig. 14.23 HFR reactor of ILL. Part of delayed neutrons

In the cases where the control rod is not extracted beyond its critical position, the dominant ω_0 is negative and $n(t)$ is a concave function that with reach a stationnary state.

It can be demonstrated that the part of delayed neutrons in the case of a subcritical reactor, with a source and in a stationnary state, is $\dfrac{\beta}{1-\rho}$. This can be seen on Fig. 14.23.

14.8.6   Rod Drop on Cabri Reactor

The integral worth of 6 BCS (control and safety rods, Hafnium) has been measured by rod drop on June 15, 2016.

After a subcritical approach, the reactor diverges at $t = 1979$ s. The power is increased until the count rate on BN1 detector reaches 700000×10^5 hits/s, and then is brought back to criticality. Thus the critical position of the 6 BCS is 417 mm, and at $t = 3000$ s, the reactor is critical, the 6 BCS are at their critical position. Then the drop of the 6 BCS is triggered at $t = 3180.8$ s.

At the end of the rod drop, the reactor is strongly undercritical, with $\rho = -8003$ pcm. Consequently, the ω_i are almost equal to $-\lambda_i$, for i in $\{0, 1, 2, 3 ..\ 7\ \}$ and ω_8 is determined by the intersection of the Nordheim curve, on its oblic asymptote, and the horizontal line $\rho = -8003$ pcm ($\omega_8 = -3391$ s^{-1}).

A parametric study of the transient has been performed, with a variation of the emissivity of the source: an arbitrary emissivity is multiplied by factors 0.0, 0.05, 0.1, 0.5, 1.0 and 5.0. The Fig. 14.24 shows that in the case $S_0 = 0$, $n(t)$ becomes a pure exponential with the dominant ω_0 after the other exponential vanish (with a logarithmic scale, it is a line). In all other cases, with $S_0 \neq 0$, the reactor reaches a stationnary state and $n(t)$ becomes constant at a level depending on the source, coherently with the Eq. 14.32.

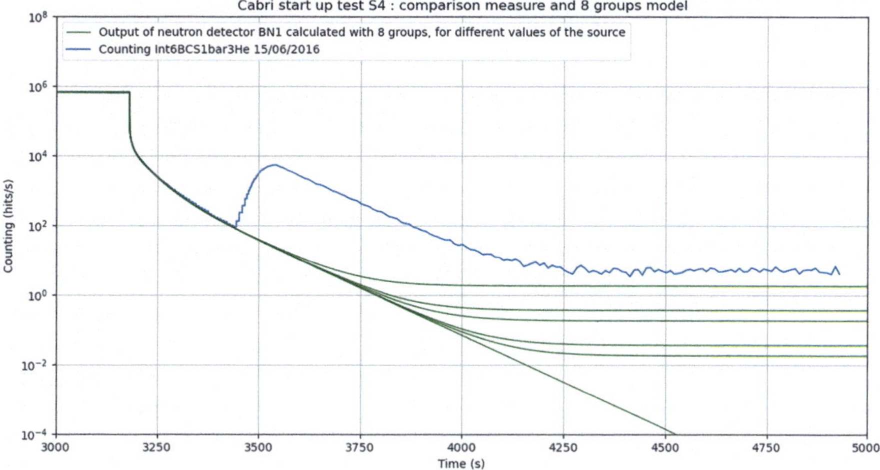

Fig. 14.24   Cabri rod drop. Measured count rate compared to calculations done with different source emissivities

The rebound of the measured count rate, visible between $t = 3440$ s and $t = 3540$ s is due to a move of the neutrons detector which is brought closer to the core by the operator, in order to counterbalance the exponential decrease of $n(t)$.

Due to the practical difficulties of accurately observing the instantaneous neutron level n_1 just after a rod drop, it is very useful to be able to determine rod worth also from the neutron decay characteristics following the drop.

14.9   The Shape Function Changes with Time During Transients: From Point Kinetics to Spatial Kinetics

14.9.1   Tilt Assumption

The instantaneous tilt assumption consists in separating the space-time problem into a space part and a time part, calculated with point kinetics. A simple way to take into account that the shape function changes with time is to make an estimation of the instantaneous shape function, based on the static flux corresponding to the instantenous values of the nuclear properties of the reactor, made critical by adjustment of some designed parameter (Radkowsky 1964, p. 958).

This replacement is referred to as the instantaneous tilt approximation, since it assumes in effect that the flux shape responds instantaneously to changes in nuclear properties. In reality, the delayed neutrons cause the shape change to lag somewhat behind the change in properties.

14.9.2 Quasi Static Method

The neutron flux is factorized in the product of an amplitude function which is only time dependent, and a shape function which is dependent on all variables including time. This approach, called quasi-static approach and introduced by Henry (1958), is based on the factorization of the neutron flux into an amplitude factor $n(t)$ which applies evenly to the entire core, and a shape functions Ψ:

$$\varphi(\boldsymbol{r}, \boldsymbol{\Omega}, E, t) = n(t)\,\Psi(\boldsymbol{r}, \boldsymbol{\Omega}, E, t)$$

The amplitude function obeys the classic point kinetics, while the shape function accounts for the spatial distortions that take place on a slower time scale and describes the time-dependent deformation of the power profile. Since the shape function varies more slowly with time, it does not need to be computed as frequently as the amplitude function.

14.9.3 Spatial Kinetics

Point kinetics may be relevant for a rector in normal operation, since a reactor in slow power transient is very close to criticality (a few pcm). However, spatial kinetics is necessary to analyse RIA (Reactivity Initiated Accidents) transients.

References

NEA Data Bank, The jef-2.2 nuclear data library. Technical report (2000)

G.I. Bell, S. Glasstone, *Nuclear Reactor Theory* (Van Nostrand Reinhold Company, 1970). https://books.google.fr/books?id=RNQmAQAAMAAJ

I. Dillmann, B. Singh, D. Abriola, Summary report of consultants' meeting on beta-delayed neutron emission evaluation (Technical report, International Atomic Energy Agency, Vienna, Austria, 2011)

A.F. Henry, The application of reactor kinetics to the analysis of experiments. Nucl. Sci. Eng. **3**(1), 52–70 (1958)

L. David, *Hetrick* (University of Chicago Press, Dynamics of Nuclear Reactors, 1971)

G.R. Keepin. *Physics of Nuclear Kinetics* (Addison-Wesley Publishing Company, 1965)

G.R. Keepin, T.F. Wimett, R.K. Zeigler, Delayed neutrons from fissionable isotopes of uranium, plutonium and thorium. J. Nucl. Energy (1954), **6**(1–2), IN2–21 (1957)

R.K. Meulekamp, S.C. Van Der Marck, Calculating the effective delayed neutron fraction with Monte Carlo. Nucl. Sci. Eng. **152**(2), 142–148 (2006)

A. Radkowsky, *Naval Reactors Physics Handbook: Selected Basic Techniques*, vol. 1 (Naval Reactors, Division of Reactor Development, United States Atomic Energy, 1964)

G. Rudstam, P. Finck, A. Filip, A D'angelo, R.D. McKnight, Delayed neutron data for the major actinides, in *A Report by the Working Party on International Evaluation Co-operation of the OECD NEA Nuclear Science Committee*, vol. 6 (2002)

W. Zwermann, N. Berner, A. Aures, K. Velkov, Sensitivities of delayed neutron fractions in the framework of scale 6.2. Ann. Nucl. Energy **153**, 108068 (2021)

Chapter 15
Subcritical Reactor

Abstract A subcritical reactor with an independent neutron source tends towards a stationary state, in which the neutron source permanently makes up the deficit of neutrons between two generations. This chapter shows how the flux shape depends on the location of the sources and the subcriticality of the reactor. It highlights the impact the location of the sources can have on the inverse multiplication curve during approach to criticality. A deep explanation of the flux shape in a subcritical reactor will be provided in the chapter Highlighting of harmonics.

15.1 Flux Shape in the Case of a Subcritical Reactor with a Source

15.1.1 Source in the Center of the Reactor

A source is introduced in the center of the reflected reactor calculated in Sect. 13.4 Flattening the radial flux. Subcriticality $\rho = -1000$ pcm is achieved by lowering the thermal utilization factor, from $f = 0.85167$ down to $f = 0.84260$. The source has a radius 0.1 cm. The shape of the flux is shown by Fig. 15.1. It can be seen that the shape of the flux is settled by the source. If the subcriticality tends to be more and more negative, the system comes down to a source in a scattering and absorbing medium.

15.1.2 Source in the External Ring of the Reactor

A source is introduced in a ring defined by internal and external radius, respectively 160 and 160.1 cm. Figure 15.2 shows that the shape of the thermal flux is determined by the source, and also by the reflector.

© The Author(s), under exclusive license to Springer Nature Switzerland AG 2026

H. Grard, *Diffusion of Neutrons in Nuclear Reactors*,

https://doi.org/10.1007/978-3-032-05088-5_15

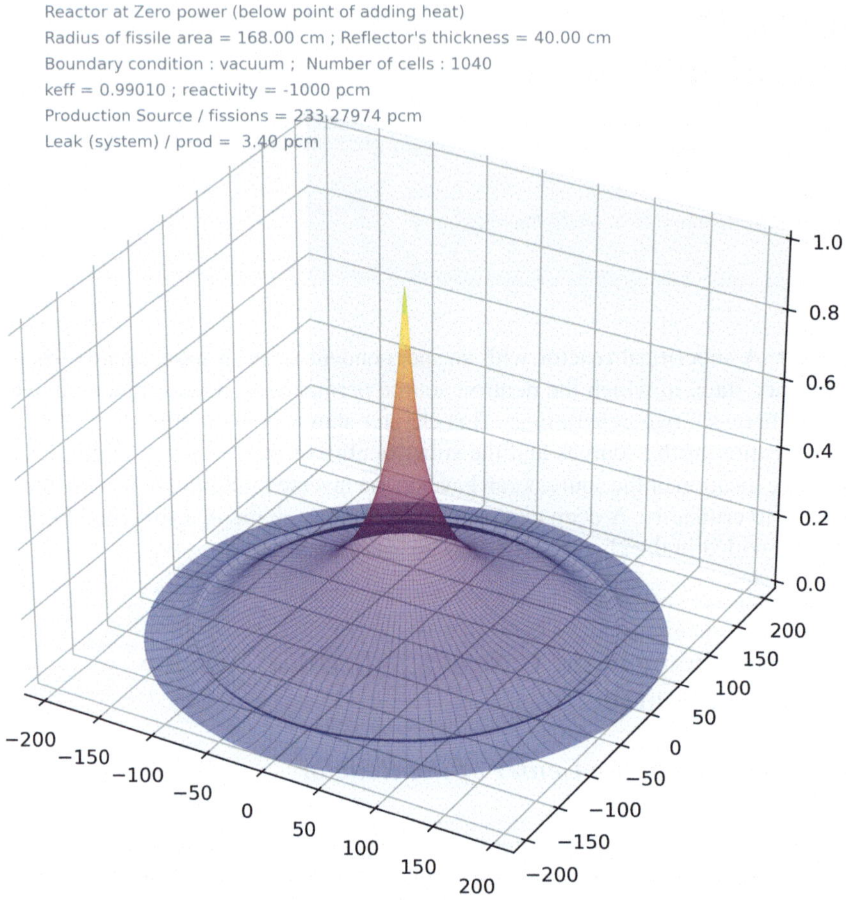

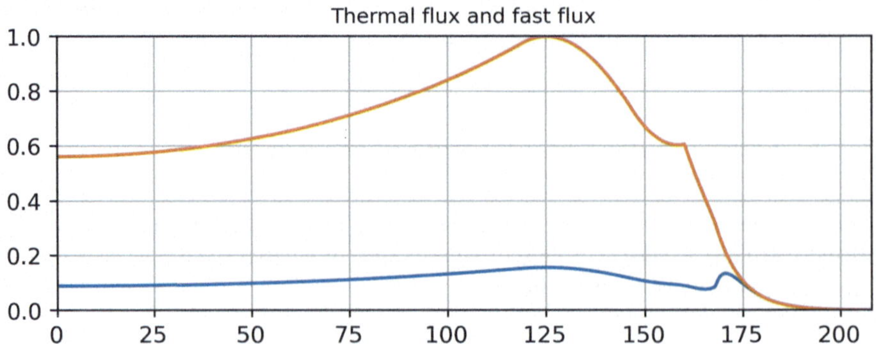

Fig. 15.1 Shape of thermal flux in the case of a source at the center of a subcritical reactor ($\rho = -1000\,$pcm)

Fig. 15.2 Shape of fast and thermal flux in the case of a source in an external ring of a subcritical reactor ($\rho = -1000\,$pcm)

15.1.3   Multiples Sources

The solution for multiple sources can be obtained by the superposition of the solutions for each source (Bell and Glasstone 1970, p. 20).

15.1.4   Homogeneous Source in the Fissile Area

If the source is homogeneous in the fissile area, the resulting flux shape is not far from the fundamental mode, and does not as much depends on the subcriticality than in the case of a narrow source.

15.1.5   Source Shape According to the Fundamental Mode

Regardless of the subcriticality of the reactor, if the source is distributed according the fundamental mode, the resulting flux shape is the fundamental mode. This is called total normal-mode multiplication, obtained in the idealized case of a source which produces neutrons with the same spatial and energy distributions as the neutrons produced by the fundamental (Keepin 1965, p. 223).

The thermal flux shape is exactly the same as the one of the critical reactor without source neutrons, as illustrated by Fig. 15.3 that illustrates the case of a subcriticality $\rho = -20\,000$ pcm.

15.1.6   Flux Shape and Subcriticality

The evolution of the flux shape is determined by the sources when the reactor is strongly subcritical. When the reactor shifts to an almost critical configuration, the shape of the flux tends to become the funtamental mode. Due to delayed neutrons, each modification of the core that increases reactivity (for example a small withdrawal of rods results in a transition to a new flux distribution determined by the new reactivity, during a few minutes, until a steady state is achieved (Fig. 15.4).

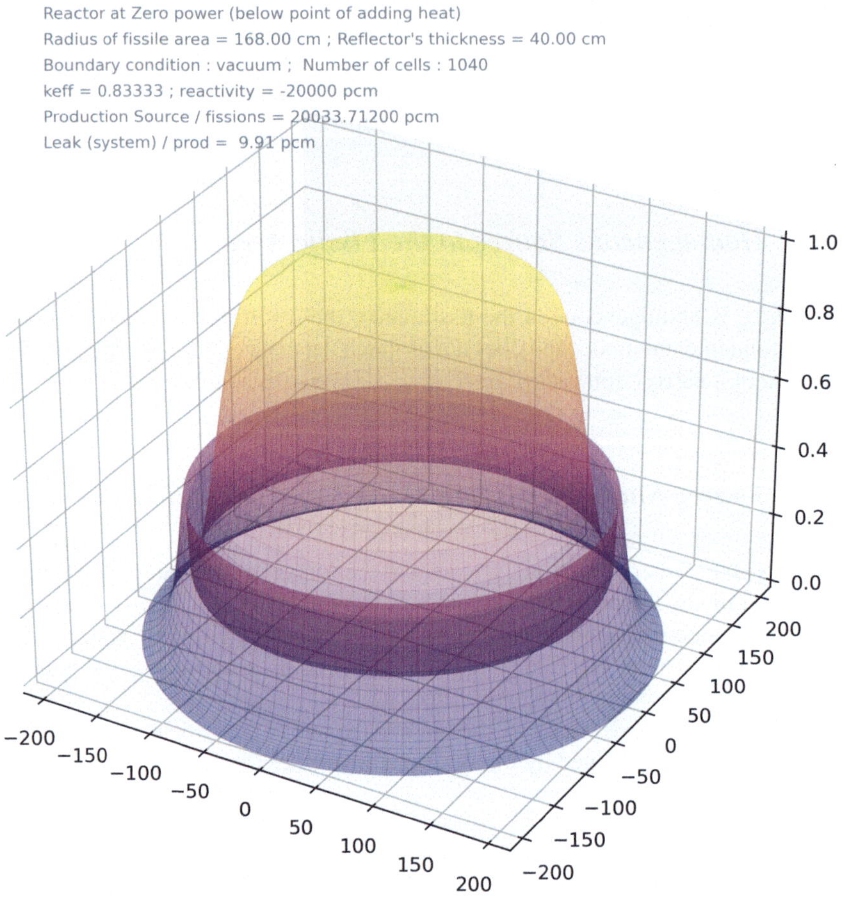

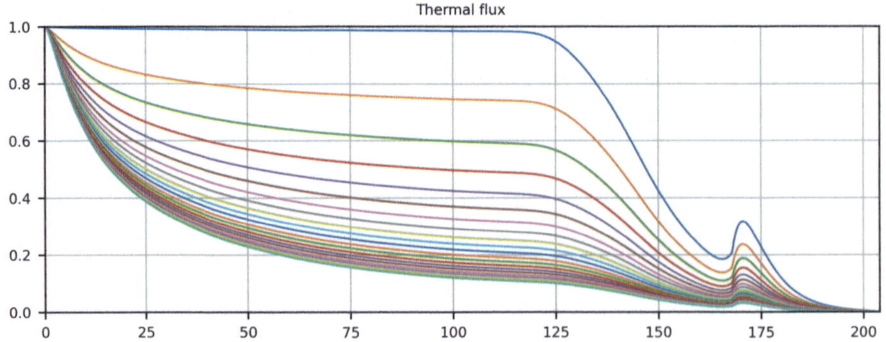

Fig. 15.3 Shape of thermal flux in the case of a source distributed according the fundamental mode, the reactor being strongly subcritical($\rho = -2.0000 \times 10^4$ pcm)

Fig. 15.4 Evolution of the shape of thermal flux in the case of a source at the center of a subcritical reactor with various reactivities (–1000 to –2 pcm)

15.2 Evaluation of the Reactivity with the Neutron Amplification Factor

15.2.1 Power and Neutron Amplification Factor

The power amplification factor PAF of a driven subcritical system is defined as the ratio of the fission power output of the blanket to the power which the driver must deliver to sustain its neutron source intensity.

Noack et al. (2013, Abstract)

Similarly we can define the neutron amplification factor of the core as the ratio of the rate of production of neutrons by neutron induced fissions to the rate of emission of source neutrons.

We define the following variables:

- R is the removal rate: number of neutrons absorbed or leaking, by unit of time
- P is the production rate: number of neutrons produced by neutron induced fissions, by unit of time.

If the source is distributed according to the fundamental mode, we can write that:

$$S_0 = \frac{(1 - k_{\text{eff}})\, n}{\theta} = (1 - k_{\text{eff}})\, R$$

$$\frac{\frac{S_0}{R}}{\frac{P}{R}} = \frac{1 - k_{\text{eff}}}{k_{\text{eff}}} = -\rho$$

And then:

$$\rho = -\frac{S_0}{P} \tag{15.1}$$

This equation is equivalent to the Eq. 14.32, which we recall below:

$$n = -\frac{S_0\, l}{\rho}$$

The subcriticality of a stationary subcritical reactor is the inverse function of the neutron amplification factor. A Monte Carlo calculation enables to calculate the global amplification factor, and thus to estimate the negative reactivity of a subcritical reactor.

15.2.2   Accuracy of the Reactivity Evaluation

The evaluation of the reactivity as the inverse function of the neutron amplification factor is accurate as far as the shape of neutrons source is close to the fundamental mode of the reactor.

Figure 15.5 compares the inverse function of the neutron amplification factor, as a function of the reactivity, for different configuration of the independent source of neutrons: a source (radius 0.1 cm) in the center of the reactor, neutrons source emitted in an external ring, and then neutrons source shaped with the fundamental mode. It can be seen that if the source is in the center of the core, reactivity is overestimated (closer to criticality), and underestimated if the source is in the external ring. That can be understood with the concept of importance. Importance may be defined as the number of neutrons that will be produced at other energies and directions as a result of fission and scattering events initiated by the original neutron (Stacey 2007,

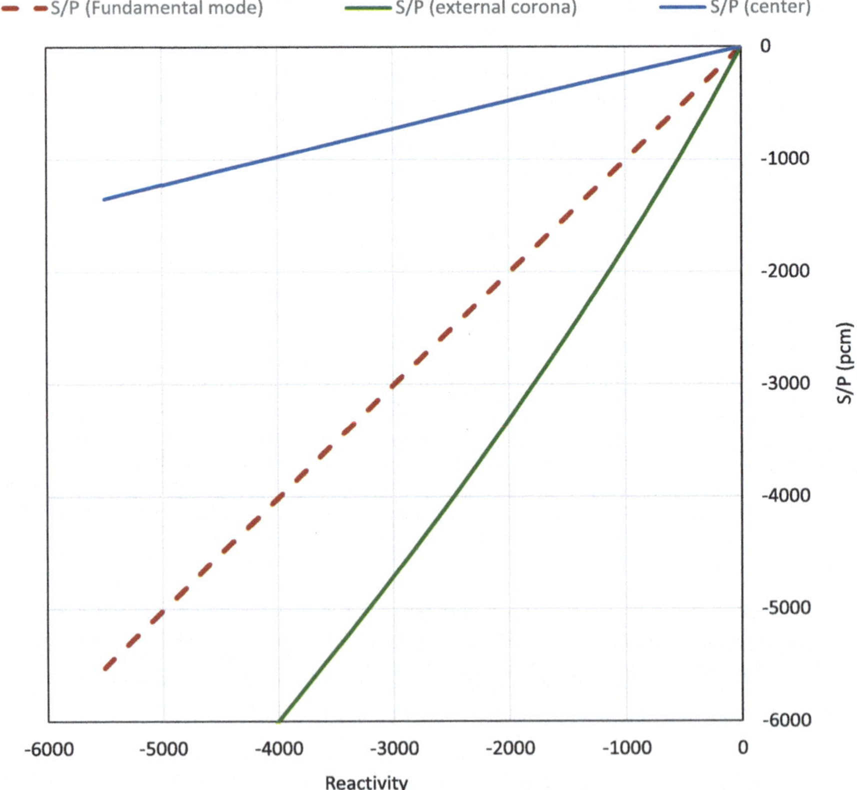

Fig. 15.5 Estimation of reactivity with the ratio source/production for different Sources configurations

p. 487). It must be mentioned that in a subcritical system, all neutrons have a finite descendancy.

> Intuitively, it is easy to understand that a neutron placed at the center of a reactor, with a good chance of causing fission, would be more "important" than a neutron placed at the surface with a high probability of escape.
>
> Reuss (2008, p. 484)

In a limit case, a neutron located at the free-surface boundary of a system, and having a velocity vector oriented outwards from the system, has no "importance" since it cannot return (Bell and Glasstone 1970, p. 257).

15.2.3 Neutron Source Efficiency

A neutrons source efficiency in a subcritical reactor can be defined by the equation (Sun et al. 2016, p. 1294):

$$\kappa = -\rho \frac{P}{S_0} \tag{15.2}$$

If the source is distributed according to the fundamental mode, then $\kappa = 1$. The efficiency may be lower or greater than unity, depending on the source location, and also on the energy and direction of the neutrons.

15.3 Approach to Criticality

In this section, the dynamic behavior of the reactor during approach to criticality, following rods withdrawals, or boric acid dilution in the reactor coolant system will not be considered. Although approach to criticality is considered as a static technique, there are transient effects. The asymptotic multiplication, following a reactivity change in a subcritical system, is observed only after all delayed neutrons have attained their equilibrium concentration. We will only focus on the successive stationary states, and on the monitoring of the core with leaking neutrons.

15.3.1 Inverse Count Rate Ratio

Excore source-range detectors give a count rate, which is an image of the neutron leakage, that is used to monitor reactivity states of reactors. A universal practice in reactor operation is that, during approach to criticality, the inverse of the counting rate is plotted as a function of the physical parameter controlling reactivity, for example the position of neutron absorbers, or the boron concentration in the case of a PWR.

After each reactivity change, the operator must wait until asymptotic multiplication is reached, i.e. the neutron population and precursors stabilize, depending on the source and reactor subcriticality. At stabilization, the operator reads out the count rate value. In the case of a boron dilution (PWR), the reactivity change is a continuous and slow process. This is why, during the dilution, the neutron population at a given instant is in quasi-equilibrium with the asymptotic solution given by the Eq. 14.32. When the inverse of the count rate approaches zero, it's good practice to reduce the dilution flow rate in stages, so as to lower the rate of change of reactivity and maintain a quasi-static equilibrium, as the asymptotic population increases in inverse proportion to reactivity.

The inverse of count rate is currently scaled at unit at the beginning of the approach to criticality, this is called the Inverse Count Rate Ratio or ICRR. As the distance from criticality reduces, the ICRR falls towards zero. A line extending from the ICRR graph down to the x-axis (ICRR equal to zero) gives a prediction of the rods position, or boron concentration, achieving criticality.

The Fig. 15.6 shows the inverse of the leaking neutron currents, out of the external surface of the water reflector, including fast and thermal neutrons. The reactor is still the reactor defined in Sect. 13.4 Flattening the radial flux, and we will consider three different kind of independent neutron sources: shaped by the fundamental mode, at the center, close to the edge of the fissile area (in the external ring).

The curvature of the inverse of the count rate, as a function of the physical parameter controlling reactivity, depends on the following parameters:

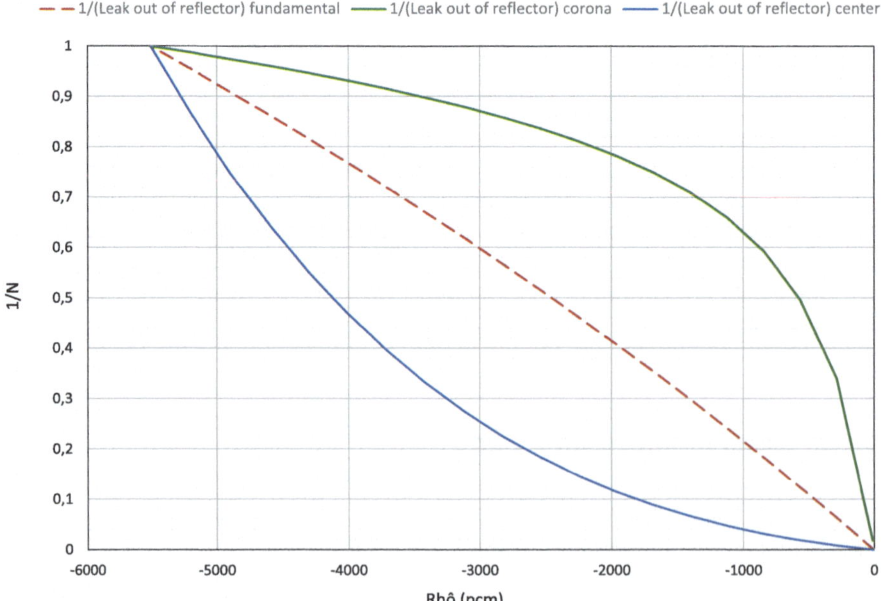

Fig. 15.6 Inverse count rate as a function of subcriticality, for different sources configurations

- The position and emissivity of the inherent and external sources of neutrons
- The fact that leak neutrons through the reflector may not be proportional to core averaged power
- The position of the detector
- The evolution of the differential worths (the one of boron concentration, or the one of rods position), is not a constant function.

Keeping illustrates typical distorsions of inverse multiplication curves in the case of experimental data obtained during startup of the Los Alamos reactor LOPO: for various detector locations, the reciprocal multiplication curves exhibit negative curvature (concave downward) or positive curvature (Keepin 1965, Fig. 8.1), depending on the detector location.

15.3.2 PWR Approach to Criticality with Reload Core Compared with Initial Core

The comparison between the ICRR graph of a PWR with initial core, and the one of a PWR with reload core is instructive.

15.3.2.1 PWR with Reload Core

In the case of a PWR with reload fuel assemblies, the inverse multiplication curve, or ICRR graph, drawn as a function of the boron concentration, is commonly close to be linear. This can be explained by the fact that spent fuel assemblies, that are the inherent neutron source, are distributed throughout the whole core, consequently we can assume that the core is supplied with neutrons by an homogeneous source. And as stated in Sect. 15.1.4, an homogeneous source results in a flux that is virtually distributed according to the fundamental mode, and therefore the ICRR graph is linear.

15.3.2.2 PWR with Initial Core

In the case of a PWR with an initial core, the inherent source coming from spontaneous fission of ^{238}U from fresh UO_2 assemblies is negligible. For example, the emissivity of a fresh fuel assembly for PWR 900 MWe, containing 461.7 kg hmi (heavy metal initial, i.e. uranium element), is about 7500 n/s. These neutrons mainly come from spontaneous fission of ^{238}U, since the branching ratio towards spontaneous fission is significantly higher for ^{238}U than ^{235}U, respectively $5.46 \times 10^{-5}\%$ and $7.2 \times 10^{-9}\%$.

That's why neutrons are provided by four ^{252}Cf sources, located on the edge of the core. The flux is like four humps centered on each source, with the neutrons

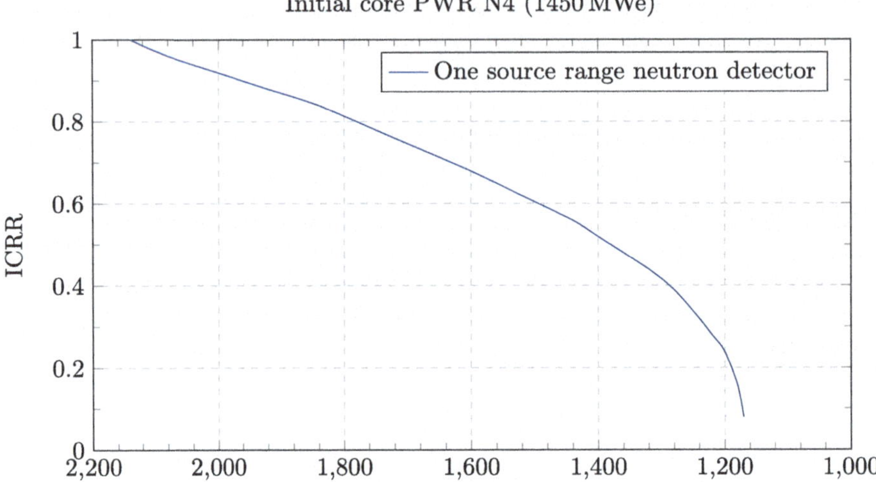

Fig. 15.7 Approach to criticality, PWR N4 loaded with initial core

they produce scattering around it and being amplified by the fissions produced in neighbouring rods. What a source-range detector sees is the activity of the source facing it, amplified by its immediate environment.

At the start of dilution, the reactivity of the overall core increases, but this does not do much for the count rate, which depends essentially on the source. The count rate is therefore not very sensitive to the increase in core reactivity and the ICRR curve varies slowly. As the core approaches criticality, the flux shape changes and the flux tends to be distributed according to the fundamental mode. An increase in reactivity translates into an increase in the amplitude of the flux over the whole core, and this has repercussions on the detector count rate, which starts to rise sharply. This is why there is a drop towards zero in ICRR (Fig. 15.7).

15.3.3 Optimal Source Position

The source position in the core is determined so as to provide a high detector response. The neutron source efficiency is an integral point of view, and the magnitude of the neutron count does not only depend on neutron production rate in the core, but also on intensity of the neutron leaks reaching the detector, which depends on the subcritical flux shape.

References

G.I. Bell, S. Glasstone, *Nuclear Reactor Theory* (Van Nostrand Reinhold Company, 1970). https://books.google.fr/books?id=RNQmAQAAMAAJ

G.R. Keepin, *Physics of Nuclear Kinetics* (Addison-Wesley Publishing Company, 1965)

K. Noack, O. Ågren, V.E. Moiseenko, A. Hagnestål, Comments on the power amplification factor of a driven subcritical system. Ann. Nucl. Energy **59**, 261–266 (2013). ISSN 0306-4549. https://doi.org/10.1016/j.anucene.2012.06.020. https://www.sciencedirect.com/science/article/pii/S0306454912002332

P. Reuss, *Neutron Physics* (EDP Sciences, 2008)

W.M. Stacey, *Nuclear Reactor Physics* (Wiley, 2007). ISBN 9783527406791. https://books.google.fr/books?id=y1UgcgVSXSkC

J. Sun, M.-S. Yahya, Y. Kim, A study on the optimal position for the secondary neutron source in pressurized water reactors. Nucl. Eng. Technol. **48**(6), 1291–1302 (2016)

Chapter 16
Highlighting of Harmonics

Abstract In this chapter we consider a bare homenegeous reactor, having an infinite cylinder geometry. At initial time, it will be assumed that a infinite line source emitting fission neutrons is introduced in the reactor. Our objective is to establish analytical equations describing the time dependant shape and magnitude of the thermal flux. The resolution of this problem combines the time-dependant diffusion equation and the age equation, and will highlight the concept of harmonics and sheds new light on criticality. This chapter is inspired by the classical book of Lamarsh (Lamarsh in Introduction to nuclear reactor theory. Addison-Wesley Publishing Company, 1966, Chap. 9). At the end of the chapter, an overture is made to the question of reactor stability to xenon-induced oscillations, although this topic is not developed in this book.

16.1   Time Dependant Equations

16.1.1   Time Dependant Equations with Delayed Neutrons

From the system of[1] Eq. 12.4, and time dependant Eq. 7.3, we can write the time dependant equation for the thermal group:

$$D_2(r)\Delta\Phi_2(r, t) - \Sigma_{r2}(r)\Phi_2(r, t) + p\Sigma_{r1}(r)\Phi_1(r, t) = \frac{\partial n(r, t)}{\partial t}$$

where $n(r, t)$ is the neutron angle and energy integrated density, over the thermal energy range $[0, E_{\text{cut off}}]$.

---

[1] John R. Lamarsh (1928, 1981), an eminent nuclear physicist and a longtime member of the New York University faculty, received his Bachelor of Science degree in General Science from the Massachusetts Institute of Technology in 1948 and attained his doctorate in Physics from this same institution in 1952. During his tenure at New York University Dr. Lamarsh published Introduction to Nuclear Reactor Theory (1966). This graduate-level textbook was based upon lectures which he gave at Cornell and New York Universities on the topic of nuclear reactor theory.

H. Grard, *Diffusion of Neutrons in Nuclear Reactors*,
https://doi.org/10.1007/978-3-032-05088-5_16

The flux can be expressed as a function of the neutron energy integrated density and the average neutron velocity:

$$\Phi_2(r, t) = n(r, t)\,\overline{v}$$

We remind that $p\Sigma_{r1}\Phi_1 = p\,q(r, \tau_{th}, t) = \Sigma_{s1\to2}\Phi_1$ and $q(r, \tau_{th}, t)$ is the slowing down current without resonant absorption.

Then we divide the equation by $\Sigma_{r2}(r)$, and:

$$L_2^2(r)\Delta\Phi_2(r) - \Phi_2(r) + \frac{p}{\Sigma_{r2}(r)}q(r, \tau_{th}, t) = \frac{1}{\overline{v}\Sigma_{r2}(r)}\frac{\partial\Phi_2(r, t)}{\partial t}$$

When age theory and theory of diffusion applied to thermal neutrons are combined, it's as if the neutrons undergo a slowing down to a cut-off energy, beyond which they are transferred to the thermal neutron group. At this cut-off energy, the age of the neutrons is τ_{th} (Bussac and Reuss 1985, p. 210).

We assume that $\Sigma_{r2}(r) \approx \Sigma_{a2}(r, \overline{v})$, and this enables to introduce the thermal diffusion time t_d (see Sect. 1.8).

Three more equations can be added: the diffusion equation, a balance of fast neutrons, and an equation for precursors with an index i.

$$
\left\{
\begin{array}{ll}
L_2^2(r)\Delta\Phi_2(r) - \Phi_2(r) + \dfrac{p}{\Sigma_{r2}(r)}q(r, \tau_{th}, t) = t_d\,\dfrac{\partial\Phi_2(r, t)}{\partial t} & (16.1a) \\[3ex]
\Delta q(r, \tau, t) = \dfrac{\partial q(r, \tau, t)}{\partial \tau} & (16.1b) \\[3ex]
q(r, 0, t) = S_0\delta(r) + \dfrac{k_\infty(1 - \sum_i \beta_i)}{p}\Sigma_{r2}(r)\Phi_2(r, t) + \displaystyle\sum_i \lambda C_i(r, t) & (16.1c) \\[3ex]
\dfrac{\partial C_i(r, t)}{\partial t} = \dfrac{k_\infty\beta_i}{p}\Sigma_{r2}(r)\Phi_2(r, t) - \lambda C_i(r, t) & (16.1d)
\end{array}
\right.
$$

Equation 16.1c is a balance of fast neutrons, of age equal to 0. The production of fast neutrons by thermal and fast fissions is represented, at criticality, by the term (Eq. 12.6):

$$\frac{k_\infty}{p}\Sigma_{r2}(r)\Phi_2(r, t) = \nu_1\Sigma_{f1}\Phi_1 + \nu_2\Sigma_{f2}\Phi_2$$

The neutrons produced by the source, by fissions and by (β^-, n) decay of the precursors are all assumed to be at the same energy. This is not rigorously true, since the energy spectrum of delayed neutrons occupies a lower energy interval than the prompt neutrons spectrum.

$C(r, t)$ is the density of precursors, and $S_0\delta(r - r_s)$ is the Dirac source function. The source is a 1D line within the infinite cylinder, located at the position r_s. If $r \neq r_s$, then $\delta(r - r_s) = 0$.

16.1.2 Time Dependant Equations Without Delayed Neutrons

$$
\begin{cases}
L_2^2(\boldsymbol{r})\Delta\Phi_2(\boldsymbol{r}) - \Phi_2(\boldsymbol{r}) + \dfrac{p}{\Sigma_{r2}(\boldsymbol{r})}q(\boldsymbol{r},\tau_{th},t) = t_d\,\dfrac{\partial\Phi_2(\boldsymbol{r},t)}{\partial t} & (16.2a) \\[2ex]
\Delta q(\boldsymbol{r},\tau,t) = \dfrac{\partial q(\boldsymbol{r},\tau,t)}{\partial\tau} & (16.2b) \\[2ex]
q(\boldsymbol{r},0,t) = S_0\delta(\boldsymbol{r}) + \dfrac{k_\infty}{p}\Sigma_{r2}(\boldsymbol{r})\Phi_2(\boldsymbol{r},t) & (16.2c)
\end{cases}
$$

16.2 Resolution in an Homogeneous Cylinder, Without Delayed Neutrons

16.2.1 Decomposition Onto an Orthogonal Basis

We define an infinite and homogeneous cylinder of radius R that is characterized by $k_\infty > 1$, consequently the positive eigen values of the Laplacian will be considered. Since the cylinder is homogeneous, Σ_{r2} and L_2^2 are constant and do not depend on $\boldsymbol{r}$. In an infinite cylinder, we can express $q(\boldsymbol{r},\tau,t)$ and $\Phi_2(\boldsymbol{r},t)$ in polar coordinates:

$$
q(\boldsymbol{r},\tau,t) = q(r,\theta,\tau,t) \quad \text{and} \quad \Phi_2(\boldsymbol{r},t) = \Phi_2(r,\theta,t)
$$

Appendix E shows how a function expressed in polar coordinates can be decomposed onto an orthogonal basis $(g_{m,n})$:

$$
\Phi_2(r,\theta,t) = \sum_{m=1}^{\infty}\sum_{n=0}^{\infty} f_{m,n}(t)g_{m,n}(r,\theta) \tag{16.3}
$$

$$
q(r,\theta,\tau,t) = \sum_{m=1}^{\infty}\sum_{n=0}^{\infty} h_{m,n}(\tau,t)g_{m,n}(r,\theta) \tag{16.4}
$$

And also:

$$
\frac{\partial q(r,\tau,t)}{\partial\tau} = \sum_{m=1}^{\infty}\sum_{n=0}^{\infty} g_{m,n}(r,\theta)\frac{\partial h_{m,n}(\tau,t)}{\partial\tau} \tag{16.5}
$$

16.2.2   Calculation of the Time Dependant Functions $f_{m,n}(t)$

$g_{m,n}(r, \theta)$ is an eigen function of the Laplacian in a disk, or infinite cylinder, with Dirichlet boundary conditions. Its associated eigen value is $(\frac{\mu_{m,n}}{R})^2$. Thus:

$$\Delta g_{m,n}(r, \theta) = -\frac{\mu_{m,n}^2}{R^2} g_{m,n}(r, \theta) \tag{16.6}$$

The substitution of (16.6) into Eqs. 16.2a and 16.2b involving Laplacians enables us to write:

$$\Delta \Phi_2(r) = \sum_{m=1}^{\infty} \sum_{n=0}^{\infty} f_{m,n}(t)(-\frac{\mu_{m,n}^2}{R^2} g_{m,n}(r, \theta)) \tag{16.7}$$

$$\Delta q(r, \tau, t) = \sum_{m=1}^{\infty} \sum_{n=0}^{\infty} h_{m,n}(\tau, t) \times (-\frac{\mu_{m,n}^2}{R^2} g_{m,n}(r, \theta)) \tag{16.8}$$

Since $(g_{m,n}(r, \theta))$ is an orthogonal basis, the corresponding terms in Eqs. 16.5 and 16.8 are equal:

$$\frac{\partial h_{m,n}(\tau, t)}{\partial \tau} = -\frac{\mu_{m,n}^2}{R^2} h_{m,n}(\tau, t) \tag{16.9}$$

The solution of the time dependant differential Eq. 16.9 is:

$$h_{m,n}(\tau, t) = T_{m,n}(t) e^{-\frac{\mu_{m,n}^2}{R^2} \tau} \tag{16.10}$$

The coefficients $T_{m,n}$ will be determined later.

Now, from Eqs. 16.4 and 16.10 we can write again the slowing down current:

$$q(r, \tau, t) = \sum_{m=1}^{\infty} \sum_{n=0}^{\infty} T_{m,n}(t) e^{-\frac{\mu_{m,n}^2}{R^2} \tau} g_{m,n}(r, \theta) \tag{16.11}$$

The neutrons source is mathematically represented by $S_0 \times \delta(r - r_s)$, where S_0 ncm/cm³/s is the emissivity of the source. The Dirac function in polar coordinates is decomposed onto the basis:

$$S_0 \delta(r - r_s) = \sum_{m=1}^{\infty} \sum_{n=0}^{\infty} S_{m,n} g_{m,n}(r, \theta)$$

The values $S_{m,n}$ depend on the position of the source r_s.

In (16.2c), we substitute $q(r, \tau, t)$ et $\Phi_2(r, \theta, t)$ by the expression (16.11) et (16.3):

$$q(\mathbf{r}, 0, t) = \sum_{m=1}^{\infty} \sum_{n=0}^{\infty} T_{m,n}(t) g_{m,n}(r, \theta)$$

$$= S_0 \delta(\mathbf{r} - \mathbf{r}_s) + \frac{k_\infty}{p} \Sigma_{r2} \sum_{m=1}^{\infty} \sum_{n=0}^{\infty} f_{m,n}(t) g_{m,n}(r, \theta)$$

$$= \sum_{m=1}^{\infty} \sum_{n=0}^{\infty} S_{m,n} g_{m,n}(r, \theta) + \frac{k_\infty}{p} \Sigma_{r2} \sum_{m=1}^{\infty} \sum_{n=0}^{\infty} f_{m,n}(t) g_{m,n}(r, \theta)$$

Given that $(g_{m,n})$ is a basis, we can write:

$$T_{m,n}(t) = S_{m,n} + \frac{k_\infty}{p} \Sigma_{r2} f_{m,n}(t) \tag{16.12}$$

We calculate the inner product vector in polar coordinates between the source function and the function $g_{m,n}$, which is a vector of the basis:

$$\langle S_0 \delta(\mathbf{r} - \mathbf{r}_s), g_{m,n} \rangle = \underbrace{\int_0^R r \int_0^{2\pi} S_0 \delta(\mathbf{r} - \mathbf{r}_s) g_{m,n}(r, \theta) \, d\theta \, dr}_{= S_0 g_{m,n}(\mathbf{r}_s) \text{ property of Dirac delta function}}$$

Since $S_{m,n}$ is the component[2] of the function $S_0 \delta(\mathbf{r} - \mathbf{r}_s)$ in the basis $(g_{m,n})$:

$$S_{m,n} = \langle S_0 \delta(\mathbf{r} - \mathbf{r}_s), g_{m,n} \rangle$$

This enables to calculate:

$$S_{m,n} = S_0 \, g_{m,n}(\mathbf{r}_s) \tag{16.13}$$

And then, from (16.12) and (16.13):

$$T_{m,n}(t) = S_0 \, g_{m,n}(\mathbf{r}_s) + \frac{k_\infty}{p} \Sigma_{r2} f_{m,n}(t) \tag{16.14}$$

The expression of $T_{m,n}(t)$ given by (16.14) can be introduced in Eq. 16.4:

$$q(\mathbf{r}, \tau, t) = \sum_{m=1}^{\infty} \sum_{n=0}^{\infty} \left(S_{m,n} + \frac{k_\infty}{p} \Sigma_{r2} f_{m,n}(t) \right) e^{-\frac{\mu_{m,n}^2}{R^2} \tau} g_{m,n}(r, \theta) \tag{16.15}$$

$$= S_{m,n} \sum_{m=1}^{\infty} \sum_{n=0}^{\infty} e^{-\frac{\mu_{m,n}^2}{R^2} \tau} g_{m,n}(r, \theta)$$

$$+ \frac{k_\infty}{p} \Sigma_{r2} \sum_{m=1}^{\infty} \sum_{n=0}^{\infty} f_{m,n}(t) e^{-\frac{\mu_{m,n}^2}{R^2} \tau} g_{m,n}(r, \theta) \tag{16.16}$$

---

[2] See Appendix B, Eq. B.5.

Equation 16.16 can now be inserted in (16.1b) if we set $\tau = \tau_{th}$:

$$t_d \sum_{m=1}^{\infty} \sum_{n=0}^{\infty} g_{m,n}(r,\theta) \frac{d\,f_{m,n}(t)}{dt} =$$

$$L_2^2 \sum_{m=1}^{\infty} \sum_{n=0}^{\infty} -f_{m,n}(t)\frac{\mu_{m,n}^2}{R^2}\,g_{m,n}(r,\theta) - \sum_{m=1}^{\infty} \sum_{n=0}^{\infty} f_{m,n}(t)g_{m,n}(r,\theta)$$

$$+ \frac{p\,S_{m,n}}{\Sigma_{r2}(r)} \sum_{m=1}^{\infty} \sum_{n=0}^{\infty} e^{-\frac{\mu_{m,n}^2}{R^2}\tau_{th}}\,g_{m,n}(r,\theta) + k_\infty \sum_{m=1}^{\infty} \sum_{n=0}^{\infty} f_{m,n}(t)e^{-\frac{\mu_{m,n}^2}{R^2}\tau_{th}}\,g_{m,n}(r,\theta)$$

Since $(g_{m,n})$ is an orthogonal basis, we can conclude that:

$$t_d \frac{d\,f_{m,n}(t)}{dt} =$$

$$- L_2^2 f_{m,n}(t)\frac{\mu_{m,n}^2}{R^2} - f_{m,n}(t) + \frac{p}{\Sigma_{r2}}S_{m,n}e^{-\frac{\mu_{m,n}^2}{R^2}\tau_{th}} + k_\infty f_{m,n}(t)e^{-\frac{\mu_{m,n}^2}{R^2}\tau_{th}} \quad (16.17)$$

We set the following new variables:

$$t_{m,n} = \frac{t_d}{1 + \dfrac{\mu_{m,n}^2}{R^2}L_2^2} \quad (16.18)$$

$$k_{m,n} = \frac{k_\infty e^{-\frac{\mu_{m,n}^2}{R^2}\tau_{th}}}{1 + \dfrac{\mu_{m,n}^2}{R^2}L_2^2} \quad (16.19)$$

And then we can write (16.17) again:

$$t_{m,n}\frac{d\,f_{m,n}(t)}{dt} = (k_{m,n} - 1)f_{m,n}(t) + \frac{k_{m,n}}{k_\infty}S_{m,n}\frac{p}{\Sigma_{r2}} \quad (16.20)$$

Since the source is introduced in the infinite cylinder at $t = 0$, $f_{m,n}(0) = 0$. The solution of the differential Eq. 16.20 is:

$$f_{m,n}(t) = \frac{k_{m,n}}{k_\infty}S_{m,n}\frac{p}{\Sigma_{r2}}\frac{1}{k_{m,n} - 1}\left[e^{(k_{m,n}-1)\frac{t}{t_{m,n}}} - 1\right] \quad (16.21)$$

16.2.3   *Description of* $\Phi_2(r,\theta,t)$ *Solution of the Problem*

$\Phi_2(r,\theta,t)$ is an infinite sum of harmonics, given by (16.3):

$$\Phi_2(r, \theta, t) = \sum_{m=1}^{\infty} \sum_{n=0}^{\infty} f_{m,n}(t) g_{m,n}(r, \theta)$$

$(g_{m,n})$ is an orthogonal basis, and the functions $g_{m,n}$ are established in the Appendix E:

$$g_{m,n}(r, \theta) = J_n\left(r \frac{\mu_{m,n}}{R}\right) \cos(n\theta)$$

The functions $g_{m,n}$ are dumped by the time decreasing functions $f_{m,n}(t)$

16.2.4   Case of a Slab Geometry

We consider here an infinite bare slab homogeneous reactor of thickness L. If the radial leakage of a finite cylinder is not taken into account, L can be seen as the height of the core.

In this case, the eigen values are $\cos B_n x$, with n an odd integer (see Sect. 10.2.1) and the time dependant thermal flux is:

$$\Phi_2(t) \sum_{n \text{ odd}} f_n(t) \cos B_n x \tag{16.22}$$

Multiplication factor k_n, and t_n can be introduced:

$$k_n = \frac{k_\infty e^{-B_n^2 \tau_{th}}}{1 + B_n^2 L_2^2} \tag{16.23}$$

$$t_n = \frac{t_d}{1 + B_n^2 L_2^2} \tag{16.24}$$

This enables to write $f_n(t)$:

$$f_n(t) = \frac{k_n}{k_\infty} \frac{2 S_0 \, p}{L \, \Sigma_{r2}} \frac{1}{k_{m,n} - 1} \left[e^{(k_n - 1)\frac{t}{t_n}} - 1 \right] \tag{16.25}$$

16.2.5   Mutiple Sources

In the case of mutiple sources, one has just to add the $S_{m,n}$ values calculated for each source.

16.2.6   Physical Sense Illustrated by an Example

16.2.7   Considered Reactor

We consider the bare homogeneous reactor described in Sect. 13.1, Table 13.1. This reactor, defined as an infinite cylinder, is critical for a radius $R = 168$ cm.

16.2.8   Multiplication Factors and Diffusion Times

The less negative eigen value of the Laplacian in an infinite cylinder is $\mu_{1,0}$. The eigen values up to $n = 9$ and $m = 10$ are given in the Table E.1. As shown on Table E.1, for a given value of n, the eigenvalues $\mu_{m,n}$ increase monotonically with m. Therefore, from Eq. 16.19, $k_{m,n}$ decrease for a given value of n and they decrease also for a given value of m.

Tables 16.1 and 16.2 give the values of the multiplication factors and diffusion times of the harmonics.

If $k_{1,0}$ is less than unity, in the case of a subcritical reactor, then all other $k_{m,n}$ are smaller, they are also less than unity. if $k_{1,0}$ is greater than unity, in practical it's not possible that a few additional $k_{m,n}$ may also be greater than unity, given the huge gap between $k_{1,0}$ and $k_{1,1}$. The reactivity difference is 1833 pcm, consequently $k_{1,1}$ could be greater than unity only in the case of a prompt critical reactor.

Table 16.1 Multiplication factors of harmonics $k_{m,n}$ in an homogeneous critical cylinder

	$m = 1$	2	3	4	5	6	7	8	9	10
$n = 0$	1.000	0.952	0.872	0.768	0.651	0.530	0.416	0.314	0.228	0.160
1	0.982	0.917	0.824	0.712	0.591	0.473	0.364	0.270	0.192	0.132
2	0.960	0.879	0.774	0.656	0.535	0.419	0.317	0.230	0.161	0.109
3	0.933	0.837	0.723	0.601	0.480	0.370	0.274	0.195	0.134	0.089
4	0.902	0.794	0.672	0.548	0.429	0.324	0.236	0.165	0.111	0.072
5	0.868	0.748	0.621	0.496	0.382	0.283	0.202	0.139	0.092	0.059
6	0.832	0.702	0.571	0.447	0.338	0.245	0.172	0.116	0.075	0.047
7	0.793	0.655	0.522	0.401	0.297	0.212	0.145	0.096	0.061	0.038
8	0.752	0.608	0.475	0.358	0.260	0.182	0.122	0.080	0.050	0.030
9	0.711	0.562	0.430	0.318	0.226	0.155	0.103	0.065	0.040	0.024

Table 16.2 Diffusion times $\times 10^{-5}$ s of harmonics $t_{m,n}$ in an homogeneous critical cylinder

	$m = 1$	2	3	4	5	6	7	8	9	10
$n = 0$	1.595	1.587	1.572	1.552	1.525	1.494	1.458	1.419	1.376	1.332
1	1.592	1.581	1.563	1.539	1.510	1.477	1.439	1.398	1.354	1.309
2	1.589	1.574	1.553	1.526	1.495	1.459	1.420	1.377	1.333	1.287
3	1.584	1.566	1.542	1.513	1.479	1.441	1.400	1.356	1.311	1.264
4	1.578	1.557	1.530	1.499	1.463	1.423	1.380	1.336	1.289	1.242
5	1.572	1.547	1.518	1.484	1.446	1.404	1.360	1.315	1.268	1.220
6	1.565	1.537	1.505	1.469	1.428	1.386	1.340	1.294	1.246	1.198
7	1.557	1.526	1.491	1.453	1.411	1.367	1.320	1.273	1.225	1.176
8	1.548	1.515	1.477	1.437	1.393	1.347	1.300	1.252	1.203	1.155
9	1.539	1.502	1.463	1.420	1.375	1.328	1.280	1.231	1.182	1.134

16.2.8.1 Functions $f_{m,n}(t)$

The functions $f_{m,n}(t)$ define the time behavior of the harmonics, the total flux being an infinite sum of harmonics. As an example, Fig. 16.1 represents the functions $f_{m,n}(t)$ for $n = 0$ and $m = 1, 2, 3, 4, 5, 6, 7, 8 and 9$ and helps to grasp the time evolution of the magnitude of the harmonics.

In the critical case, function (16.21) is not defined. However,

$$\lim_{k_{m,n} \to 1} \frac{1}{k_{m,n} - 1} \left[e^{(k_{m,n} - 1)\frac{t}{t_{m,n}}} - 1 \right] = \frac{t}{k_{m,n}}$$

This means than in a critical reactor, the fundamental mode characterized by the multiplication factor $k_{1,0} = 1$ grows linearly. The Sect. 14.8.3 gives the time evolution of a critical point reactor with a source.

All the subcritical harmonics reach a steady state, and do not vanish even if the exponential is negative. Indeed:

$$\lim_{t \to \infty} \frac{e^{\frac{(k-1)t}{t_d}}}{k - 1} = \frac{1}{1 - k} \quad \text{if } k < 1$$

In the case $k_{1,0} > 1$, $f_{1,0}$ grows exponentially and the fundamental mode dwarfs the harmonics. When the harmonics become negligible, the flux is shaped by the fundamental mode.

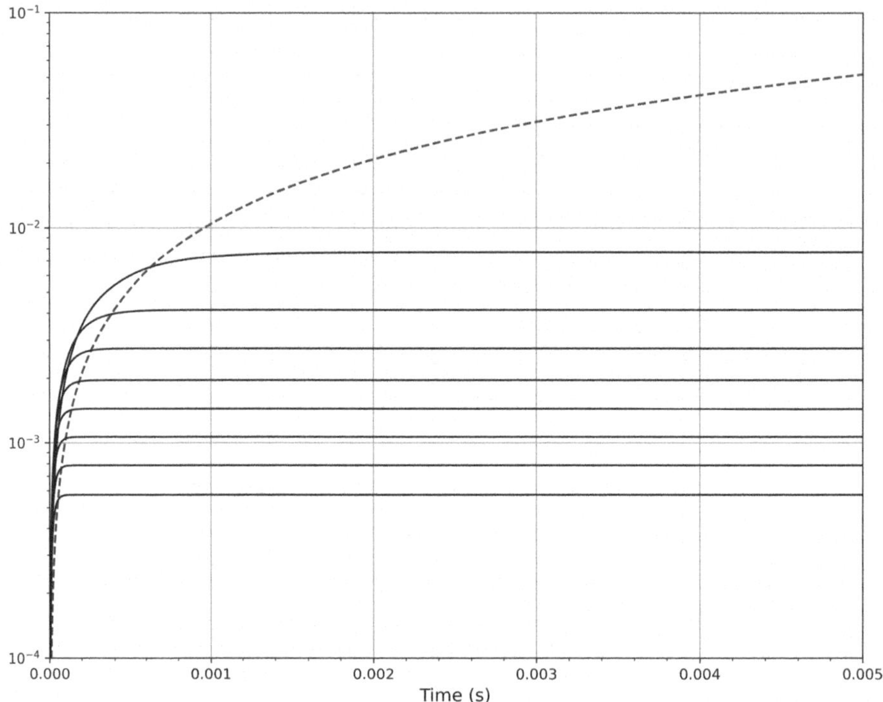

Fig. 16.1 Function $f_{m,0}$ for $m = 2, 3, 4, 5, 6, 7, 8 \, and \, 9$, and $f_{1,0}$ the critical fundamental mode, in the case of an homogeneous cylinder

16.2.8.2   Fundamental Mode

The fundamental mode corresponds to $(m = 1, n = 0)$. The multiplication factor of the fundamental mode is $k_{1,0}$. Let's introduce again the buckling, and write that the buckling squared is the absolute value of the less negative eigen value: $B_g^2 = \dfrac{\mu_{1,0}^2}{R^2}$.

From Eq. 16.19 we can write:

$$k_{1,0} = \frac{k_\infty e^{-B_g^2 \tau_{th}}}{1 + B_g^2 L_2^2} \tag{16.26}$$

$k_{1,0}$ is the multiplication factor for the fundamental mode, and is also the effective multiplication factor defined in (13.10):

$$k_{\text{eff}} = \frac{k_\infty e^{-B_g^2 \tau_{th}}}{1 + B_g^2 L_2^2} \approx \frac{k_\infty}{(1 + \tau_{th} B_g^2)(1 + L_2^2 B_g^2)}$$

And indeed, since $\tau_{th} B_g^2$ is close to zero:

$$e^{-B_g^2 \tau_{th}} \approx \frac{1}{1 + \tau_{th} B_g^2}$$

The probability of non leakage during the process of thermalization is quantified by $e^{-B_g^2 \tau_{th}}$. The age a thermal neutrons is related to the crow flight distance between the fission point and the point at which neutrons are thermalized. A higher of τ_{th}, as it is the case in graphite moderated reactors, has to be offset by a larger dimension of the core, that diminushes B_g^2. An expression of the non leakage probability, valid also for a non critical reactor, is:

$$P_{NL,f} = e^{-B_g^2 \tau_{th}} \tag{16.27}$$

From Eq. 16.18 we can write:

$$t_{1,0} = \frac{t_d}{1 + B_g^2 L_2^2} \tag{16.28}$$

Taking account of $t_d = \frac{1}{\overline{v} \Sigma_{r2}}$ and $D_2 = \Sigma_{r2} L_2^2$ enables us to write:

$$t_{1,0} = \frac{1}{\overline{v}(\Sigma_{r2} + B_g^2 D_2)}) \tag{16.29}$$

The Eq. 16.29 is consistent with (14.3).
$\frac{1}{\overline{v} \Sigma_{r2}}$ is the diffusion time in an infinite reactor.

16.3 Simulations Results

16.3.1 Case of a Source Located in the Center

The location in the center of the cylinder has a major consequence on the calculation of the source function in the basis. In the Eq. 16.13, the numerator $g_{m,n}(r_s)$ becomes $g_{m,n}(r = 0)$. The Appendix B shows that the Bessel functions of the first kind are null at the origin, for all the integer orders $n > 0$. This implies that $g_{m,n \neq 0}(r = 0) = 0$, and finally $f_{m,n \neq 0}(t) = 0$.

The centerline source excites only the zero-order eigenmodes, which have a concentric shape. By analogy, we can say that a shock to the drum surface excites only the zero-order vibrations. The harmonics $(m = 1, n = 0)$ and $(m = 2, n = 0)$ can be seen on Fig. E.1.

16.3.2   *Critical Reactor*

The inputs for calculations are given in Chap. 13, Table 13.1. The radius of the infinite homogeneous cylinder is the critical radius 168 cm.

At $t = 0$, four 1D line source are introduced in the reactor, at 150 cm from the center, and in the directions 0, $\frac{pi}{2}$, π, and $\frac{3\pi}{2}$. A numerical calculation of Eq. 16.21 enables us to draw the thermal flux in the cylinder radial plane, as a function of time.

Figure 16.2 shows the evolution of the shape of the thermal flux after the introduction of the sources in the critical reactor. The values on the scales reveal the amplitude increase of the thermal flux. The harmonics tend to be negligible compared with the fundamental function $g_{1,0}$, thus the flux tends to be in the fundamental mode. If

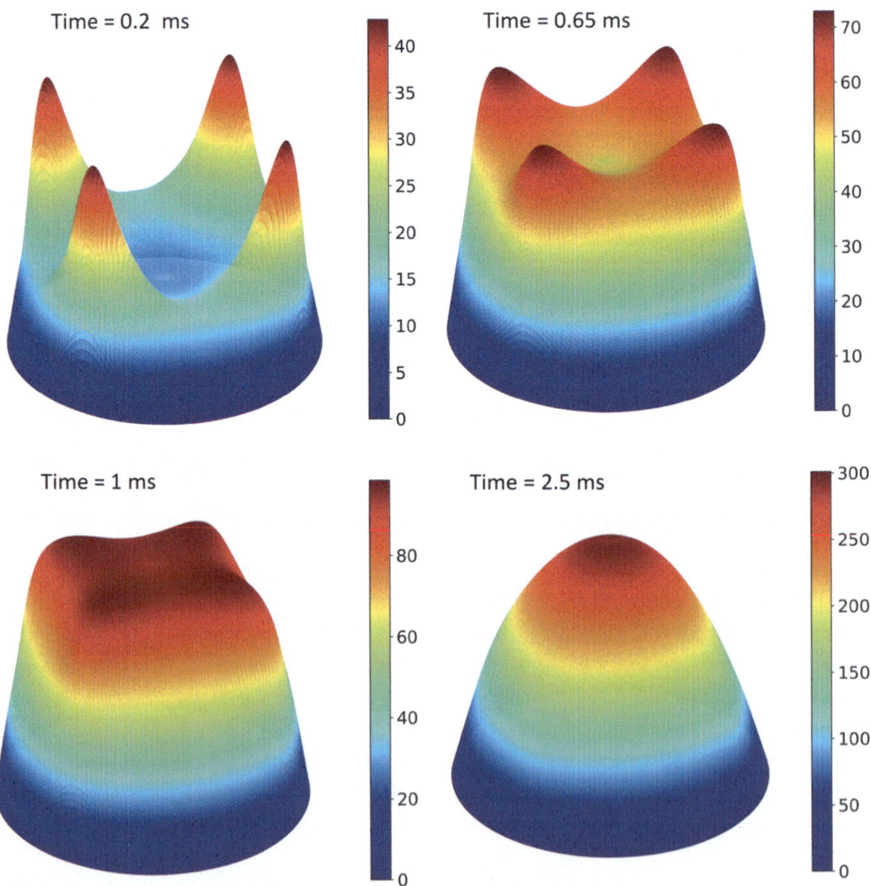

Fig. 16.2 The stacking of harmonics making the thermal flux, at different times after introduction of the source into a critical reactor, calculated without delayed neutrons

the sources were removed from the core, then the fundamental mode would sustain itself, and the harmonics would fade away, since they are associated to multiplication factors $k_{m,n\neq0} < 1$.

16.3.3 Subcritical Reactor

The reactor, that is critical with a thermal utilization factor $f = 0.85167$, is made subcritical by changing the thermal utilization factor: with $f = 0.815$, the reactivity is $\rho = -4225\,\text{pcm}$. After the introduction of four sources at $t = 0$, the flux tends to a steady state, as described in the Chap. 15. The sources are 1D lines located at 150 cm from the center at $t = 0$. Figure 16.3 shows the intersection of a vertical plane containing the centerline of the cylinder and one 1D line source. It reveals how the flux is a compound of harmonics $J_{m,n}$ that are ponderated by the convergence $\lim_{t\to\infty} f_{m,n}(t)$.

If the reactor is subcritical and close to criticality, then the harmonics tend to be negligible and the flux is almost in the fundamental mode. As an example, let's take $k_{1,0} = 0.9999$ which corresponds to $\rho = -10$ pcm. Thus $k_{1,1} = 0.9999 \times 0.982$.

$$\lim_{t\to\infty} \frac{f_{1,0}(t)}{f_{1,1}(t)} = \frac{k_{1,0}\,(1 - k_{1,1})}{k_{1,1}\,(1 - k_{1,0})} = 184.3$$

The magnitude of the second harmonic ($m = 1$ and $n = 1$) tends to be worth 0.5% the magnitude of the fundamental mode ($m = 1$ and $n = 0$).

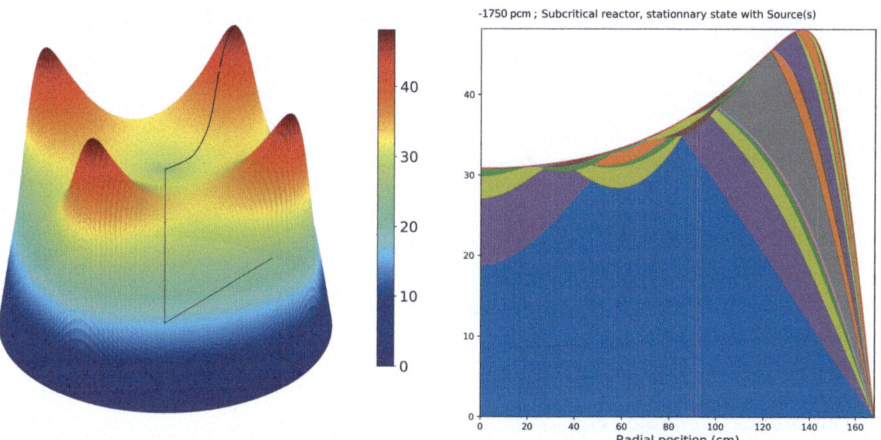

Fig. 16.3 The stacking of harmonics making the thermal flux, in a stationary subcritical reactor with source

16.4   Resolution with Delayed Neutrons

A common mathematical procedure for studying nonseparable kinetics problems consists of expansion of the neutron flux, source, and precursor densities in terms of a set of orthogonal eigenfunctions. Perhaps the most familiar eigenfunctions are the space-dependent flux modes or harmonics

Keepin (1965), p. 167

16.4.1   Additional Terms Due to Delayed Neutrons

The following term in introduced in the Eq. 16.1c:

$$C(\boldsymbol{r}, t) = C(r, \theta, t) = \sum_{m=1}^{\infty} \sum_{n=0}^{\infty} B_{m,n}(t)\, g_{m,n}(r, \theta)$$

The unit of $B_{m,n}(t)$ is cm^{-3}, the same than $C(r, \theta, t)$. A new equation for $T_{m,n}(t)$ is written, replacing (16.12):

$$T_{m,n}(t) = S_{m,n} + \frac{k_{\infty}(1 - \beta)}{p} \Sigma_{r2} f_{m,n}(t) + \lambda B_{m,n}(t) \tag{16.30}$$

The expression (16.30) is introduced into Eq. 16.4, and then, following the same steps than in section, we write a similar equation to (16.20). We combine it with the delayed neutrons equation, and we obtain the system of coupled equations:

$$\begin{cases} \dfrac{d\, f_{m,n}(t)}{dt} = \dfrac{k_{m,n}(1 - \sum_i \beta_i) - 1}{t_{m,n}} f_{m,n}(t) + \dfrac{k_{m,n}}{k_{\infty}} \dfrac{p}{\Sigma_{r2} t_{m,n}} \sum_i \lambda_i B_{i,m,n}(t) + \dfrac{k_{m,n}}{k_{\infty}} \dfrac{p}{\Sigma_{r2} t_{m,n}} S_{m,n} \\[2ex] \dfrac{d\, B_{i,m,n}(t)}{dt} = \dfrac{k_{\infty}}{p} \beta_i\, \Sigma_{r2} f_{m,n}(t) - \lambda_i B_{i,m,n}(t) \end{cases}$$

We define a new function $C_{m,n}(t)$:

$$C_{i,m,n}(t) = B_{i,m,n}(t) \frac{p\, k_{m,n}}{\Sigma_{r2}\, k_{\infty}\, t_{m,n}}$$

$C_{i,m,n}(t)$ and $f_{m,n}(t)$ have the same unit $\mathrm{n/cm}^3$.

The system of coupled equations becomes:

$$
\begin{cases}
\dfrac{d\,f_{m,n}(t)}{dt} = \dfrac{k_{m,n}(1 - \sum_i \beta_i) - 1}{t_{m,n}}\, f_{m,n}(t) + \sum_i \lambda_i C_{i,m,n}(t) + \dfrac{k_{m,n}}{k_\infty}\,\dfrac{p}{\Sigma_{r2} t_{m,n}}\, S_{m,n} \\[3mm]
\dfrac{d\,C_{i,m,n}(t)}{dt} = \beta_i\,\dfrac{k_{m,n}}{t_{m,n}}\, f_{m,n}(t) - \lambda_i C_{i,m,n}(t)
\end{cases}
$$

$$(16.31)$$

In the case $m = 1$, $n = 0$, the system (16.31) is equivalent to the system (14.13) established in the frame of point kinetics theory, which assumes that the flux is distributed according to the fundamental mode. And indeed, $f_{1,0}(t)$ is the fundamental mode. The elegant solution described by the system of Eq. 16.31 enables to work out the theory of diffusion coupled to kinetics in analytical fashion.

The calculation of steady states with this approach, which consists in decomposing functions onto harmonic basis, would consume computer processing time with no advantage, and a modelization with finite differences is much more efficient in this case. However, a solution to the system (16.31) can be calculated with implicit numerical method in the case the reactivity is time dependent. And in the case the reactivity is not time dependent (that is true for a step function at initial time), the Eq. D.11 established in Appendix D can be used:

$$
\boldsymbol{u}_{m,n}(t) = e^{A_{m,n}(t-t_0)}\boldsymbol{u}_{m,n}(t_0) + \int_{t_0}^{t} e^{A_{m,n}(t-s)}\boldsymbol{b}_{m,n}(s)\,ds
\qquad (16.32)
$$

With

$$
\boldsymbol{u}_{m,n}(t) = \begin{bmatrix} f_{m,n}(t) & C_{i,m,n}(t) \end{bmatrix}
$$

$$
\boldsymbol{b}_{m,n}(s) = \begin{bmatrix} S_{m,n}(t)\dfrac{k_{m,n}}{k_\infty}\,\dfrac{p}{\Sigma_{r2} t_{m,n}} & 0 \cdots\cdots 0 \end{bmatrix}
$$

$$
\begin{aligned}
A_{m,n}[0,0] &= \frac{k_{m,n}(1 - \sum_i \beta_i) - 1}{t_{m,n}} && \text{first line, first column of the matrix} \\[3mm]
A_{m,n}[0,j] &= \lambda_i && \text{first line, column } j \text{ of the matrix} \\[3mm]
A_{m,n}[i+1,0] &= \beta_i\,\frac{k_{m,n}}{t_{m,n}} && \text{line } i+1, \text{ first column of the matrix} \\[3mm]
A_{m,n}[i+1,i+1] &= -\lambda_i && \text{diagonal of the matrix, with } i = 1 \text{ to } 6
\end{aligned}
$$

16.5 Spatial Kinetics with Harmonics

We still consider the same homogeneous reactor.

The system is made subcritical by an appropriate thermal utilization factor. Two neutron sources are defined: the first source provides neutrons according to the fundamental function $g_{m,n}$, and the second source provides what we could call anti neutrons, which means that the components $S_{m,n}$ are multiplied by -1 in order to produce constant absorbing effect. The geometry of the second source is a vertical cylinder of radius 15 cm that represents a central control rod.

The components $S_{m,n}$ of the cylinder onto the basis are calculated by the integral defining the inner product in polar coordinates (cf. Eq. G.2):

$$S_{m,n} = \int_0^R \int_0^{2\pi} \Gamma(r, \theta)\, g(r, \theta)\, r\, dr d\theta$$

where $\Gamma(r, \theta)$ is defined up to a multiplicative constant:

$$\Gamma(r, \theta) = \begin{cases} 1, & \text{The point } (r, \theta) \in \text{ the cylinder} \\ 0, & \text{The point } (r, \theta) \notin \text{ the cylinder} \end{cases}$$

The combination of a fundamental mode distributed source and an intrinsically subcritical system allows us to create a stationary state, in the presence of neutron absorption by the central cylinder. This represents the state of a critical reactor. The precursors, represented by the functions $C_{i,m,n}(t)$, are calculated at equilibrium with neutrons, which means that:

$$C_{i,m,n}(t) = \frac{\beta_i \dfrac{k_{m,n}}{t_{m,n}} f_{m,n}(t)}{\lambda_i}$$

The worth of the cylindric rod is calculated as the ratio between its absorption and the total absorption of the non perturbated core. This worth in our example is evaluated to be 519.68 pcm. At $t = 0$, while the reactor is initially at equilibrium, we assume that the absorbing cylinder is instantaneously extracted.

Figure 16.5 enables us to observe the evolution of the thermal flux shape as a function of time, and the evolution of the concentration of the fourth group of precursors. The fourth group is chosen for plotting since it's the main producer of delayed neutrons, as shown by the Fig. 14.12. The Chap. 14 highlighted that, since the production of precursors evolves without a prompt jump, unlike neutron density, it lags behind the latter. The spatial aspect is also worth considering. Figure 16.5 shows that the shape of the precursors concentration evolves more slowly than the shape of the thermal flux. The thermal flux depression in the center, which is due to the initial insertion of the absorbing cylinder, vanishes in $10 \times t_{1,0}$. Concerning the delayed neutrons, the depression vanishes in roughly one second. The fact that the shapes of thermal neutron flux and precursors concentrations are not consistent as they are in a static situation means that the reactivity is affected.

This example illustrates the need for a static-dynamic correction during the DRWM test, which is the subject of the Chap. 19 (Fig. 16.4).

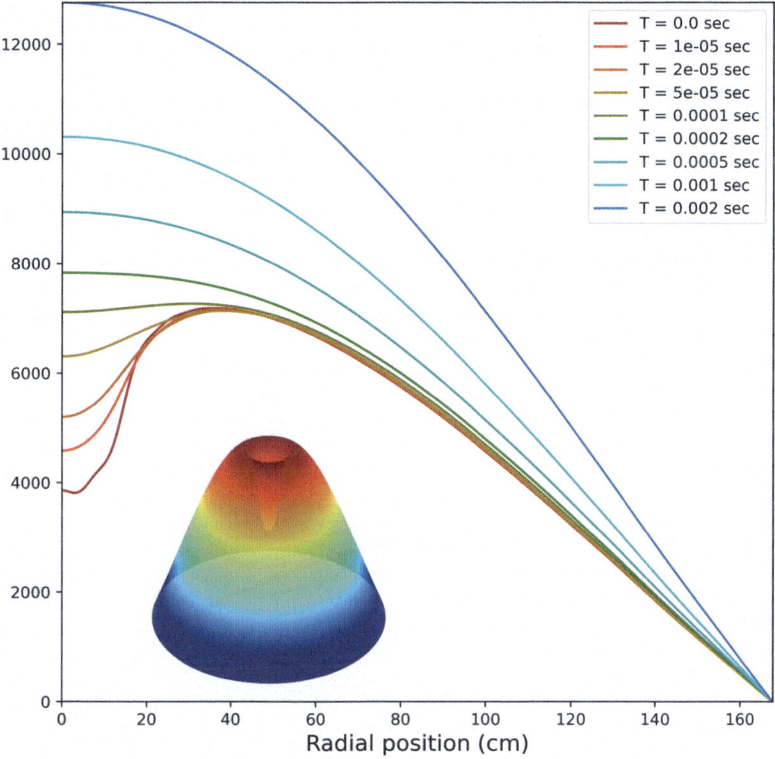

Fig. 16.4 Thermal flux shape after rod extraction

16.6 Migration Area, Size of the Reactor and Sustainability of Harmonics

16.6.1 Harmonics on the Disk

Equation 16.19, that was established for an homogeneous infinite cylinder of radius R, shows that harmonics tend to vanish, by comparison with the fundamental mode, if τ_{th}/R^2 increases enough. The harmonics having the highest multiplication factor $k_{m,n}$ within an homogeneous cylinder are $g_{1,1}$, $g_{1,2}$, $g_{2,0}$ and $g_{1,3}$.

We can easily highlight that harmonics tend to weaken in a subcritical reactor if τ_{th}/R^2 increases enough. Let's consider the same subcritical reactor than in Sect. 16.3.3, with a source distributed according a linear combination of basis vectors, of zero order ($n = 0$). In this section, we choose $f = 0.7$, and thus $\rho = -20000 \times 10^4$ pcm. The combination of harmonics defining the source is:

$$J_0\left(r\frac{\mu_{1,0}}{R}\right) + 0.4 \times J_0\left(r\frac{\mu_{2,0}}{R}\right) + 1.5 \times J_0\left(r\frac{\mu_{3,0}}{R}\right) + 0.6 \times J_0\left(r\frac{\mu_{4,0}}{R}\right)$$

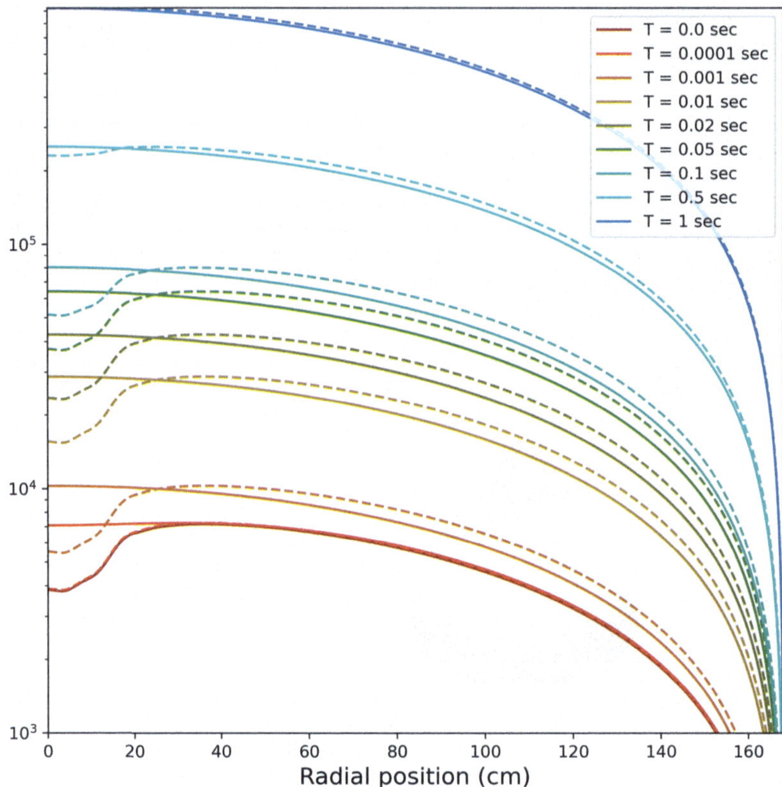

Fig. 16.5 Thermal flux shape after rod extraction, and precursors concentration shape (scaled for comparison, plotted with a dotted line)

By construction, the source excites the first radial harmonics ($n = 0$), and not the azimutal harmonics ($n \geq 1$). Figure 16.6 shows this function source, compared with the function $J_0\left(r\frac{\mu_{1,0}}{R}\right)$ alone, for the same volume integration on the disk. Figure 16.7 enables to compare the thermal flux shapes for various fissile radius, 210, 126, 105 and 42 cm, with the same medium and thus the same thermal age τ_{th}. We can see a tendancy, related to the ratio $R/\sqrt{\tau_{th}}$, although a sharp threshold cannot be revealed.

When $R/\sqrt{\tau_{th}} < 20$, the radial harmonics are strongly weakened. This is consistent with the assertion of Duderstadt and Hamilton, quoted in Sect. 16.6.3.

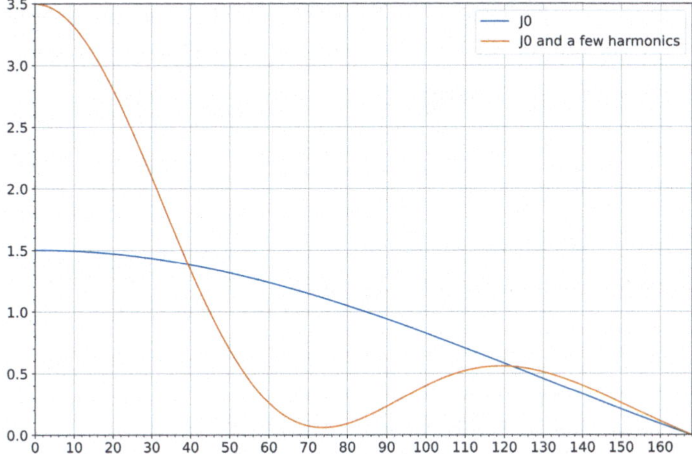

Fig. 16.6 Source function combining the fundamental mode shape and a few harmonics

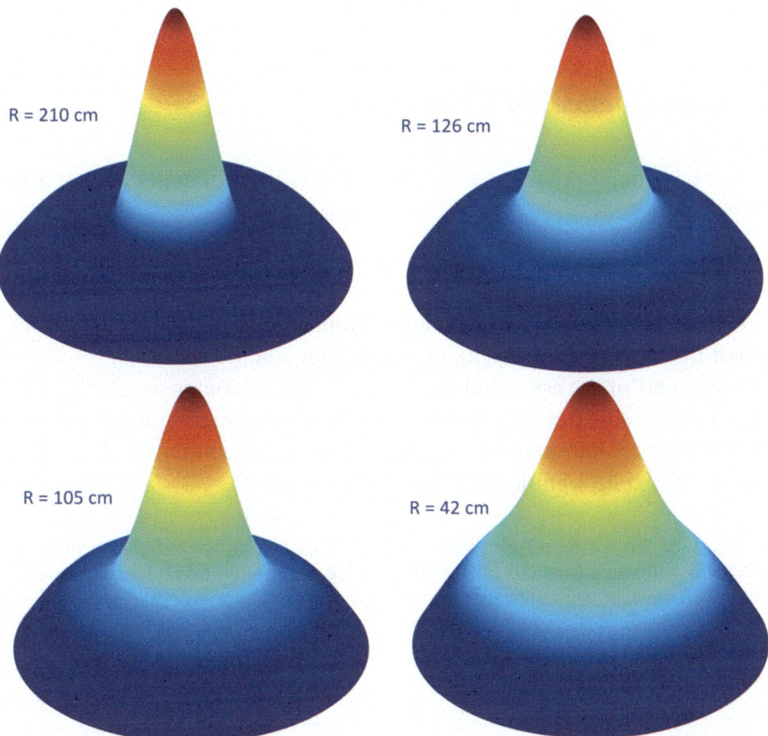

Fig. 16.7 Subcritical homogeneous reactor with a source combining a few harmonics, for different sizes of homothetic systems

Table 16.3 Comparison of the leakages associated with axial and radial/azimutal harmonics

(m, n)	(1, 0)	(1, 1)	(1, 2)	(2, 0)	(1, 3)
$e^{-\frac{\mu_{m,n}^2}{R^2}\tau_{th}}$	0.989 807 15	0.974 325 61	0.954 351 01	0.947 449 99	0.930 425 55
n	1	3	5	7	9
$e^{-B_n^2\tau_{th}}$	0.997 284 43	0.975 823 71	0.934 277 78	0.875 251 78	0.802 310 2

16.6.2   *Comparison Between Slab and Disk*

The active length of a PWR 1300 fuel assembly is 426 cm. Thus we choose the value $L = 426$ cm as an example, in order to calculate the values k_n associated to the harmonics. n are odd integers and the first harmonic is $n = 3$.

Table 16.3 compares the leakages associated with axial and radial/azimutal harmonics and shows there aren't any significant differences. In all cases, the higher the mode, the greater is the damping due to leakage.

16.6.3   *Consequences of Reactor Size*

The fact that harmonics are likely to be excited only in sufficiently large reactors has practical consequences:

- The tilting effect due to control rods insertion has a much higher magnitude in a large core. This tilting effect is illustrated from a radial point of view in Sect. 17.3. This effect is well known in the axial direction, and the insertion of control rods in the top of the core produces an axial flux change, the flux tilts and become higher in the lower half of the core, and the Axial Offset[3] becomes negative.
- A large reactor is sensitive to xenon induced flux harmonic instability, especially of the first and higher harmonics of the flux, since the variations of xenon poisoning are capable to excite significantly these harmonics.

The stability of the spatial power distribution with respect to xenon-induced oscillations will decrease with increasing core size or decreasing neutron migration length. As a rule of thumb, xenon oscillations will be a problem if the reactor core is over 30 migration lengths in size.

Duderstadt and Hamilton (1976), p. 580

---

[3] The Axial Offset is defined from P_T and P_B, which are respectively the power produced in the upper half and lower half of the reactor. AO is usually defined in % when multiplied by 100

$$AO = \frac{P_T - P_B}{P_T + P_B}$$

In the example of PWR 1300, the active length squared over thermal age (H^2/τ_{th}) is 6 times the radius squared over thermal age (R^2/τ_{th}). We see that, for such a reactor, the first radial harmonic is less likely to be excited than the first axial harmonic, so that the threshold flux[4] for radial oscillation lies above that for an axial oscillation.

References

J. Bussac, P. Reuss, *Traité de neutronique: Physique et calcul des réacteurs nucléaires avec Application aux réacteurs à eau pressurisée et aux réacteurs à neutrons Rapides*, vol. 25 (Hermann, 1985)

J.J. Duderstadt, L.J. Hamilton, *Nuclear Reactor Analysis* (Wiley, 1976)

G.R. Keepin, *Physics of Nuclear Kinetics* (Addison-Wesley Publishing Company, 1965)

J.R. Lamarsh, *Introduction to Nuclear Reactor Theory*, in Addison-Wesley Series in Nuclear Engineering (Addison-Wesley Publishing Company, 1966). ISBN 9780201041262. https://books.google.fr/books?id=by5RAAAAMAAJ

---

[4] In order to sustain a significant xenon oscillation, the average thermal flux of the reactor should have a minimal value. This threshold values of the flux for oscillations are function of reactor size and degree of flux flattening.

Chapter 17
Effect of Control Rods

Abstract This chapter allows the reader to start thinking about the effects that control rods have on the overall power distribution in the core, and the coupling between different banks of rods, known as the shadowing and anti-shadowing effects. Our approach is based on a radial plane, and not from the axial point of view, as is usually the case.

17.1 Control Rods

The term control rod refers to rods, plates, or tubes containing a neutron absorbing material such as boron carbide B_4C, hafnium, AIC alloy (AgInCd) that can be removed from or inserted into the reactor core to increase or decrease the reactivity.

In this chapter, considering our simplified infinite cylinder geometry, control rods are represented by rings of additional absorbing material.

17.2 Absorbing Worth and Spatial Effect

17.2.1 Calculation Without Thermalhydraulics Feedback Effects

The reactor is supposed to be at constant power, which is the case if a diminution of the boric acid concentration compensates for the insertion of a rod cluster control assembly bank. Although the total power of the core remains constant, radial variations of the flux shape are held back by Doppler feedback effect : for example, if the flux tends to diminish in some part of the core, then the fuel rods tend to cool down which entails a lower resonant absorption countering the flux diminution. The calculations done in this subsection are all things being equal, and thus without feedback effects: that's the case of a reactor at zero power. A reactor is said to be at zero power when it operates below the point of adding heat, which is at the order of magnitude 0.1 %PN (percent of nuclear rated power) (Grard 2014, p. 46, 98, 181). If the fission

© The Author(s), under exclusive license to Springer Nature Switzerland AG 2026   277
H. Grard, *Diffusion of Neutrons in Nuclear Reactors*,
https://doi.org/10.1007/978-3-032-05088-5_17

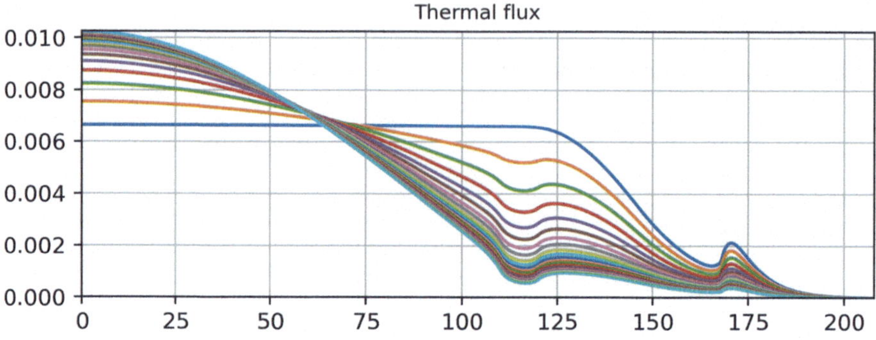

Fig. 17.1   Shape of thermal flux as a function of absorber's worth

reaction rate is too low, the fuel rods do not heat up, and fresh fuel assemblies are at thermal equilibrium with water: in this situation, a flux variation does not produce any temperature change in the fuel, therefore no change in resonant absorption. The insertion of rod cluster control assemblies is represented by the addition of a supplementary macroscopic absorption cross section, in a ring. This will affect the radial shape of flux.

In our example, the ring is delimitated by the radius 110 and 120 cm.

The absorption cross section is increased from 0 by arithmetic steps of $0.01\,\mathrm{cm}^{-1}$, and 20 steps are calculated.

Figure 17.1 shows the shapes of thermal flux for the different absorber's worths. The flattened flux curve is for an non absorbing ring. Two observations can be done: the flux is dished in the vicinity of the absorbing ring, the overall shape of the flux is affected.

Figure 17.2 shows the reactivity effect due to the absorbing ring, as a function of the absorbing worth which is on a linear scale. It allows us to see that adidtional antireactity has a dressing yield: this can be explained by spatial self-shielding, which is presented in the Sect. 5.4 in the context of absorption by fuel rods. In our studied configuration, we can see that the leak are affected, as a consequence of the variation of the flux shape, which is scaled down in the outlying area of the core. The variation of the leak is involved in the reactivity effect.

17.2.2   Calculation with Doppler Feedback Effect

The Doppler feedback effect diminishes the antitrap factor, also coined resonance escape probability. The feedback is calculated according to the following simplifying assumptions:

- the local increase of resonant absorption is a linear function of the local linear power density in the fuel rods

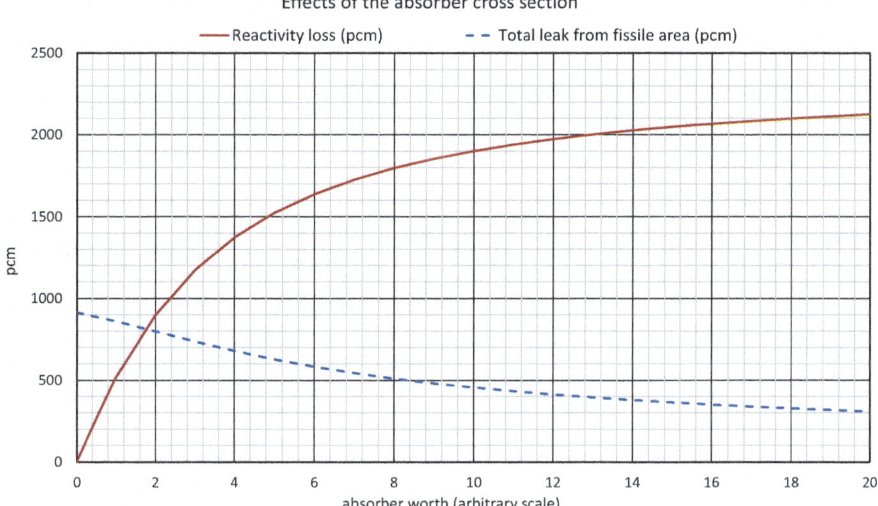

Fig. 17.2 Reactivity effect as a function of absorber macroscopic cross section

- if the power is perfectly uniform in the core (which is of course not the case), then the feedback effect is worth -12 pcm/%PN, where %PN is a percentage of nuclear rated power.

During the power build-up from 0 to 100 %PN, in this case the reactivity loss due to Doppler effect is 1200 pcm (Grard 2014, Figs. 2.19 and 2.20, p. 56).

Figure 17.3 enables to compare the radial disruption due to an absorbing ring, of increasing worth, calculated at zero power and at rated power. It an be seen that the spatial disruption is dampened by the Doppler feedback, which opposes itself to local reactivity changes: if the local power diminishes, then the fuel tends to cool down which leads to reactivity gain partially compensating for the reactivity loss. This is the reason why Doppler feedback effect attenuates the phenomenon of xenon oscillations.

17.3 Effect of the Location

In this section, we add an absorbing ring with a macroscopic cross section of 0.01 cm between the internal radius 110 cm and external radius 120 cm. Internal and external radius are both decreased by steps of 5 cm, and 20 configurations are calculated. The shapes of thermal flux are shown on Fig. 17.4 . When the ring is at its initial position, it tends to create a redistribution of the flux in the central part of the core. Gradually as the absorbing ring shifts towards the center of the core, the flux in the central part

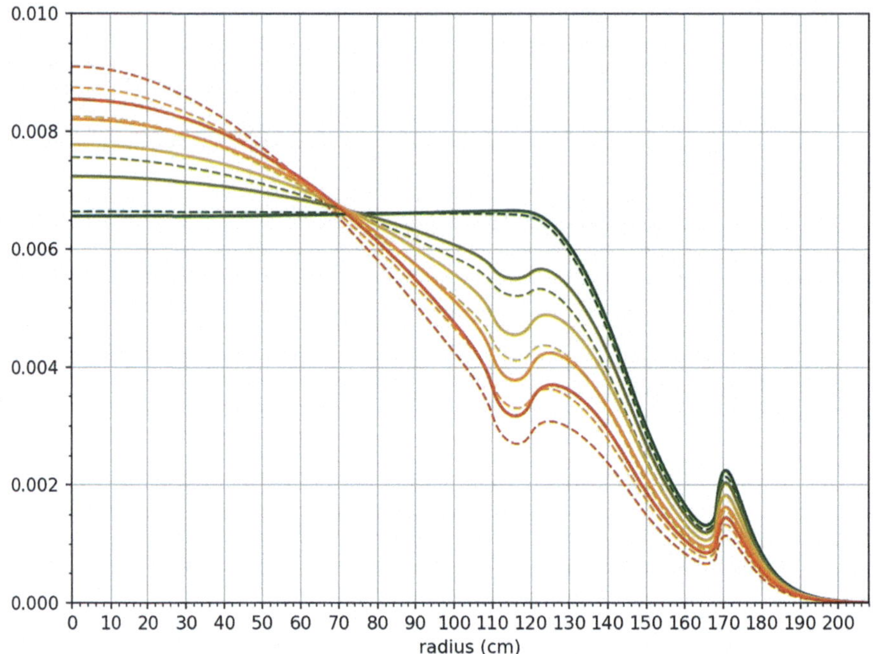

Fig. 17.3 Radial disruption due to an absorbing ring, at zero power (dashed line) and rated power

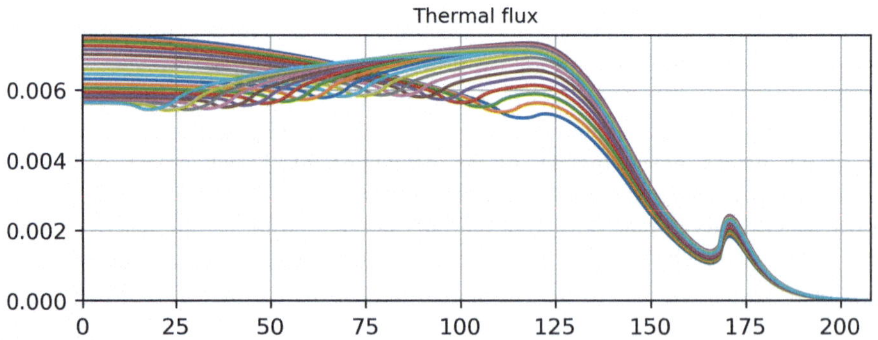

Fig. 17.4 Effect of the position of an absorbing ring on the flux shape

of the core delimitated by the ring tends to decrease, and the flux beyond the ring tends to increase. At some point, the flux tilts towards the peripheral part of the core.

The phenomenon we have just described echoes the axial behaviour of neutron flux in the case of Part Length rods in Westinghouse PWRs. Part Length rods were designed with an AIC part absorbing neutrons, corresponding to 25% of the total length, i.e. 91 cm in the case of 900 MWe reactor. They were dedicated to control the axial offset with manual adjustment, with a slow motion governed by screw mechanisms. It has been found out that Part Length rods could lead to penalizing axial distributions occuring when the flux tilts significantly on one side of the absorbing zone, making the protection system uncapable of protecting the core from the boiling crisis. Their use has been stopped in the mid-1970s (Grard 2014, p. 20).

The axial effects of Total Length rods, which is widely known, is different: the more a control rods bank is inserted in the upper half part of the core, the more the shape of flux is unbalanced with higher values in the lower half part (negative axial offset), the assymmetry is maximam for a bank located in the middle of the core, and when the bank is below the middle, the asymmetry steps back.

17.4 Collective Effect

It is necessary to be attentive to the fact that the antireactivity values of different control rods banks are not additive, that is to say that the antireactivity of all the banks inserted in the core differs from the sum of the antireactivities of each of each bank, evaluated alone in the core (Bussac and Reuss 1985, p. 456).

Shadowing and anti-shadowing effects appear in the case of more than one bank is introduced in the core. They consist in the fact that the worth of a set of banks is less than (shadowing effect), or more than (antishadowing effect), the addition of each bank's worth considered alone in the core (Bussac and Reuss 1985, p. 460).

This may be apprehended with our 1D diffusion solver by introducing different absorbing rings.

17.4.1 Spatial Shadowing Effect

> (..) the worth of the two rods is less than twice the worth of the two rods acting independently. This reduction in effectiveness per rod is due to the fact that the flux at each rod is reduced by the presence of the other, an effect known as *shadowing*.

> Lamarsh (1966), p. 516

The shadowing effect is intuitively easy to grasp. If a bank of absorbing rods were inserted in the immediate vicinity of another bank that is already inserted, the insertion would take place in a locally depressed flux, resulting in a reduction in the worth of the bank.

Table 17.1 Antishadowing effect with three absorbing rings

Inserted rings	Antireactivity (pcm)	Sum of the antireactivities of each ring (pcm)	Antishadowing (> 0) or shadowing effect (pcm)
(C_1)	238	238	0
(C_2)	604	604	0
(C_3)	896	896	0
(C_1) and (C_2)	789	$238 + 604 = 842$	-53
(C_1) and (C_3)	1349	$238 + 896 = 1134$	215
(C_2) and (C_3)	1680	$604 + 896 = 1500$	180
(C_1) and (C_2) and (C_3)	2102	$238 + 604 + 896 = 1738$	364

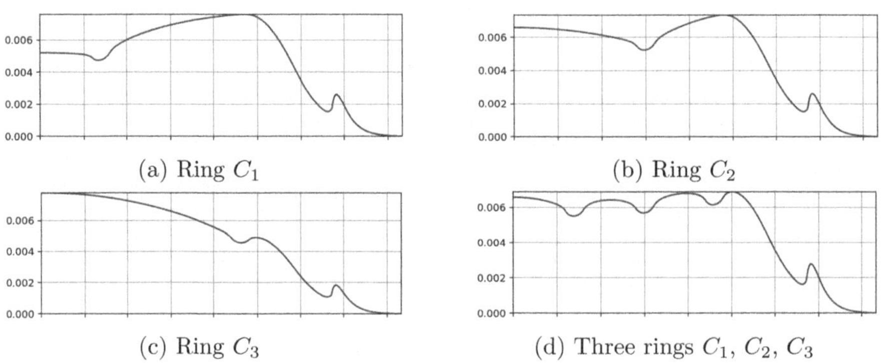

(a) Ring C_1

(b) Ring C_2

(c) Ring C_3

(d) Three rings C_1, C_2, C_3

Fig. 17.5 Flux shape tilts due to absorbers insertion and antishadowing effect

17.4.2 Antishadowing Effect

Three different absorbing rings are considered:

1. $30\,\mathrm{cm} < r < 40\,\mathrm{cm}$ (C_1)
2. $70\,\mathrm{cm} < r < 80\,\mathrm{cm}$ (C_2)
3. $110\,\mathrm{cm} < r < 120\,\mathrm{cm}$ (C_3)

Each ring is weighed out alone, then rings are weighed by pairs, and then the three rings together (Table 17.1).

Figure 17.5 shows that when the absorbing rings are inserted alone, the flux tilts: it decreases in the area affected by the ring, and migrates towards the opposite area of the core (internal versus external areas from a radial point of view, of bottom versus top from an axial point of view). This freedom to tilt does limitate the worth of the absorbing rings. Consequently, when the three rings are inserted together, they cover radially the whole core and the flux has no room for tilting and is squashed: the worth of the three rings together is higher than the sum of each worth.

(..) the worth of the two rods is greater than the sum of the worths of each rod acting alone. This effect, which is known as *antishadowing*, is caused by the increase in flux at each rod due to the other.

Lamarsh (1966), pp. 516–517

It is interesting to have in mind what was shown in Sect. 16.6: this enables to understand that the greater the size of the core compared with the migration area, the greater the amplitude of the tilting phenomenon.

17.5 Worth of Absorbers and Neutronic Parameters of the Core

The control rods worth is all the more high that (Bussac and Reuss 1985, p. 458):

- the diffusion area is high: there are more thermal neutrons that may enter into the absorbing material,
- the reactor is small: the area under the influence of the absorbing rods represents a higher relative proportion of the volume of the reactor,
- the radius of the absorbing rod is high: in the case of a black absorbing rod, the absorbing worth is approximatively proportional to the external surface of the absorbing rod, and in the case of a grey absorbing rod, the absorbing worth is an increasing function of the volume.

The slowing area affects the absorbing worth provided that the material absorbs epithermal neutrons.

References

J. Bussac, P. Reuss, *Traité de neutronique: Physique et calcul des réacteurs nucléaires avec Application aux réacteurs à eau pressurisée et aux réacteurs à Neutrons Rapides*, vol. 25 (Hermann, 1985)

H. Grard, *PWRs Physics, Operation and Safety: The Producing Reactor; Physique, fonctionnement et sûreté des REP. Le réacteur en Production* (EDP Sciences, 2014)

J.R. Lamarsh, *Introduction to Nuclear Reactor Theory*, in Addison-Wesley Series in Nuclear Engineering (Addison-Wesley Publishing Company, 1966). ISBN 9780201041262. https://books.google.fr/books?id=by5RAAAAMAAJ

Chapter 18
Poisoning of the Reactors

Abstract The short-term time dependence of two fission products, ^{149}Sm and ^{135}Xe, which also have strong absorption cross sections of thermal neutrons, introduces some interesting reactivity change during reactor power transients. The xenon's behaviour during power changes makes it specially important and interesting. This chapter illustrates the dynamics of xenon and samarium poisoning with the examples of a PWR and a high flux experimental reactor. The phenomenon of xenon induced spatial oscillations is not studied in this book, and the reader will find an extensive bibliography on this subject.

18.1 ^{135}Xe Poisoning

18.1.1 Main Features of ^{135}Xe

Of all the isotopes present in a reactor core, the isotope ^{135}Xe of xenon is the strongest thermal neutron absorber. It has an exceptional thermal absorption cross-section of 2.6520×10^6 b (at $2200\,\mathrm{m \cdot s^{-1}}$), far greater than those of gadolinium and boron, since the effective neutron absorption cross-sections at $2200\,\mathrm{m \cdot s^{-1}}$ are 2.5400×10^5 b for ^{157}Gd and 3840 b for ^{10}B.

The fact that ^{135}Xe has 81 neutrons and that the additional neutron completes the layer at 82, the magic number, explains why ^{135}Xe is particularly greedy for an extra neutron.

The cross section for thermal absorption is visible on Fig. 3.4, and at $306\,^{\circ}$C, the averaged value of the ^{135}Xe cross section over a Maxwellian distribution is $\overline{\sigma_{c,th}} = 2.4290 \times 10^6$ b (see Sect. 3.2.7).

© The Author(s), under exclusive license to Springer Nature Switzerland AG 2026

H. Grard, *Diffusion of Neutrons in Nuclear Reactors*,

https://doi.org/10.1007/978-3-032-05088-5_18

18.1.2   The Production of ^{135}Xe

^{135}Xe is produced directly by fission in two forms: ^{135}Xe and ^{135m}Xe (metastable). ^{135m}Xe undergoes a more than 99% transition to ^{135}Xe with a half-life of $T_{\frac{1}{2}} =$ 15.29 min (917.4 s).

Part of the ^{135m}Xe comes from the β^- decay of ^{135}I

- β^- decay of ^{135}I, branching ratio 16.509% towards ^{135m}Xe
  $Q = 2.1215$ MeV
- β^- decay of ^{135}I, branching ratio 83.491% towards ^{135}Xe
  $Q = 2.648$ MeV.

It is interesting to note that the cross section of ^{135m}Xe is significantly higher than the one of ^{135}Xe, although it is not well known.

We consider the isobar 135 of fission products:

$$^{135}_{49}\text{In} \xrightarrow{\beta^-} {}^{135}_{50}\text{Sn} \xrightarrow{\beta^-} {}^{135}_{51}\text{Sb} \xrightarrow{\beta^-} {}^{135}_{52}\text{Te} \xrightarrow{\beta^-} {}^{135}_{53}\text{I} \xrightarrow{\beta^-} {}^{135}_{54}\text{Xe}$$

Table 18.1 gives an overview of the different fission yields for the thermal fission of ^{235}U, involved in the final production of ^{135}Xe. A small part of the fissions are fast fissions , and the thermal fissions of ^{239}Pu and ^{241}Pu also contribute to the production of ^{135}Xe.

The following calculation illustrates how cumulative fission yields are calculated. We multiply the cumulative fission yield of ^{135}I by the β^- decay branching ratio towards ^{135m}Xe, and then add the independent fission yield of ^{135m}Xe:

$0.063853 \times 0.165086 + 0.1669 = 1.2210\%$; that is the cumulative fission yield of ^{135m}Xe.

Table 18.1 Fission yields of isobar 135 for the thermal fission of U-235 (0.0253 eV). *Source* JEFF 3.1.1

FP	Independent fission yield (%)	Cumulative fission yield (%)	Half-life $T_{\frac{1}{2}}$ (s)	Branching ratio β^- (%)
^{135}Sn	6.8852×10^{-4}	6.886×10^{-4}	0.53	
^{135}Sb	0.178	0.1785	1.74	84.3 (else β^-, n)
^{135}Te	3.6828	3.8367	19	100
^{135}I	2.5486	6.3853	2.3652×10^4	$83.49 + 16.51$ (^{135m}Xe)
^{135}Xe	0.069 118	6.614	3.2904×10^4	100
^{135m}Xe	0.1669	1.221		

18.1.3 Equations Governing ^{135}Xe and ^{135}I

There are two possible modeling options:

- Ignore ^{135m}Xe and assume that the branching ratio of iodine to ^{135}Xe (non-metastable) is 100%.
- Take ^{135m}Xe into account since its capture cross-section is significantly higher than that of ^{135}Xe. In this case, a third equation is added, with ^{135m}Xe produced directly by fission and by 16% of ^{135}I, and with its own capture cross-section.

In the remainder of this chapter, we will disregard ^{135m}Xe.

We can observe on Table 18.1 that, along the isobar, the lifetimes tend to be higher after each β^- that bring the nuclei closer to the valley of stability. Because the lifetime of ^{135}Te is very short compared to the one of ^{135}I, it can be assumed that fissions produce ^{135}I directly, with a cumulative yield 6.38%. This enables to represent the chain of reactions that produce and destroy ^{135}Xe on Fig. 18.1.

We adopt the following notations:

$$\lambda_I = \frac{\ln 2}{T_{\frac{1}{2}}(^{135}\mathrm{I})}, \quad \lambda_X = \frac{\ln 2}{T_{\frac{1}{2}}(^{135}\mathrm{Xe})}, \quad \sigma_X = \overline{\sigma_{c,th}}(^{135}\mathrm{Xe})$$

And Σ_f is the macroscopic fission cross section for thermal neutrons. Thus we can write the system of differential Eq. 18.1 describing the volumic concentration of ^{135}I and ^{135}Xe as a function of time:

$$\begin{cases} \dfrac{dX(t)}{dt} = \gamma_X\,\Sigma_f\phi + \lambda_I I - (\lambda_X + \sigma_X\phi)X \\[2mm] \dfrac{dI(t)}{dt} = \gamma_I\,\Sigma_f\phi - \lambda_I I \end{cases} \tag{18.1}$$

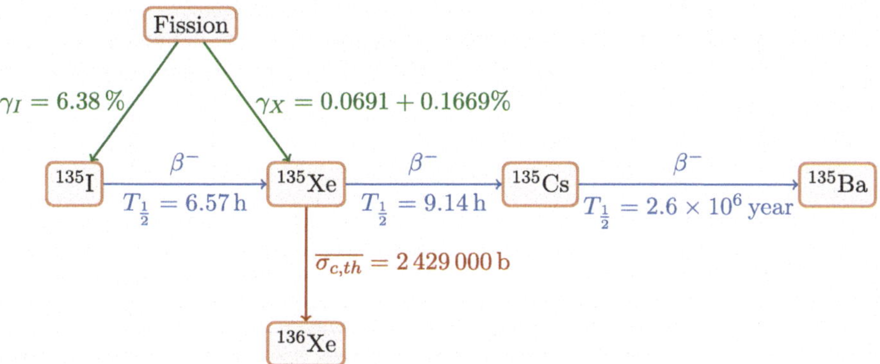

Fig. 18.1 Series of reactions for xenon poisoning

The fission reaction rate $\Sigma_f \phi$ (see (2.3)) is multiplied by the fission yield so as to give the number of nuclei produced per unit volume per unit time.

$\lambda_I I$, in the xenon equation, represents the production of ^{135}Xe by decay of ^{135}I.

$-\sigma_X \phi X$ represents neutron capture by xenon.

If the flux is constant, there is a stationary solution:

$$\begin{cases} X_{eq} = \dfrac{(\gamma_x + \gamma_I)\,\Sigma_f \phi}{\lambda_X + \sigma_X \phi} \\ I_{eq} = \dfrac{\gamma_I\,\Sigma_f \phi}{\lambda_I} \end{cases} \tag{18.2}$$

For high flux values, in practice higher than 1×10^{15} neutrons/cm^{2}s, the xenon concentration depend less on flux, and tends to an asymptotic behaviour. Iodine concentration is proportional to the power density of the reactor.

18.1.4   Estimation of the Xenon Poisoning

If we note that the fission appear and remain within the fuel, and that their capture, if it has to be taken into consideration, occurs essentially in the thermal domain, the we can see that the fission products will affect the thermal utilization factor slightly (..), and will mainly affect the reproduction factor η because if affects it directly via this same cross section $\Sigma_{a,f}$. By convention, poisoning is defined as the effect on η only, expressed to the first order and as an absolute value. Reuss (2008, p. 301)

The reactivity difference (Marguet 2018, p. 841) between a core named core 1 without xenon, and the same core with xenon, named core 2, is:

$$\rho_2 - \rho_1 = \frac{k_2 - k_1}{k_1 \times k_2}$$

We admit that xenon only affects the reproduction factor η, defined as the ratio of the number of fast neutrons produced by thermal fission to the number of thermal neutrons absorbed in the fuel (see Sects. 13.1.6 and 13.1.3.5). Then we can write:

$$\rho_2 - \rho_1 = \frac{\eta_2 - \eta_1}{\eta_2} \times \frac{1}{k_1}$$

For reactors 1 and 2, we can write the reproduction factor:

$$\eta_1 = \frac{\nu \Sigma_f}{\Sigma_{a,\,\text{fuel}}} \quad \text{and} \quad \eta_2 = \frac{\nu \Sigma_f}{\Sigma_{a,\,\text{xenon}} + \Sigma_{a,\,\text{fuel}}}$$

$$\frac{\eta_2 - \eta_1}{\eta_2} = -\frac{\Sigma_{a,\,\text{xenon}}}{\Sigma_{a,\,\text{fuel}}}$$

And the xenon poisoning is:

$$\rho_{Xe} = \left| \frac{\eta_2 - \eta_1}{\eta_2} \right| = \frac{\Sigma_{a,\,\text{xenon}}}{\Sigma_{a,\,\text{fuel}}}$$

The approximation for the calculation of xenon antireactivity can be written:

$$\rho_X = \frac{X \sigma_X}{\Sigma_{a,\,\text{fuel}}} \tag{18.3}$$

And we define ρ_I as the antireactivity that would occur in the core if all the iodine nuclei were turned into xenon nuclei. Thus:

$$\rho_I = \frac{I \sigma_X}{\Sigma_{a,\,\text{fuel}}} \tag{18.4}$$

Taking into account the reproduction factor according to Eq. 13.16, we are now able to calculate ρ_X and ρ_I:

$$\rho_X = \frac{(\gamma_x + \gamma_I)\, \Sigma_f \phi\, \sigma_X}{(\lambda_X + \sigma_X \phi) \Sigma_{a,\,\text{fuel}}}$$

$$\rho_X = \frac{\eta}{\nu}\, \frac{(\gamma_x + \gamma_I)\, \sigma_X \phi}{(\lambda_X + \sigma_X \phi)} \tag{18.5}$$

$$\rho_I = \rho_X \frac{I_{\text{eq}}}{X_{\text{eq}}}$$

$$\rho_I = \rho_X \frac{\gamma_I\, \Sigma_f \phi \times (\lambda_X + \sigma_X \phi)}{\lambda_I \times (\gamma_x + \gamma_I)\, \Sigma_f \phi}$$

$$\rho_I = \rho_X \frac{\gamma_I\, (\lambda_X + \sigma_X \phi)}{\lambda_I\, (\gamma_x + \gamma_I)} \tag{18.6}$$

18.2 ^{149}Sm Poisoning

18.2.1 Main Features of ^{149}Sm

^{149}Sm has two major differences from xenon: it is stable and does not decay, and is not directly produced by fission.

The averaged value of the ^{149}Sm cross section over a Maxwellian distribution is $\overline{\sigma_{c,th}}$=6.3800 $\times$ 10^4 b.

18.2.2   The Production of ^{149}Sm

18.2.3   Equations Governing ^{149}Pm and ^{149}Sm

Figure 18.2 represents the chain of reactions that produce and destroy ^{149}Sm. Because the lifetime of ^{149}Nd is significantly short compared to the one of ^{149}Pm, it can be assumed that fissions produce ^{149}Pm directly, with a cumulative yield 6.38%. The system of differential Eq. 18.7 describing the volumic concentration of ^{149}Pm and ^{149}Sm as a function of time is:

$$\begin{cases} \dfrac{dS(t)}{dt} = \lambda_P P - \sigma_S \phi S \\ \dfrac{dP(t)}{dt} = \gamma_P \, \Sigma_f \phi - \lambda_P P \end{cases} \tag{18.7}$$

The concentrations at equilibrium are:

$$\begin{cases} P_{\text{eq}} = \dfrac{\gamma_P \, \Sigma_f \phi}{\lambda_P} \\ S_{\text{eq}} = \dfrac{\gamma_P \, \Sigma_f}{\sigma_S} \end{cases} \tag{18.8}$$

This equilibrium is also known as samarium reservoir since all of the promethium must undergo decay to samarium. The equilibrium ^{149}Sm concentration is independent of neutron flux and power level. It will be observed that although the equilibrium samarium is independent of the flux, the post-shutdown samarium increases

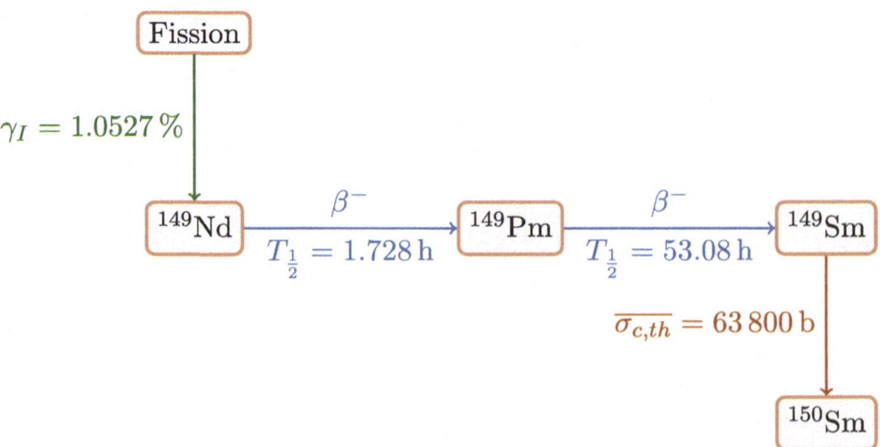

Fig. 18.2   Series of reactions for samarium poisoning

Table 18.2 Fission yields of Nd-149 and Pm-149 for the thermal fission of U-235 (0.0253 eV). Source: JEFF 3.1.1

FP	Independent fission yield (%)	Cumulative fission yield (%)	Half-life $T_{\frac{1}{2}}$ (s)
^{149}Nd	6.8489×10^{-3}	1.0527	6220.8
^{149}Pm	$\approx 4.7 \times 10^{-6}$		1.9109×10^{5}

with increasing flux, since the equilibrium promethium is proportional to the flux (Lamarsh 1966, p. 474) (Table 18.2).

18.2.4 *Estimation of the Samarium Poisoning*

With the concentrations at equilibrium (18.8), we can calculate the poisoning due to samarium ρ_S, and also ρ_P that is defined in the same way as ρ_I.

$$\rho_S = \frac{\gamma_P \, \Sigma_f}{\Sigma_{a, \text{fuel}}} \quad \text{and} \quad \rho_P = \rho_S \frac{\sigma_S \phi}{\lambda_P}$$

In the case of a PWR 1300 MWe modelized as an homogeneous cylinder in the Chap. 13, we have estimated the four factors (see Sect. 13.1.6), and in particular the reproduction factor $\eta = 1.5355$.

In order to calculate an estimation of the ratio fission over absorption in the fuel in the case of a PWR, we have to divide the reproduction factor by the average number of neutrons produced by fission. We know that, in a PWR, fissions are mainly divided between thermal fissions of ^{235}U, ^{239}Pu and fast fissions of ^{238}U (see Sect. 14.1.5). Taking into account the average number of neutrons produced by thermal fissions of ^{235}U, ^{239}Pu and by fast fissions of ^{238}U (see Sect. 3.3.2), we can write:

$$\overline{\nu}_{\text{PWR}} = 0.67 \times 2.4355 + 0.26 \times 2.8836 + 0.07 \times 2.819 = 2.5789$$

$$\frac{\Sigma_f}{\Sigma_{a, \text{fuel}}} = \frac{\eta}{\overline{\nu}_{\text{PWR}}} = \frac{1.5355}{2.5789} = 0.595\,41$$

And then we can calculate an approximation of ρ_S:

$$\rho_S = 0.010\,527 \times 0.595\,41 \times 1 \times 10^{5} \text{pcm} = 626.79 \, \text{pcm}$$

The same calculation is now carried out for the high flux reactor of the institute ILL. We admit that the ratio fission over absorption in the fuel is 0.757 (à 14 EFPD). This value is not very different from the ratio calculated in the case of a PWR, and for the high flux reactor we can estimate $\rho_S = 0.010\,527 \times 0.757 \times 1 \times 10^5\,\text{pcm} = 796.89\,\text{pcm}$.

It is interesting to calculate an estimation of the samarium poisoning after a reactor trip. We know that:

$$\frac{\rho_P}{\rho_S} = \frac{\sigma_S \phi}{\lambda_P}$$

We admit that the thermal flux, within the fuel plates of the ILL high flux reactor, is $\phi = 1.44 \times 10^{14}\,\text{n/cm}^2/\text{s}$. The averaged value of the capture cross section will be higher than the value calculated at $306\,°\text{C}$, since the temperature of heavy water between the plates of the fuel assembly is much lower than in a PWR. We take the value $\overline{\sigma_{c,th}} = 7.1671 \times 10^4\,\text{b}$. As a result:

$$\frac{\rho_P}{\rho_S} = \frac{7.1671 \times 10^4 \times 1 \times 10^{-24} \times 1.44 \times 10^{14} \times 1.9109 \times 10^5}{\ln 2} = 2.8452$$

We can see that, at equilibrium, the potential of poisoning stored in promethium represents almost 3 times the samarium poisoning. After a sufficient number of days following a trip of the considered reactor (4 weeks for example), promethium has almost entirely decayed into samarium and an extra amount of poisoning $2.8452 \times \rho_S$ is added to the initial poisoning ρ_S. The poisoning after promethium decay is $(1 + 2.8452) \times \rho_S = 3064.2\,\text{pcm}$.

In this situation, the high flux reactor may not be able to diverge anymore with the samarium poisoned fuel element.

18.3   Evolution of the Poisons Dynamics Throughout an Operating Cycle

Throughout the operating cycle, the number of fissile nuclei decreases, even if we take into account the conversion of ^{238}U to fissile ^{239}Pu known as fuel breeding, that partially replaces fissile ^{235}U. It is interesting to observe that the average fission reaction rate remains constant for the same nuclear power, and this explains why the thermal flux does naturally increase as the quantity of fissile material decreases. In consequence, the dynamics of xenon and iodine changes all along the cycle.

18.3.1   Example of the ILL High Flux Reactor

The fuel assembly of the RHF is highly enriched in ^{235}U at 93%.

The rated power of the reactor, 58 MeV, enables to calculate the fission rate:

$$\frac{58 \times 10^6}{1.6022 \times 10^{-19} \times 2.00 \times 10^8} = 1.81 \times 10^{18} \text{ fissions/s}$$

Thus the rate of fission of ^{235}U is 61.026 g/EFPD.

The probability that neutron absorption leads to a fission is:

$$P_{\text{F/A}} = \frac{\sigma_f}{\sigma_f + \sigma_{n,\gamma}}$$

where $\sigma_f = 570.79$ b is the fission cross section of ^{235}U for thermal neutrons, and $\sigma_{n,\gamma}$ is the radiative capture cross section. This enables to calculate:
$\frac{61.026}{0.8538} = 71.475$ g of ^{235}U are destroyed each day at rated power. Since the initial mass of ^{235}U in the fuel assembly is 8568 g, the daily loss of fissile material at rated power is 0.834 21% of the initial mass. At 46 EFPD, which is the end of the cycle, 38.374% of the initial mass of ^{235}U is destroyed.

The macroscopic cross section of fission is:

$$\Sigma_f = N_{^{235}\text{U}} \times \sigma_f$$

where $N_{^{235}\text{U}}$ is the volumic concentration of ^{235}U in the fuel plates (nuclei/cm^3).

We can assume that Σ_f does linearly decrease along with the mass of ^{235}U.

Thus $\Sigma_f(0\,\text{EFPD}) = 1.6291$ and $\Sigma_f(46\,\text{EFPD}) = 1.0040$.

$R_f = \Sigma_f \times \Phi_{\text{th}}$ and consequently, the thermal flux does increase: $\Phi_{\text{th}}(0\,\text{EFPD}) = 1.4443 \times 10^{14}$ n/cm^2/s and $\Phi_{\text{th}}(46\,\text{EFPD}) = 2.3436 \times 10^{14}$ n/cm^2/s.

The reproduction factor is defined as defined as the ratio of the number of fast neutrons produced by thermal fission to the number of thermal neutrons absorbed in the fuel (see Sect. 13.1.3.5):

$$\eta = \bar{\nu} \frac{N_{^{235}\text{U}}\sigma_f(^{235}\text{U})}{N_{^{235}\text{U}}\sigma_a(^{235}\text{U}) + N_{^{238}\text{U}}\sigma_f(^{238}\text{U}) + \Sigma_{a,\text{FP}}}$$

We admit the following estimations:

$\eta(0\,\text{EFPD}) = 1.9$ and $\eta(46\,\text{EFPD}) = 1.6909$.

At the beginning of the cycle, the fuel is fresh and η is almost equal to the reproduction factor of the fissile isotope ^{235}U: $\eta_{^{235}\text{U}} = \bar{\nu}P_{\text{F/A}} = 2.066$. Our estimations of η, Σ_f, Φ_{th} as functions of the position in the operating cycle enable us to calculate the xenon and samarium poisonings at different positions in the cycle.

The operating cycle of the reactor has a duration of 46 EFPD. 8 reactor trips are calculated, at 5, 10, 15, 20, 25, 30, 35 and 40 EFPD. Xenon are Samarium poisoning are calculated for each transient, and results are shown in Figs. 18.3 and 18.4. The height of the peaks becomes greater as the cycle progresses, due to an equilibrium iodine concentration proportional to increasing flux during the cycle. The xenon concentration decreases during the cycle, because at constant power, X_{eq} exhibits a constant numerator proportional to fission reaction rate and Φ at the denominator

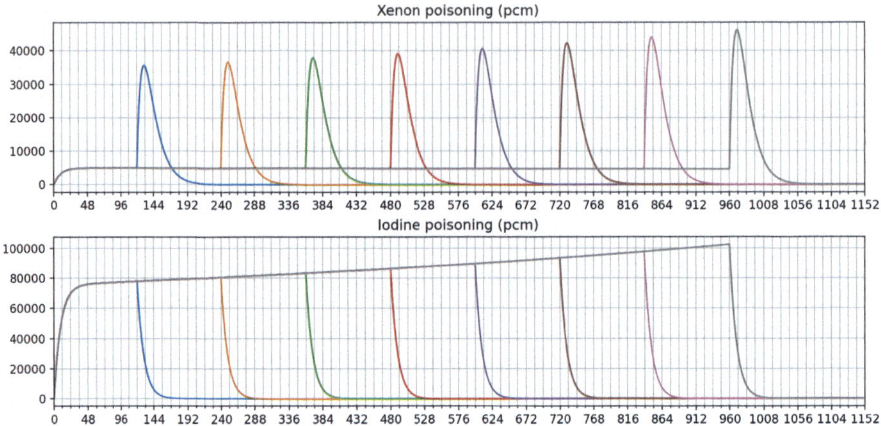

Fig. 18.3 Xenon peaks for ILL high flux reactor trips calculated at different positions in the cycle

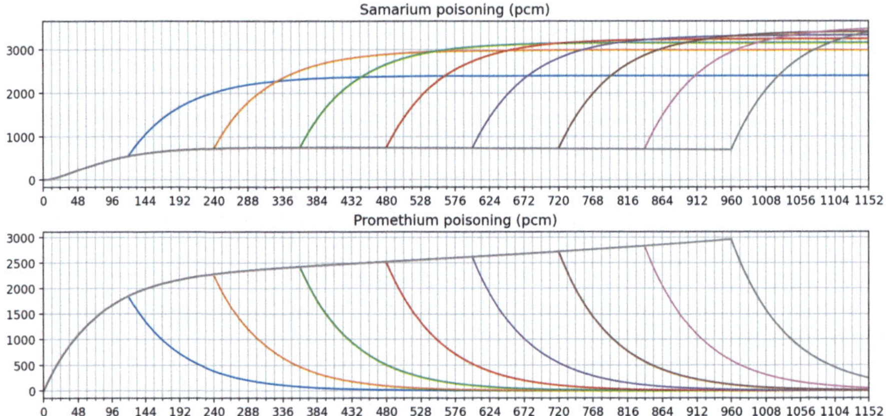

Fig. 18.4 Samarium poisoning for ILL high flux reactor trips calculated at different positions in the cycle

(see (18.2)). However, xenon poisoning varies little, since flux is at the numerator and denominator in Eq. 18.5.

It is also essential to remember that xenon peaks are considerably higher on a high-flux reactor than on a PWR.

18.3.2   Case of a PWR

The considered reactor is a PWR 1300 MWe, with a fuel enriched at 4%. This last precision is necessary, since, for a given reactor, the enrichment determines the

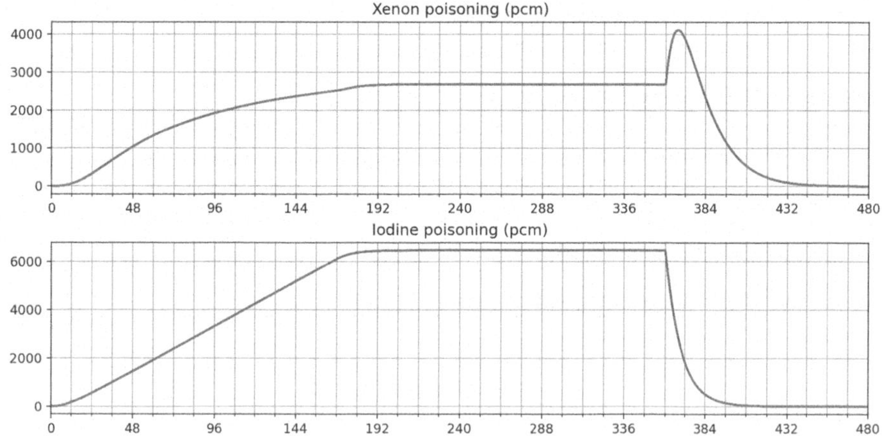

Fig. 18.5 Xenon build up at the beginning of the cycle, followed by a reactor trip at 15 EFPD

thermal flux and thus the dynamics of xenon. With former enrichments, e.g. 3.1%, the xenon peaks used to be notably higher.

This value $\eta = 1.5355$ is at the beginning of the operating cycle, and we will admit that when the boron concentration reaches $BC = 10$ ppm (for example at 370 EFPD), before stretch out operation,[1] the reproduction factor has decreased to $\eta = 1.3734$. Figure 18.5 shows the evolution of xenon after the core refueling. We assume that the power increases linearly during 7 d,[2] until it reaches rated power. When the reactor reaches rated power, xenon is still not yet at equilibirum, roughly 48 h at stabilized power are necessary to achieve xenon equilibrium. The reactor trip at 15 EFPD produces a peak in xenon (4100 pcm). At the end of the cycle, the peaks are higher, a few hundreds pcm more.

---

[1] Stretch-out (SO) operation is a common and approved mode of operation that may be used at "natural end of the cycle", when all control rods are at the top of the reactor (except the RCCA controlling the average coolant temperature) and boron concentration is close to 0 ppm (in practical 10 ppm). The reactivity balance is achieved by a tolerated cooling of the primary water, since boron concentration cannot anymore be reduced by dilution. During stretch out operation, small electrical power reductions are periodically performed, in such a way as to respect the turbine limit: the average temperature is allowed to fall only below a specific power supplied to the turbine. The SO operation of nuclear power plants improves the flexibility of outage schedule, and also increases the depth of burnup and improves the economy of fuel.

[2] The ramping up of reactor power takes in the order of 7 and is, of course, not linear. For example, a power step at 8%Pn enables to couple the generator to the grid, and intermediate steps enable to perform some flux mappings.

18.4   Load Following by a PWR

Parametric study with various low power levels

A parametric study with load following transients are calculated. A reactor, initially at rated power, performs a load reduction during 2 h and remains at the low plateau during 12 h. Then power is increased during 2 h up to rated power again. The low power levels are 30, 40, 50, 60, 70, 80 and 90%Pn (percentage of rated power) (Figs. 18.6, 18.7, 18.8, 18.9 and 18.10).

Parametric study with various low power level durations

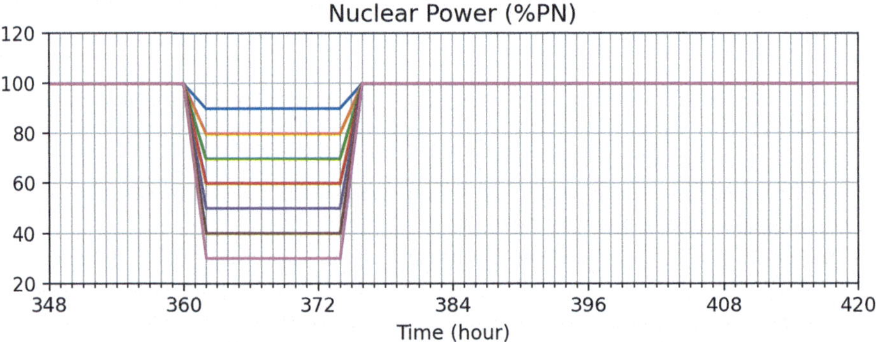

Fig. 18.6 Parametric study: load following with various low power levels

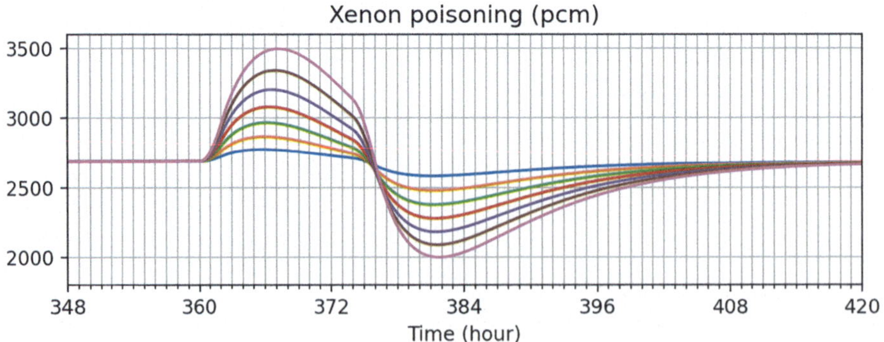

Fig. 18.7 Xenon poisoning and iodine in the load following parametric study

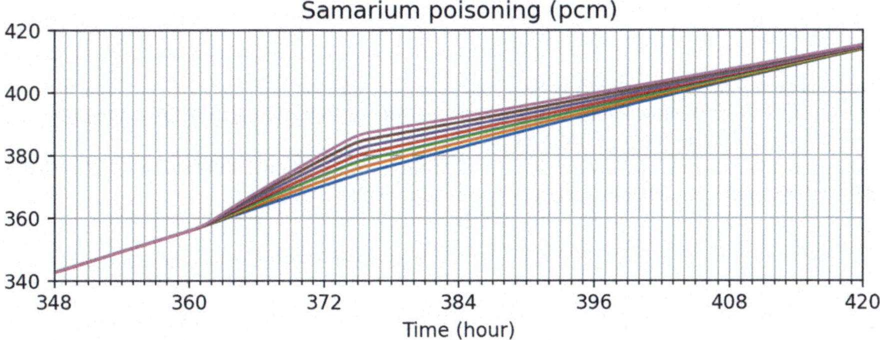

Fig. 18.8 Samarium poisoning and promethium in the load following parametric study

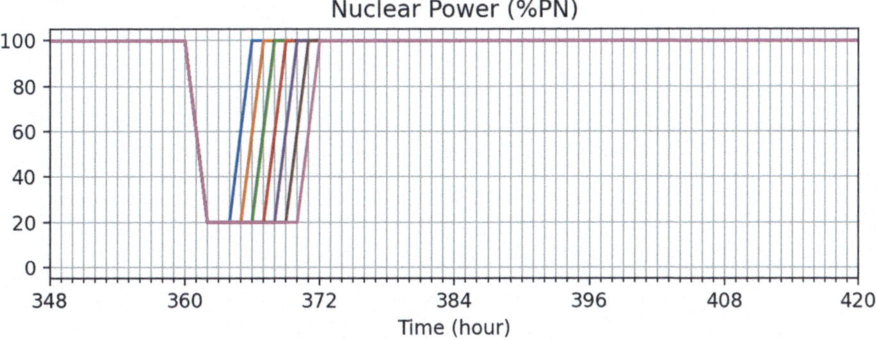

Fig. 18.9 Parametric study: load following with various low power level durations

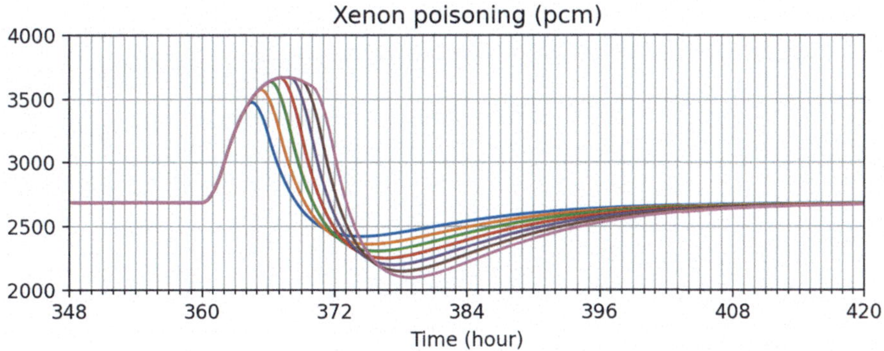

Fig. 18.10 Xenon poisoning and iodine in the load following parametric study (duration)

References

J.R. Lamarsh, *Introduction to Nuclear Reactor Theory*, in Addison-Wesley Series in Nuclear Engineering (Addison-Wesley Publishing Company, 1966). ISBN 9780201041262. https://books.google.fr/books?id=by5RAAAAMAAJ

S. Marguet, *The Physics of Nuclear Reactors* (Springer International Publishing, 2018). ISBN 9783319595603. https://books.google.fr/books?id=9DlODwAAQBAJ

P. Reuss, *Neutron Physics* (EDP Sciences, 2008)

Chapter 19
Dynamic Rod Weight Measurement

Abstract The short study of dynamic rod worth method provides an interesting illustration of various topics discussed in other chapters, notably the separation of time, and space-energy in point kinetics, the radial effects of rod banks. A DRWM test performed on a reactor is analysed.

19.1   Introduction

One of the objectives of zero power physics testing is to measure the integral worth of each bank and to compare it with calculated static bank worth.

The Dynamic Rod Weight Measurement has been used for over 20 years as the method used to measure the worths of control or shutdown banks during zero power physics testing (ZPPT) at numerous reactor startups with excellent results. It was first developed and applied by JSI[1] and then independently developed and commercialized by the Westinghouse Electric Company, under the trademark DRWM™(Dynamic Rod Worth Measurement). This method replaces the boron dilution and rod swap methods (RSM) that are no longer in use.

The boron dilution method enabled obtain the worth of heaviest bank called reference bank. It is a very slow but accurate process during which the rod bank had to be held in its position after each insertion (a few steps) into the core while boron was diluted to bring back the reactivity to a positive value, in preparation for the following insertion. In this way, reactivity was maintained within a range of $[-20, 20\,\text{pcm}]$ for example. The duration of measurement of the reference bank is determined by the dilution process.

For example, we assume that the initial boron concentration is $BC_i = 1700\,\text{ppm}$, the boron differential worth is $\epsilon_d = -8\,\text{pcm/ppm}$, the reactor coolant mass is $M_{RCS} = 270\,\text{t}$ and the worth of the reference bank is $\Delta\rho = 1000\,\text{pcm}$. The total injected mass of demineralized water compensating for the insertion of the reference bank is:

---

[1] The Jožef Stefan Institute in Slovenia operates a TRIGA research nuclear reactor built in 1966 by General Atomics, and reconstructed in 1991.

H. Grard, *Diffusion of Neutrons in Nuclear Reactors*,
https://doi.org/10.1007/978-3-032-05088-5_19

$$m = M_{RCS} \log_e \frac{BC_i}{BC_i - \frac{\Delta\rho}{\varepsilon_d}} = 20.6\,\mathrm{t}$$

If the zero power physics testing rules a dilution flow rate of $10\,\mathrm{t} \cdot \mathrm{h}^{-1}$, the duration of the worth measurement is 2 h. It is important to note that the zero power physics testing is on the critical path of the reactor start up after refueling.

During the rod swap process, a test bank is step-wisely inserted into the core while the reference bank is pulled in small steps to compensate for reactivity loss. The test bank worth is determined as the partial worth of reference bank. A typical duration for all the banks is 6 h.

The DRWM method offers three major advantages over the boron dilution and rod swap methods:

- Safety: Safety gain, since banks worths measurements are carried out with a single bank in the core, which is essentially subcritical.
- Waste: Reduction of the volume of effluent generated, as measurements are carried out at a constant boron concentration.
- Availability: Reduction of the testing duration, with a potential saving of around 8 h on the critical path.

What's more, since each banks are measured individually, banks worth measurements with no interferences with another rod bank in the core, by shade or antishade effects, which used to be the case with the RSM method.

19.2 DRWM Test Sequence

This section presents a sequence performed in a PWR 1300 and its results.

The first bank to be inserted is the R bank, dedicated to the control of the average coolant temperature in power operation. The reactor is initially in a critical state with the R bank partially inserted, so that its extraction results in an over-criticality of between 50 and 75 pcm, and the corresponding doubling time remains greater than 40 s. The R bank is then withdrawn to the top of the core, i.e., the allrod-out (ARO) position at 260 steps. As a consequence, in the considered example, the flux level is increased to a value of 6 times the flux corresponding to the point of adding heat (of the order of 0.05%PN). However, before inserting a bank for proceeding to the measurement, the flux level has to be less than 10 times the point of adding heat.

Banks are weighed from lightest to heaviest. Each bank to be weighed is inserted in manual mode at maximum stepping rate (60 steps/min for G1, G2, N1 and N2, or 72 steps/min for R resulting in a duration of 3 min 40 s) from allrod-out down to 5 steps. Measurements are acquired during insertion. At the end of insertion, the bank is withdrawn in manual mode at maximum stepping rate to the allrod-out (ARO) position. When the same initial flux conditions for measurement are once again achieved, the next bank is inserted.

There are also weak currents induced by gamma-rays, both from spontaneous gamma-ray sources in the reactor and neutron-material reaction induced gamma-ray sources near the detector. Especially, at the time a bank is fully inserted into the core, ex-core power range detector currents can decrease below 1 nA (0.1 nA durings the weighings analysed in the next section), thus the contribution of gamma-ray to detector currents cannot be neglected (Lee et al. 2005, p. 1466).

19.3 Analysis of a DRWM

19.3.1 Presentation of the Test

Figure 19.1 shows an example of a DRWM performed in a PWR 1300. The banks were weighed in the following order: G1, SD, G2, SA, N2, SB, SC, N1 and R. The scale on the left is the reactivity estimated by the digital reactivity meter whose input measurement is the current of a power range neutron detector. The scale on the right is the intensity measured by one of the four power range neutron detectors. The average coolant temperature is plotted in green, and not scaled.

The weighs of the banks, calculated by the DRWM method, are indicated on purple.

The pale red vertical rectangles in Fig. 19.1 represents time intervals during which the flux is above the point of adding heat. Withdrawal of a bank prior to its insertion for weighing brings positive reactivity and the flux increases up to 10 times the point of adding heat; then the banks are inserted for weighing. As long as the flux is within the range $[I_{\text{POAH}} 10 \times I_{\text{POAH}}]$, the fuel is generating heat, which is reflected in the mean primary temperature, which rises in waves.

During the DRWM weighing, the reactor is significantly subcritical, and this lasts until the weighed bank has risen again to its critical position. As a result, the flux drops by about 4 decades. Flux is restored by complete extraction of the bank. Reactivity is then around 60 pcm. At 60 pcm, the doubling time is about 60 s. An increase of the flux by 4 requires a duration between 13 and 14 doublings times ($2^{13} = 8192$), i.e. a duration of just over 13 min, as shown in the Fig. 19.1. Given that the flux can drop by up to 4 decades, it is worthwhile to start the insertion above the POAH, so as not to fall below permissive P6[2] at the end of the insertion. In hot stand by below permissive P6, the reactor is considered converged. Starting insertion with an over critical reactor helps to limit the time during which the core is subcritical, and results in a higher flux at the end of insertion.

---

[2] The source range trip provides protection against reactivity excursions and startup accidents. This trip may be manually blocked, for example in the context of a reactor start up, when at least one of the two intermediate range channels exceeds the P6 setpoint value. When source range trip is manually blocked, power is removed from the source range neutron detectors, de-energizing both channels. The trip function is automatically reinstated when both intermediate range channels decrease below the P6 setpoint. If the core falls below P6 outside the ZPPT framework, then the reactor is considered to be converged.

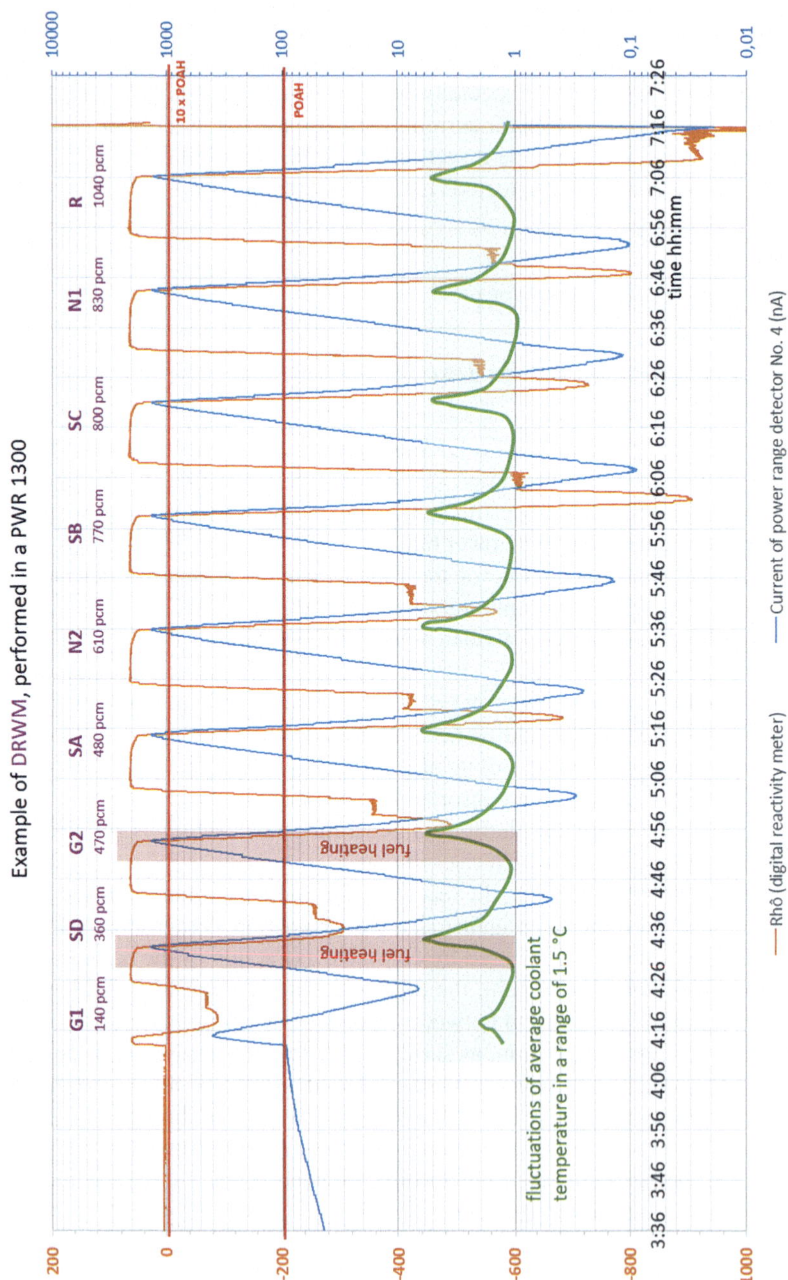

Fig. 19.1 DRWM performed in a PWR 1300

During banks withdrawals, some stops of 3 min are achieved to enable the cooling down of control rod drive mechanisms.

A low insertion speed would be penalizing in two ways: the duration of the test is extended, and the lower the group's insertion speed, the greater the flux reduction at the end of insertion is, since the time during which the core is subcritical is then greater. This can lead to the core falling below permissive P6. This is why insertions are carried out at maximum stepping rate.

Figure 19.2 shows the dozen or so decades of neutron flux between refueling and rated power (100%PN), and the measurement ranges of the different types of instrumentation (source, intermediate and power level detectors). The flux range in which the DRWM takes place is shown in gray; the range in which the source neutrons is negligible and the range in which nuclear heating occurs (and therefore negative feedback take place) are also shown.

The common principle of these analog and digital reactivity meters is to solve an inverse problem of the conventional point kinetics equation, where a spatial shape of neutron flux distribution is assumed to remain unchanged by any perturbations. In a large thermal power reactor and a loosely coupled-core research reactor, however, a local and large reactivity perturbation such as an insertion of control rod must significantly change the spatial shape and consequently lead to a fatal failure of this inverse kinetics analysis. Sano et al. (2018, Introduction)

The core averaged power and the ex-core detector signal are far from being linearly dependent during the weighing of a bank, because of the radial and axial spatial redistribution of the neutron flux. Important spatial as well as temporal effects occur, which change the spatial distribution of the prompt and delayed neutron population. Since a detector measures the local flux at the detector location, the proportionality factor changes with time, because the neutron population density distribution also changes (Merljak et al. 2018, p. 97).

We can see on Fig. 19.1 that the reactivity given by the reactivity meter can be very different from the reactivity resulting from the DRWM method.

The Fig. 19.3 enables us to visualize the modifications to the radial power shape produced by the insertion of G1, with the result that the ex-core detectors overestimate the core averaged power. The same phenomenon happens during the weighing of SD (shutdown banks are not represented on Fig. 19.3). This results in an underestimation of G1, SD weights by the reactivity meter, by comparison with the accurate DRWM method, a shown by Fig. 19.1.

Given their locations in the core, the banks SA, SB and G2 will reduce the flux in the crown on the outskirts of the core, facing the ex-core detectors. The underestimation of the core averaged power results in an overestimation of the integral worths of SA, SB and G2 by the reactivity meter.

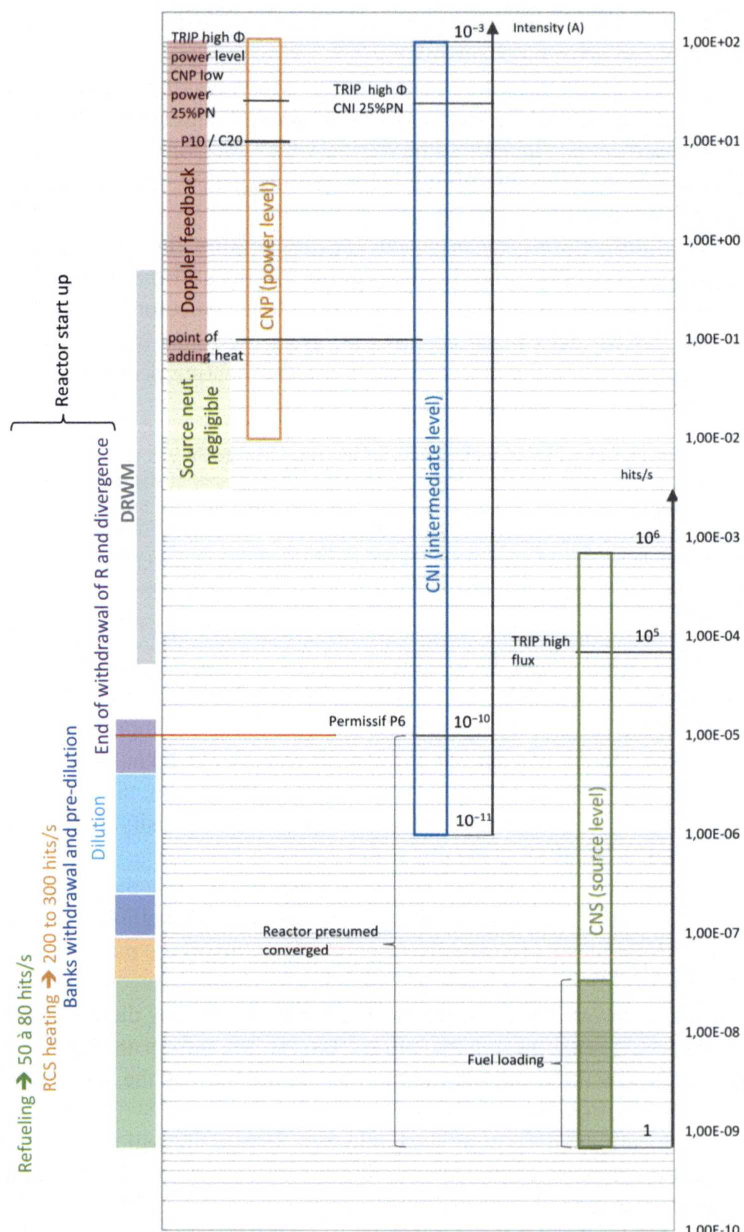

Fig. 19.2 Excore neutron detectors for source, intermediate and power level ranges

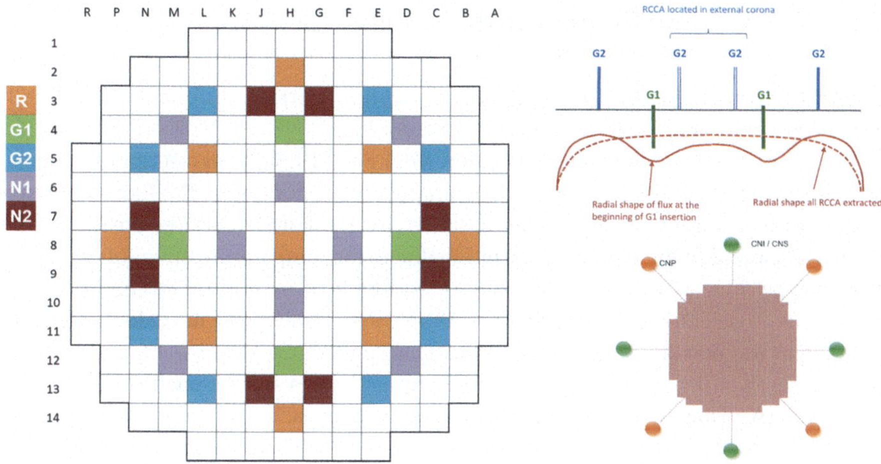

Fig. 19.3 Position of the banks; radial effect of G1 and G2; radial position of detectors

19.4 Spatial and Dynamic Effects of the DRWM Method

Higher modes are, of course, sustained for much longer times by the delayed neutrons. However, the relative intensity of theses modes is extremely low and they can usually be neglected. A notable exception is the "rod-drop" method or shutdown experiment where suddenly the prompt-neutron population is greatly diminished, and the modal distribution of delayed-neutron emission (amplified by residual multiplication of the system) predominates until an equilibrium is reached. Keepin (1965, p. 166)

The insertion of a bank during weighing produces spatial and dynamic effects:

- A static spatial effect due to the insertion of the bank, that occurs immediately for prompt neutrons. The prompt fission neutron source yields rapidly to the inserted absorber, while the delayed neutron precursors decay more slowly and keep the shape close to that from a previous time step.
- A dynamic effect, since the spatial distribution of delayed neutrons evolves in a distorted and delayed manner compared to that of prompt neutrons: everything happens as if the shape of the delayed flux is trying to join that of the prompt flux in constant evolution.

A static correction due to the redistribution of the neutron flux, and a dynamic correction due to the temporal delay (the delayed neutron distribution is trailing behind the distribution of the prompt neutrons) are applied (Merljak et al. 2018, p. 97) in order to calculate a time dependent reactivity from the ex-core measurements, using the inverse kinetics equations. Static correction may be applied before the inverse point kinetics calculation, and the dynamic correction after the inverse point kinetics calculation. Corrective factors are obtained by high-accuracy three-dimensional static and transient simulations carried out according to a specific bank insertion.

19.5   DRWM and Temperature Effects

In order to estimate the weight of a bank, some final corrections enable to account for the contribution of fuel and primary fluid cooling.

As we have seen in our example of DRWM in Sect. 19.3, the measurement begins above the point of adding heat. We can estimate that the cooling of the fuel following the drop in neutron flux brings a gain in reactivity estimated at around 5 pcm.

An assessment of the reactivity variation due to the temperature effect of the moderator can be done as follows: The point of adding heat is supposed to be visible at a power range detector current $100\,\text{nA} \approx 0.05\%\text{PN}$. To simplify the calculation, we assume that the power extracted from the steam generators is constant, and we can integrate the heating power transferred to primary water during each time interval above $100\,\text{nA}$. That is roughly $4000\,\text{MJ}$. The heat capacity of the reactor cooling system is estimated in the range [$1400\,\text{MJ}/\,^\circ\text{C}$ for "fast" transients[3]; $2600\,\text{MJ}/\,^\circ\text{C}$ for "slow" transients], the later taking into account the exchange of thermal energy between the primary fluid and the metal structures. Thus we can estimate that each heating phase is the range [$1.5\,^\circ\text{C}$; $2.8\,^\circ\text{C}$]. This is consistent with the range of average coolant temperature fluctuations observed during the considered test (see Fig. 19.1). Cooling begins during a bank insertion when the core operates below the point of adding heat, and continues after the bank is fully inserted until the power extracted from the steam generators returns to equilibrium with the primary power. The moderator temperature coefficient is $-6.6\,\text{pcm}/^\circ\text{C}$ during ZPPT, with a boron concentration around $1740\,\text{ppm}$. Thus the reactivity gain due to primary cooling during this DRWM test remains less than $10\,\text{pcm}$.

References

G.R. Keepin, *Physics of Nuclear Kinetics* (Addison-Wesley Publishing Company, 1965)

E.K. Lee, H.C. Shin, S.M. Bae, Y.K. Lee, New dynamic method to measure rod worths in zero power physics test at PWR startup. Ann. Nucl. Energy **32**(13), 1457–1475 (2005)

V. Merljak, M. Kromar, A. Trkov, Rod insertion method analysis-a methodology update and comparison to boron dilution method. Ann. Nucl. Energy **113**, 96–104 (2018)

T. Sano, K. Hashimoto, H. Taninaka, U. Hironobu, Significant spatial dependence observed in inverse kinetics analysis for a loosely coupled-core system of the Kyoto university critical assembly. J. Nucl. Sci. Technol. **55**(11), 1355–1361 (2018)

---

[3] During a fast transient, one can neglect the heat exchange between coolant and structures, thus the heat capacity of the reactor cooling system is the heat capacity of the water. If the mass of water in the loops (pressurizer excluded) is $246\,\text{t}$, at an average temperature $306\,^\circ\text{C}$, then $MC_p \approx 5.6172 \times 246 = 1380\,\text{MJ} \cdot \,^\circ\text{C}^{-1}$. In the case of a slow transient, the heat capacity of the structures (pipes, vessel etc.) is added, assuming the thermal equilibrium between water and structures.

Chapter 20
Adjoint Flux and Perturbation Method

Abstract In this chapter, consideration will be given to the equation which is adjoint to the diffusion equation, in the cases of one and two energy groups. The solutions of adjoint equations have a physical significance as the "importance" of neutrons within a system. The adjoint function can be used to obtain the response to the small perturbation, and a simplified overview is provided in this chapter, with the first order perturbation theory.

20.1 The Importance Function

20.1.1 Iterated Fission Probability, Progeny of Neutrons

Henry Hurwitz, Jr. defined the iterated fission probability as an asymptotic power originating in a neutron injected in a critical core at zero power.

> The effect on reactivity of an arbitrary change in an assembly can be described in terms of a function $F(r, E)$ which has the following definition. Let a neutron be introduced in the assembly (which is assumed to be just critical) at point r and with energy E. This neutron will, on the average, produce a certain number of fissions with a certain spatial distribution. Neutrons from these fissions will then produce further fissions, etc., each succeeding generation having a distribution closer to the actual power distribution in the operating assembly. Furthermore, since the assembly is at critical, the number of fissions produced in the nth generation will approach a limit as n approaches infinity, and this limit is defined as $F(r, E)$. Hurwitz (1948, p. 4)

The iterated fission probability $F(r, E)$ must not be confused with the probability $P(r, E)$ that a neutron introduced at point r and with energy E will produce a fission, although both seem similar. For example, the fact that the considered initial neutron is close to the edge of the fissile area does not affect the probability P, but the iterated fission probability $F(r, E)$ is lowered: indeed the neutron introduced near the edge produces fissions near the edge of the core, and these fissions will have less than the average chance of producing further fissions. Another confusion should also be avoided: the total number of fissions induced by a neutron and its progeny is not proportionnal to the iterated fission probability (Nauchi and Kameyama 2010, p. 978).

H. Grard, *Diffusion of Neutrons in Nuclear Reactors*,
https://doi.org/10.1007/978-3-032-05088-5_20

The set of neutrons produced from the initial neutrons is coined the progeny.

> In reactor physics we can refer to one neutron at position r at time t and to its progeny at some time later. The progeny are all those neutrons that trace their life back to the original neutron through fission, scattering, etc., or even the original neutron itself after it moved away from location r. We make no distinction between this original neutron and the "new" neutrons in counting the progeny. (..) a particle is as important as its progeny. Lewins (1965, pp. 20, 22)

20.1.2 A Physical Approach to the Adjoint Flux

The direct equation, also known as the forward equation, establishes a balance between neutron density at a given point in space (but also at an energy E and in an Ω direction, in the case of the transport equation): it is the balance is between departures due to local motion or reaction rate, and arrivals.

The earliest discussion of the physical meaning of the adjoint flux was given by Hurwitz (Hurwitz 1948, p. 978). He described that the adjoint flux is proportionnal to the iterated fission probability. The adjoint equation establishes a balance on the importance of neutrons. The adjoint flux is often called the importance function.

> When the current is approximated by a Fick's law of expression, we may draw a balance of probabilities for either density or importance. In this balance the diffusion equation adopts the Eulerian view-point of observing local rate of change rather the Lagrangian viewpoint of transport theory following the neutron path. Lewins (1965, p. 57)

Thus, the adjoint flux has the physical significance of being proportionnal to the overall power level which will result asymptotically if neutrons of lethargy u having direction Ω are introduced at point r in a critical reactor initially at zero power (Radkowsky 1964, p. 858).

If we consider a critical core of zero power without external sources, and let a neutron be injected. The fission neutron emission produced by the progeny converges to the fundamental mode.

The formal identification between the neutron importance and the adjoint flux enables to compute the importance function with Monte Carlo calculation codes, and thus estimate the adjoint neutron flux for multiplying systems.

20.2 Calculation of Adjoint Flux

20.2.1 Adjoint in the Case of One Group

The critical equation, in the case of one neutron group, cas be written as:

$$\nu\Sigma_f(r)\Phi(r) - \Sigma_a(r)\Phi(r) + \mathrm{div}(D(r) \cdot \overrightarrow{\mathrm{grad}}\Phi(r)) = 0$$

If we assume that $D(r)$ is constant within the reactor, then we can write:

$$\nu\Sigma_f(r)\Phi(r) - \Sigma_a(r)\Phi(r) + D\Delta\Phi(r)) = 0$$

Let H be the operator defined by $(\nu\Sigma_f - \Sigma_a)$. The critical equation is then $H\Phi = 0$.

The direct flux is the eigenfunction of H associated with the eigen value 0.

In an inner vector space over $\mathbb{R}$, a linear operator and its adjoint have the same eigenvalues, thus 0 is also an eigenvalue of H^*. The adjoint flux is the eigenfunction of H^* associated with the eigenvalue 0 and $H^*\Phi^* = 0$.

The Laplacian operator Δ and the multiplication by a function operator $(\nu\Sigma_f(r) - \Sigma_a(r))$ are self-adjointoperators (Appendix G, Sect. G.5.1). The operator H is self-adjoint, thus $H^* = H$.

It can be concluded that the direct flux $\Phi(r)$ and the adjoint flux $\Phi^*(r)$ are proportionnal.

If $D(r)$ is not constant within the reactor, then a similar conclusion is reached: Let u and v be any two functions which vanish at the boundary of the system, on the surface of V. It can be demonstrated that (Lamarsh 1966, p. 527):

$$\int_V u(r)\mathrm{div}(D(r)\cdot\overrightarrow{\mathrm{grad}}\,v(r))d\tau = \int_V v(r)\mathrm{div}(D(r)\overrightarrow{\mathrm{grad}}\,u(r))d\tau$$

Which shows that the operator $\mathrm{div}(D(r)\cdot\overrightarrow{\mathrm{grad}})$, in which $D(r)$ is an arbitrary function of position r, is self-adjoint.

Hence, it can be concluded that one-group operators are self-adjoint in nonuniform, reflected, or multiregion reactor. As a consequence, the direct flux $\Phi(r)$ and the adjoint flux $\Phi^*(r)$ are proportionnal.

20.2.2 Adjoint in a Bare Homogeneous Reactor with Two Groups

We write again the coupled Eq. 12.7:

$$\begin{cases} D_1\Delta\Phi_1(r) - \Sigma_{r1}\Phi_1(r) + \dfrac{k_\infty}{p}\Sigma_{r2}\Phi_2 = 0 \\[2ex] D_2\Delta\Phi_2(r) - \Sigma_{r2}\Phi_2(r) + p\Sigma_{r1}\Phi_1(r) = 0 \end{cases}$$

Since the system is assumed to be homogeneous, D_1, D_2, Σ_{r1}, Σ_{r2} are constants and not functions of r.

20.2.2.1   Critical Condition Calculated with Energy Matrix

The coupled equations can be written in matrices in vector form (Barjon et al. 1993, p. 415):

$$\left(\underbrace{\begin{bmatrix} D_1\Delta & 0 \\ 0 & D_2\Delta \end{bmatrix}}_{D} + \underbrace{\begin{bmatrix} -\Sigma_{r1} & 0 \\ 0 & -\Sigma_{r2} \end{bmatrix}}_{A} + \underbrace{\begin{bmatrix} 0 & \frac{k_\infty}{p}\Sigma_{r2} \\ p\Sigma_{r1} & 0 \end{bmatrix}}_{P} \right) \begin{pmatrix} \Phi_1(r) \\ \Phi_2(r) \end{pmatrix} = 0$$

$p\Sigma_{r1}$ represents the arrival in the thermal group from the slowing down of fast neutrons, and $\frac{k_\infty}{p}\Sigma_{r2}$ represents the production of fast neutrons from thermal fissions.

$$\Delta \begin{pmatrix} \Phi_1(r) \\ \Phi_2(r) \end{pmatrix} + \underbrace{\begin{bmatrix} -\frac{\Sigma_{r1}}{D_1} & \frac{k_\infty}{pD_1}\Sigma_{r2} \\ p\frac{\Sigma_{r1}}{D_2} & -\frac{\Sigma_{r2}}{D_2} \end{bmatrix}}_{E \text{ energy matrix}} \begin{pmatrix} \Phi_1(r) \\ \Phi_2(r) \end{pmatrix} = 0 \tag{20.1}$$

The eigen values of the energy matrix E, λ_a and λ_b, are the solutions of the equation:

$$\det(E - \lambda I) = 0$$

$$\left(\frac{\Sigma_{r1}}{D_1} + \lambda \right) \left(\frac{\Sigma_{r2}}{D_2} + \lambda \right) - k_\infty \frac{\Sigma_{r1}\Sigma_{r2}}{D_1 D_2} = 0$$

$$\frac{\Sigma_{r1}\Sigma_{r2}}{D_1 D_2} + \lambda \left(\frac{\Sigma_{r1}}{D_1} + \frac{\Sigma_{r2}}{D_2} \right) + \lambda^2 - k_\infty \frac{\Sigma_{r1}\Sigma_{r2}}{D_1 D_2} = 0$$

$$(\tau_{th} L_2^2)\lambda^2 + \lambda(\tau_{th} + L_2^2) + (1 - k_\infty) = 0 \tag{20.2}$$

The discriminant of Eq. 20.2 is $(\tau_{th} + L_2^2)^2 - 4\tau_{th} L_2^2(1 - k_\infty)$.

If $k_\infty < 1$, criticality is not achievable.

If $k_\infty = 1$, $\lambda_a = 0$ and $\lambda_b = -\frac{\tau_{th} + L_2^2}{\tau_{th} L_2^2}$.

If $k_\infty > 1$, $\lambda_a > 0$ and $\lambda_b < 0$.

The eigenvalues λ_a and λ_b of the energy matrix are also the eigenvalues of the Laplacian, each eigen value being the same for each neutron energy group as shown by Eq. 20.1. In the case $k_\infty > 1$, the positive eigenvalue of the Laplacian λ_a is associated with an eigenfunction $f_0(r)$ that does not change of sign and equal to zero at the boundary of the system. $f_0(r)$ defines the fundamental mode.

To emphasize that $\lambda_a > 0$, we write $\lambda_a = B^2$. The flux fulfils the equations:

$$\begin{cases} \Delta\Phi_1(r) + B^2\Phi_1(r) = 0 \\ \\ \Delta\Phi_2(r) + B^2\Phi_2(r) = 0 \end{cases} \tag{20.3}$$

Where $\Phi_1(r)$ is the eigenfunction $f_0(r)$ times an arbitrary constant Φ_1: $\Phi_1(r) = \Phi_1 f_0(r)$. Similarly $\Phi_2(r) = \Phi_2 f_0(r)$, and the ratio $\frac{\Phi_2}{\Phi_1}$ can be expressed as a function of k_∞ and of the energy matrix coefficients.

$$B^2 \frac{\Phi_2}{\Phi_1} = p \frac{\Sigma_{r1}}{D_2} - \frac{\Sigma_{r2}}{D_2} \frac{\Phi_2}{\Phi_1}$$

$$\frac{\Phi_2}{\Phi_1} = \frac{p \Sigma_{r1}}{D_2 B^2 + \Sigma_{r2}} = \frac{\frac{\Sigma_{s1\to2}}{\Sigma_{r2}}}{1 + L_2^2 B^2}$$

20.2.2.2 Criticality and Direct Flux

The critical equation can be written as:

$$([D] + [A] + [P]) \begin{pmatrix} \Phi_1(r) \\ \Phi_2(r) \end{pmatrix} = 0$$

It is also possible to say that the direct critical flux is solution of the following eigen equation:

$$([D] + [A] + [P]) \begin{pmatrix} \Phi_1(r) \\ \Phi_2(r) \end{pmatrix} = \lambda \begin{pmatrix} \Phi_1(r) \\ \Phi_2(r) \end{pmatrix}$$

With an eigen value $\lambda = 0$.

20.2.2.3 Criticality and Adjoint Flux

An operator and its adjoint (see Appendix G, Sect. G.4) in a inner product space over $\mathbb{R}$ have the same eigen values. Thus $\lambda = 0$ is also an eigen value of the adjoint equation:

$$([D] + [A] + [P])^* \begin{pmatrix} \Phi_1^*(r) \\ \Phi_2^*(r) \end{pmatrix} = \lambda \begin{pmatrix} \Phi_1^*(r) \\ \Phi_2^*(r) \end{pmatrix}$$

Taking into account the properties of adjoint linear operators, we write the adjoint critical equation:

$$([D^*] + [A^*] + [P^*]) \begin{pmatrix} \Phi_1^*(r) \\ \Phi_2^*(r) \end{pmatrix} = 0$$

Given that $[D]$ and $[A]$ are diagonal matrices, $[D^*] = [D]$ and $[A^*] = [A]$.

$$[P]^* = \begin{bmatrix} 0 & p\Sigma_{r1} \\ \frac{k_\infty}{p}\Sigma_{r2} & 0 \end{bmatrix}$$

Thus the adjoint critical Eq. 20.2.2.3 becomes:

$$\Delta \begin{pmatrix} \Phi_1^*(r) \\ \Phi_2^*(r) \end{pmatrix} + \underbrace{\begin{bmatrix} -\frac{\Sigma_{r1}}{D_1} & p\frac{\Sigma_{r1}}{D_1} \\ \frac{k_\infty}{pD_2}\Sigma_{r2} & -\frac{\Sigma_{r2}}{D_2} \end{bmatrix}}_{E^* \text{ energy matrix of adjoint equation}} \begin{pmatrix} \Phi_1^*(r) \\ \Phi_2^*(r) \end{pmatrix} = 0 \qquad (20.4)$$

It should be observed that the energy matrix of adjoint Eq. **is not** the adjoint of the energy matrix of the direct equation. However, the equations $\det(E - \lambda I) = 0$ and $\det(E^* - \lambda I) = 0$ are identical (20.2), which means that energy matrix of adjoint equation and the energy matrix of the direct equation have the same eigenvalues. As a consequence, the critical condition for the direct equation and the adjoint equation is the same.

One important consequence is that, in an homogeneous reactor, the direct and adjoint flux have the same shape $f_0(r)$ in both energy groups.

$$\Phi_1(r) = \Phi_1 f_0(r) \quad \text{and} \quad \Phi_1^*(r) = \Phi_1^* f_0(r)$$

$$\Phi_2(r) = \Phi_2 f_0(r) \quad \text{and} \quad \Phi_2^*(r) = \Phi_2^* f_0(r)$$

However, the fast over thermal ratio of the adjoint flux will be different from the ratio of the direct flux.

We can introduce B^2, that is both the positive eigen value of the direct and the adjoint critical equation:

$$\begin{bmatrix} -\frac{\Sigma_{r1}}{D_1} & p\frac{\Sigma_{r1}}{D_1} \\ \frac{k_\infty}{pD_2}\Sigma_{r2} & -\frac{\Sigma_{r2}}{D_2} \end{bmatrix} \begin{pmatrix} \Phi_1^*(r) \\ \Phi_2^*(r) \end{pmatrix} = B^2 \begin{pmatrix} \Phi_1^*(r) \\ \Phi_2^*(r) \end{pmatrix} \qquad (20.5)$$

$$\frac{-1}{\tau_{th}}\Phi_1^* + p\frac{\Sigma_{r1}}{D_1}\Phi_2^* - B^2\Phi_1^* = 0$$

$$\frac{\Phi_1^*}{\Phi_2^*} = p\frac{1}{1 + B^2\tau_{th}} \qquad (20.6)$$

It cas also be easily established that:

$$\frac{\Phi_2^*}{\Phi_1^*} = \frac{k_\infty}{p(1 + B^2 L_2^2)} > 1$$

$\frac{1}{1 + L_2^2 B^2}$ is the thermal non leakage factor.

As an example, we consider the bare homogeneous critical reactor presented in Sect. 13.1, Table 13.1, with Dirichlet boundary conditions. The fissile radius is 168 cm. The values from Table 13.1 enable to calculate:

$$\tau_{th} = \frac{D_1}{\Sigma_{r1}} = 50\,\text{cm}^2$$

$$p = \frac{\Sigma_{s12}}{\Sigma_{r1}} = 0.723\,22$$

$$L_2^2 = \frac{D_2}{\Sigma_{r2}} = 6\,\text{cm}^2$$

$$B_g^2 = \frac{2.4048^2}{168^2} = 2.0490 \times 10^{-4}\,\text{cm}^2$$

$$k_\infty = (1 + L_2^2 B_g^2)(1 + \tau_{th} B_g^2) = 1.0115$$

And then:

$$\frac{\Phi_2^*}{\Phi_1^*} = \frac{1 + \tau_{th} B_g^2}{p} = \frac{k_\infty}{p(1 + L_2^2 B_g^2)} = 1.3969$$

$$\frac{\Phi_1}{\Phi_2} = \frac{1 + L_2^2 B_g^2}{\frac{\Sigma_{s12}}{\Sigma_{r2}}} = 6.2918$$

Neutrons introduced into the slow group contribute more to the chain reaction than neutrons introduced into the fast group, since neutrons introduced into the fast group have to escape fast leakage (probability $\frac{1}{1+B^2\tau_{th}}$) and to escape resonances during slowing down (probability p) before becoming thermalized neutrons. This is equivalent to saying that neutrons added to the thermal flux are more important than neutrons added to the fast group. Importance of neutrons is a concept introduced in the next section.

20.2.3 Amplification of a Source and Ajoint Flux

We whish to highlight the fact that the adjoint fast flux describes the importance of fast neutrons, i.e. the stationary population of neutrons that is established for a fission neutron introduced into the core. To do this, we first modify the mutiple areas and reflected reactor defined in Table 13.6, and increase $\nu_2 \Sigma_{f2}$ in order to change the

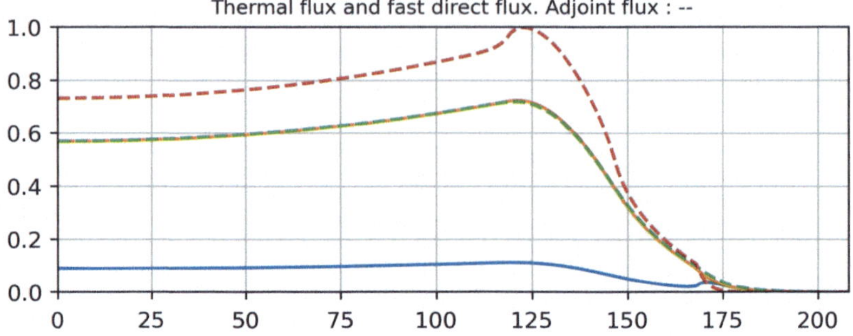

Fig. 20.1 Calculation of a direct and adjoint thermal and fast flux, with a deformed shape

shape of the direct and adjoint fluxes. The calculation of the fundamental adjoint and direct neutron fluxes is given in Fig. 20.1.

In order to calculate the stationnary state of the reactor with a source and made subcritical, the thermal utilization factor affecting the entire core is lowered. Four different negative reactivities are considered. The source is a radial ring of width 1 cm and 20 calculations are performed with various radial positions of the source, for each reactivity. For each source position, the ratio between the production rate of neutrons by fission and the emission rate of the source is calculated, that is the amplification factor defined in Sect. 15.2.

The Fig. 20.2 shows that the amplification factor estimates the adjoint fast flux, which is the importance function of fast neutrons. It's important to note that the iterated fission probability, proportionnal to the adjoint flux, is defined for an almost critical reactor. The subcritical reactor, for which the amplification factor is calculated, is different from the critical reactor for which the fundamental adjoint flux is calculated. The more subcritical the reactor, the less accurate the amplification factor estimates the adjoint fast flux. The variations of the adjoint fast flux reflect the variations of the source efficiency, as a function of the source position.

We can also add that the amplification factor of a source in a subcritical reactor and the iterated fission probability are similar, and respectively consider rates of neutrons per unit time and absolute numbers of neutrons.

20.3   Adjoint in the Case of Two Groups and Multiple Areas Reactor

Two groups direct flux, or fundamental flux, and adjoint flux have the same shape in the case of the bare homogeneous reactor, as shown in Sect. 20.2.2. This is not true in a multiple areas reactor, for which direct flux and adjoint flux do not have the same shape. This is specially visible in the case of a reflected reactor. We calculate numerically the adjoint flux in the case of the reactor described in Sect. 13.4.2, with a

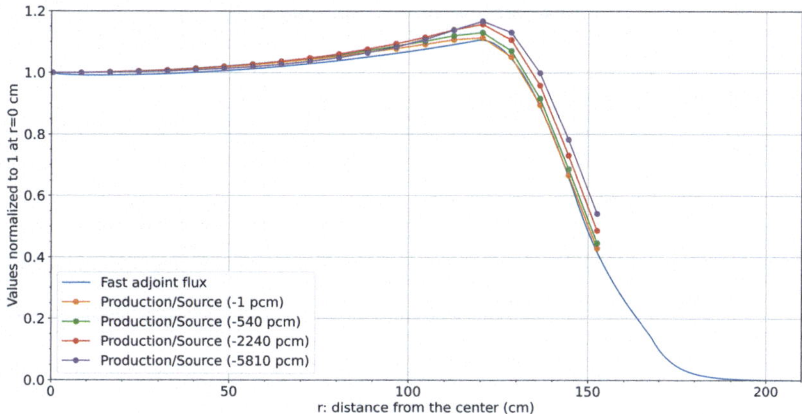

Fig. 20.2 Comparison between the fast adjoint flux and the ratio production/source for various source positions

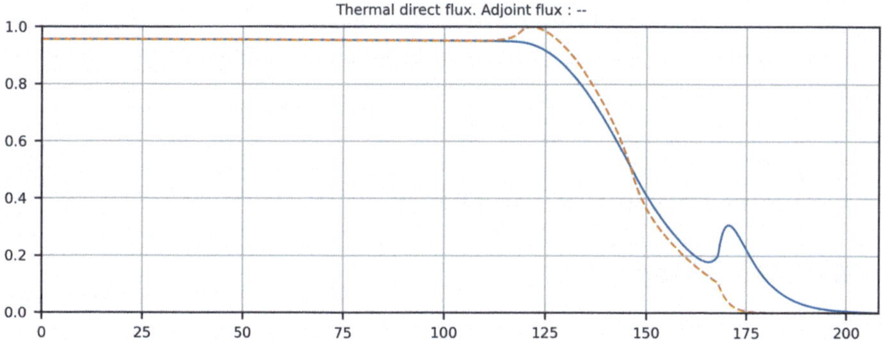

Fig. 20.3 Comparison between thermal direct and adjoint flux (rescaled) for a non homogeneous reactor

flattened thermal flux shown on Fig. 13.8. Figure 20.3 enables to compare the shapes of the direct thermal flux and the adjoint thermal flux. The two curves are scaled together.

The peaking of the thermal flux in the reflector arises from the thermalization of fast neutrons that escape from the core. Since the absorption by the reflector is small, the thermalized neutrons accumulate in this region. The neutrons reflecting back to the core from the reflector may induce nuclear fissions. The probability that a neutron is reflected to the core decreases with the distance from the core boundary. Thus, the thermal adjoint flux, that weights the progeny of thermal neutrons as a function of their location in the system, does decrease monotonically in the reflection region (Hayashi 2018).

Figure 20.4 shows four curves: thermal direct and adjoint flux, fast direct and ajoint flux. The fast direct and ajoint flux are scaled together so as to be compared

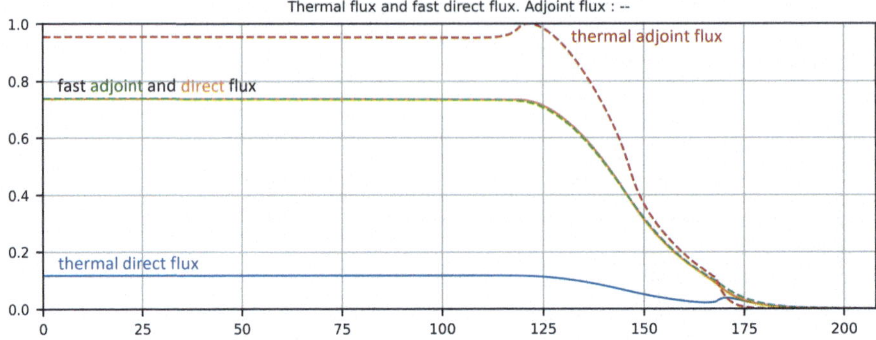

Fig. 20.4 Thermal and fast direct and adjoint flux

easily, and this scaling factor is applied to thermal direct and adjoint flux. Fast direct and ajoint flux are almost overlapped. Thermal direct flux and thermal adjoint appear with different magnitudes, this is due to the fact that ratios $\frac{\Phi_2}{\Phi_1}$ and $\frac{\Phi_2^*}{\Phi_1^*}$ are different. The fast adjoint is less than the slow adjoint in the core although the fast flux is greater than the slow flux in the same region.

20.4   Ajoint Flux in Reactor Theory

20.4.1   Perturbation

The various one-group perturbation formulas developed in this section are necessarily limited by the inadequacies inherent in the one-group method. For example, it will be recalled that the peaking of the thermal flux in the reflector is not predicted by one-group theory. Thus the effect of placing an absorber in the reflector cannot be determined accurately with one-group perturbation theory. The one-group formulas also are not applicable to problems involving change in the moderating properties of the system or the production of fast neutrons because these things simply are not included in one-group theory (..) one-group perturbation theory has very limited applications. A much wider range of problems can be handled in two group theory. Lamarsh (1966, p. 534)

20.4.1.1   Non Perturbated Diffusion Equation for Two Groups

We can write again the coupled Eq. 12.5, in which criticality is achieved by the properties of the system (matter and geometry):

$$\begin{cases} D_1(\boldsymbol{r})\Delta\Phi_1(\boldsymbol{r}) - \Sigma_{r1}(\boldsymbol{r})\Phi_1(\boldsymbol{r}) + \nu_2\Sigma_{f2}(\boldsymbol{r})\Phi_2(\boldsymbol{r}) + \nu_1\Sigma_{f1}(\boldsymbol{r})\Phi_1(\boldsymbol{r}) = 0 \\[2ex] D_2(\boldsymbol{r})\Delta\Phi_2(\boldsymbol{r}) - \Sigma_{r2}(\boldsymbol{r})\Phi_2(\boldsymbol{r}) + \Sigma_{s1\to2}(\boldsymbol{r})\Phi_1(\boldsymbol{r}) = 0 \end{cases}$$

This leads to a new different definitions of matrices $[A]$ and $[P]$:

$$\left(\underbrace{\begin{bmatrix} D_1\Delta & 0 \\ 0 & D_2\Delta \end{bmatrix}}_{D} + \underbrace{\begin{bmatrix} -\Sigma_{r1} & 0 \\ \Sigma_{s1\to2} & -\Sigma_{r2} \end{bmatrix}}_{A} + \underbrace{\begin{bmatrix} 0 & \frac{k_\infty}{p}\Sigma_{r2} \\ 0 & 0 \end{bmatrix}}_{P} \right) \begin{pmatrix} \Phi_1(r) \\ \Phi_2(r) \end{pmatrix} = 0$$

In the following equation, we assume that criticality is achieved by the multiplication factor k that enables to balance the operator of production with the other operators. By themselves, the properties of the system do not make it be critical.

$$\left([D] + [A] + \tfrac{1}{k}[P] \right) \begin{pmatrix} \Phi_1(r) \\ \Phi_2(r) \end{pmatrix} = 0$$

In an homogeneous system with two neutron groups, $[D]$, $[A]$ and $[P]$ are 2×2 matrices.

If we consider a more realistic system with different areas, in which a numerical resolution with finite-difference method is performed, then the matrices $[D]$, $[A]$ and $[P]$ become, for example:

$$P = \begin{bmatrix} [P11] & [P12] \\ [P21] & [P22] \end{bmatrix}$$

$P11$, $P12$, $P21$ and $P22$ are 4 square matrices of dimension $I + 1$ where I is the number of cells (see Appendix F).

$$[P11] = \begin{bmatrix} \nu\Sigma_{f1,1} & 0 & \cdots & & \cdots & 0 \\ 0 & \nu\Sigma_{f1,2} & & & & \\ \vdots & & \ddots & & & \vdots \\ \vdots & & & \nu\Sigma_{f1,I} & 0 \\ 0 & \cdots & & \cdots & 0 & 0 \end{bmatrix}$$

$$[P12] = \begin{bmatrix} \nu\Sigma_{f2,1} & 0 & \cdots & & \cdots & 0 \\ 0 & \nu\Sigma_{f2,2} & & & & \\ \vdots & & \ddots & & & \vdots \\ \vdots & & & \nu\Sigma_{f2,I} & 0 \\ 0 & \cdots & & \cdots & 0 & 0 \end{bmatrix}$$

And $[P21] = [P22] = [0]$.

$$[A11] = \begin{bmatrix} -\nu\Sigma_{r1,1} & 0 & \cdots & \cdots & 0 \\ 0 & -\nu\Sigma_{r1,2} & & & \vdots \\ \vdots & & \ddots & & \vdots \\ \vdots & & & -\nu\Sigma_{r1,I} & 0 \\ 0 & \cdots & \cdots & 0 & 0 \end{bmatrix}$$

$$[A22] = \begin{bmatrix} -\nu\Sigma_{r2,1} & 0 & \cdots & \cdots & 0 \\ 0 & -\nu\Sigma_{r2,2} & & & \vdots \\ \vdots & & \ddots & & \vdots \\ \vdots & & & -\nu\Sigma_{r2,I} & 0 \\ 0 & \cdots & \cdots & 0 & 0 \end{bmatrix}$$

$$[A21] = \begin{bmatrix} \Sigma_{s1\rightarrow2,1} & 0 & \cdots & \cdots & 0 \\ 0 & \Sigma_{s1\rightarrow2,2} & & & \vdots \\ \vdots & & \ddots & & \vdots \\ \vdots & & & \Sigma_{s1\rightarrow2,I} & 0 \\ 0 & \cdots & \cdots & 0 & 0 \end{bmatrix}$$

And $[A12] = [0]$.

20.4.2 *Perturbated Diffusion Equation for Two Groups*

20.4.2.1 Example of First Order Perturbation Theory with a Perturbation of Thermal Absorption

As an example, we assume that the perturbation applies to the absorption matrix; the perturbated matrix is $[A] + \delta[A]$.

$$\delta[A] = \begin{bmatrix} [0] & [0] \\ \hline [0] & \delta[A22] \end{bmatrix}$$

So as to keep the pertubated reactor critical, the multiplication factor applying to production matrix becomes $\frac{1}{k_p}$. This enables to write the critical perturbated equation:

$$\left([D] + [A] + \delta[A] + \frac{1}{k_p}[P] \right) \begin{pmatrix} \Phi_1^p(r) \\ \Phi_2^p(r) \end{pmatrix} = 0$$

Where $\Phi_1^p(r)$ and $\Phi_2^p(r)$ are the perturbated fast and thermal flux.

$$\left([D] + [A] + \delta[A] + \tfrac{1}{k}[P] + \left(\tfrac{1}{k_p} - \tfrac{1}{k}\right)[P]\right)\begin{pmatrix}\Phi_1^p(r)\\\Phi_2^p(r)\end{pmatrix} = 0$$

$$\left([D] + [A] + \tfrac{1}{k}[P]\right)\begin{pmatrix}\Phi_1^p(r)\\\Phi_2^p(r)\end{pmatrix} = \left(-\delta[A] + \tfrac{k_p-k}{k_p \times k}[P]\right)\begin{pmatrix}\Phi_1^p(r)\\\Phi_2^p(r)\end{pmatrix}$$

The non perturbated adjoint equation is:

$$\left([D] + [A] + \tfrac{1}{k}[P]\right)^*\begin{pmatrix}\Phi_1^*(r)\\\Phi_2^*(r)\end{pmatrix} = 0$$

The following equation results the definition of the operator adjoint.

$$\left\langle \left([D] + [A] + \tfrac{1}{k}[P]\right)\begin{pmatrix}\Phi_1^p(r)\\\Phi_2^p(r)\end{pmatrix}, \begin{pmatrix}\Phi_1^*(r)\\\Phi_2^*(r)\end{pmatrix}\right\rangle =$$

$$\left\langle \begin{pmatrix}\Phi_1^p(r)\\\Phi_2^p(r)\end{pmatrix}, \left([D] + [A] + \tfrac{1}{k}[P]\right)^*\begin{pmatrix}\Phi_1^*(r)\\\Phi_2^*(r)\end{pmatrix}\right\rangle$$

Thus it can be deduced that:

$$\left\langle \left([D] + [A] + \tfrac{1}{k}[P]\right)\begin{pmatrix}\Phi_1^p(r)\\\Phi_2^p(r)\end{pmatrix}, \begin{pmatrix}\Phi_1^*(r)\\\Phi_2^*(r)\end{pmatrix}\right\rangle = 0$$

In the vector space of real functions whose domaine is enclosed within boundaries, the inner product is defined as the integration over an axis (or surface, or volume) of the product of the functions (see Appendix G).

$$\frac{k_p - k}{k_p \times k}\int_{\text{reactor}}\begin{pmatrix}\Phi_1^*(r)\\\Phi_2^*(r)\end{pmatrix}[P]\begin{pmatrix}\Phi_1^p(r)\\\Phi_2^p(r)\end{pmatrix}dV = \int_{\text{reactor}}\begin{pmatrix}\Phi_1^*(r)\\\Phi_2^*(r)\end{pmatrix}\delta[A]\begin{pmatrix}\Phi_1^p(r)\\\Phi_2^p(r)\end{pmatrix}$$

If the perturbation is limitated to an affected volume, then the integration can be limited to this affected volume and instead an integral over the reactor, we write:

$$\int_{\text{affected volume}}\begin{pmatrix}\Phi_1^*(r)\\\Phi_2^*(r)\end{pmatrix}\delta[A]\begin{pmatrix}\Phi_1^p(r)\\\Phi_2^p(r)\end{pmatrix}$$

We now assume that the perturbation is "small" and thus the perturbated flux and the non perturbated flux are almost equal. This approach, the first order perturbation theory, avoids having to calculate the direct perturbated flux. Only the flux and adjoint flux of the initial case need to be calculated (Reuss 2008, p. 488). And thus:

$$\frac{k_p - k}{k_p \times k} = \frac{\int_{\text{affected volume}} \begin{pmatrix} \Phi_1^*(\boldsymbol{r}) \\ \Phi_2^*(\boldsymbol{r}) \end{pmatrix} \delta[A] \begin{pmatrix} \Phi_1(\boldsymbol{r}) \\ \Phi_2(\boldsymbol{r}) \end{pmatrix} dV}{\int_{\text{reactor}} \begin{pmatrix} \Phi_1^*(\boldsymbol{r}) \\ \Phi_2^*(\boldsymbol{r}) \end{pmatrix} [P] \begin{pmatrix} \Phi_1(\boldsymbol{r}) \\ \Phi_2(\boldsymbol{r}) \end{pmatrix} dV}$$

$\frac{k_p - k}{k_p \times k}$ is the reactivity difference (Marguet 2018, p. 841) that we write $\Delta\rho$, between the perturbated and the non perturbated reactor.

20.4.2.2 Exact Perturbation Formula

From a formal point of view, the reactivity difference calculated with the perturbation theory in the frame of neutron transport theory and the one calculated in the frame of diffusion theory are expressed similarly.

P is the production operator, and L represents all other operators. It can be established that:

$$\Delta\rho = \frac{\langle \Phi^*, (\delta L - \frac{1}{k}\delta P)\Phi^p \rangle}{\langle \Phi^*, (P + \delta P)\Phi^p \rangle}$$

20.4.3 Example of Calculation with First Order Perturbation Theory

Fortunately, problems of this type (localized perturbation) can be handled by perturbation theory, provided the perturbation is not so large as to distort substantially the flux in the neighborhood of the perturbation. This technique is not limited to problems involving localized perturbations; the effect on a reactor of all (small) nonuniform changes as well as uniform changes can also be determined by perturbation theory. Lamarsh (1966, p. 523)

20.4.3.1 Uniform Supplementary Absorption

We consider again the reactor described in Sect. 13.4.2 already used as an example in this chapter, in Sect. 20.3. A supplementary thermal absorption of 0.01 cm is added uniformaly in the fissile area of the system, that becomes the perturbated reactor. The shape of direct thermal and fast flux is almost non affected by the supplementary absorption. Thus the first order pertubation method can be used with a satisfactory confidence level.

The non perturbated reactor is critical. The numerical calculation of eigen values for the perturbated reactor gives $k_{\text{eff}} = 0.950\,76$ and $\delta\rho = -5179\,\text{pcm}$. The effect of supplementary absorption over reactivity numerically calculated with first order

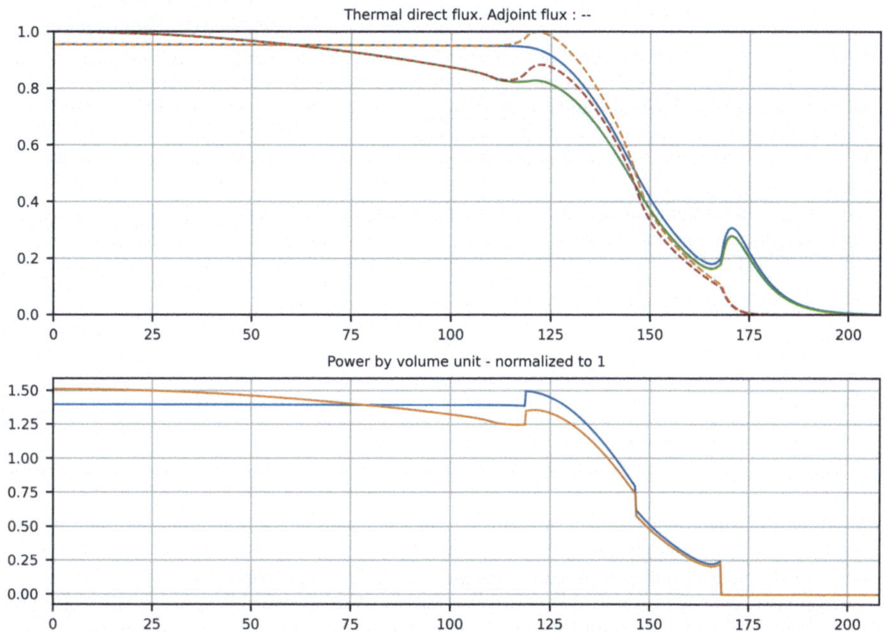

Fig. 20.5 Thermal direct and adjoint flux for non perturbated and perturbated reactor, by an absorbing ring

perturbation theory is $\delta\rho = -5490.2$ pcm. The reactivity effect estimated by the first order perturbation theory is shifted of 6 % with respect to the eigen value calculation.

20.4.3.2 Absorbing Ring

The same reactor is now perturbated by an absorbing ring: 5×10^{-3} cm between the radii 110 and 120 cm. The Fig. 20.5 shows that the absorbing ring warps the shape of thermal flux.

The numerical calculation of eigen values for the perturbated reactor gives $k_{\text{eff}} = 0.997\,24$ and $\delta\rho = -276.9$ pcm. The effect of supplementary absorption over reactivity numerically calculated with first order perturbation theory is $\delta\rho = -212.51$ pcm. Since the pertubation warps significantly the flux, the evaluation of reactivity effect with first order perturbation theory is less accurate than for the uniform supplementary absorption.

The one group adjoint function $\psi(\boldsymbol{r_0})$ is proportional to the (negative) change in reactivity of the reactor per neutron absorbed per second at $\boldsymbol{r_0}$. If, for example, the absorber is placed at a point where $\psi(\boldsymbol{r_0})$ is small, the reactivity will also be small. In other words, $\psi(\boldsymbol{r_0})$ measures the importance of the point $\boldsymbol{r_0}$ with respect to the reactivity introduced by an absorber located at that point, and it is for this reason that $\psi(\boldsymbol{r_0})$ is also called the importance function.(..)
$\psi(\boldsymbol{r_0})$ *is proportionnal to the (positive) reactivity per (net) neutron introduced per second*

by fissile material located at r_0. In view of this result and the conclusions of the preceding paragraph, it is evident that the adjoint function represents the relative importance of *any* local change at the point r_0 which either introduces or removes neutrons from the system.(..) In a similar way the physical significance of the two-group adjoint fluxes can easily be established. As would be expected, the functions $\psi_1(r_0)$ and $\psi_2(r_0)$ are proportional, respectively, to the reactivites resulting from perturbations which introduce or remove neutrons from the fast or slow groups.(..)

The group adjoint function $\psi_n(r)$ is proportional to the gain or loss in reactivity of a reactor due to the insertion or removal of one neutron per second in the group at the point r. Lamarsh (1966, pp. 541–543)

References

1. R. Barjon, L. Neel et al., *Physique des réacteurs nucléaires* (Institut des sciences nucléaires, 1993)
2. M. Hayashi, Analytical solution of diffusion adjoint equation for homogeneous bare critical reactor. Ann. Nucl. Energy **114**, 569–570 (2018)
3. H. Hurwitz, Note on a theory of danger coefficients. Technical report (1948)
4. J.R. Lamarsh, *Introduction to Nuclear Reactor Theory*. Addison-Wesley Series in Nuclear Engineering (Addison-Wesley Publishing Company, 1966). ISBN 9780201041262. https://books.google.fr/books?id=by5RAAAAMAAJ
5. J. Lewins, *Importance: The Adjoint Function: The Physical Basis of Variational and Perturbation Theory in Transport and Diffusion Problems* (Pergamon Press, 1965)
6. S. Marguet, *The Physics of Nuclear Reactors* (Springer International Publishing, 2018). ISBN 9783319595603. https://books.google.fr/books?id=9DlODwAAQBAJ
7. Y. Nauchi, T. Kameyama, Development of calculation technique for iterated fission probability and reactor kinetic parameters using continuous-energy Monte Carlo method. J. Nucl. Sci. Technol. **47**(11), 977–990 (2010)
8. A. Radkowsky, *Naval Reactors Physics Handbook: Selected Basic Techniques*, vol. 1, Naval Reactors, Division of Reactor Development (United States Atomic Energy, 1964)
9. P. Reuss, *Neutron Physics* (EDP Sciences, 2008)

Appendix A
Scalar Fields, Vector Fields and Some of Their Operators

A.1 Scalar and Vector Field

A.1.1 Definitions

In mathematics and physics, a scalar field is a function associating a single number to every point in a space. The scalar may either be a pure mathematical number (dimensionless) or a physical quantity (with units). The temperature distribution throughout space, or the pressure distribution in a fluid are examples of scalar fields.

A vector field on two (or three) dimensional space is a function F that assigns to each point (x, y) or (x, y, z) a two (or three dimensional) vector given by $F(x, y)$ or $F(x, y, z)$.

$$F(x, y, z) = P(x, y, z)i + Q(x, y, z)j + R(x, y, z)k \tag{A.1}$$

where P, Q, and R are scalar fields.

The Fig. A.1 shows a gravitational field and the velocity of a river at points on its surface. The gravitational field is exerted by two astronomical objects, such as a star and a planet or a planet and a moon. The vector associated with a given point on the river's surface gives the velocity of the water at that point.

A.1.2 Flux of a Vector Field

F is a vector field, and S is a surface.

In physics, the flux of the vector field trough S is defined by:

$$\Phi = \iint_S F.dS$$

© The Editor(s) (if applicable) and The Author(s), under exclusive license to Springer Nature Switzerland AG 2026
H. Grard, *Diffusion of Neutrons in Nuclear Reactors*,
https://doi.org/10.1007/978-3-032-05088-5

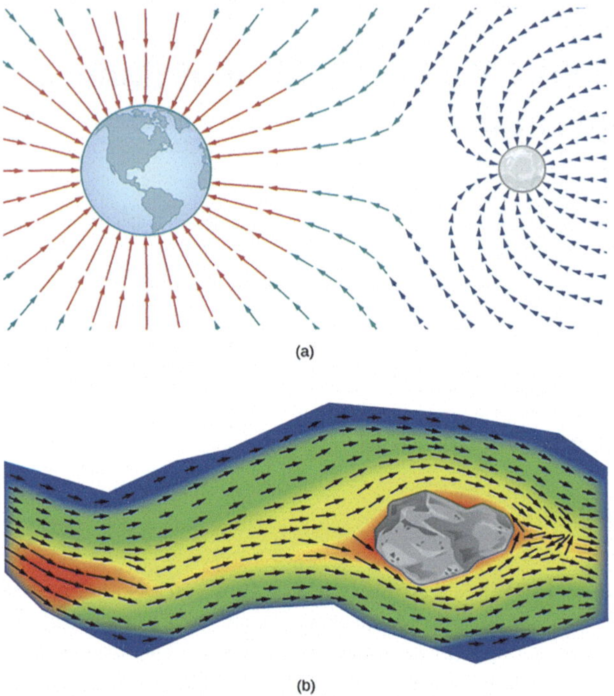

Fig. A.1 Gravitational field and surface flow velocity of a river. Access for free at https://openstax.org/books/calculus-volume-3/pages/1-introduction [Gilbert Strang, 2016, 6.1 Vector Fields]

If the surface is closed, by convention of sign, $\boldsymbol{F}.d\boldsymbol{S}$ is positive if $\boldsymbol{F}$ is oriented to the output.

A.2   Three Main Operators for Scalar and Vector Fields

$\boldsymbol{F}$ is a vector field, and $f(x, y, z)$ is a scalar field.

A.2.1   Gradient

The gradient of a **scalar** field is a **vector** field defined by:

$$\overrightarrow{\text{grad}}\, f(x_0, y_0, z_0) = \begin{bmatrix} \frac{\partial f}{\partial x}(x_0, y_0, z_0) \\ \frac{\partial f}{\partial y}(x_0, y_0, z_0) \\ \frac{\partial f}{\partial z}(x_0, y_0, z_0) \end{bmatrix} \tag{A.2}$$

Fig. A.2 A vector of the gradient vector field

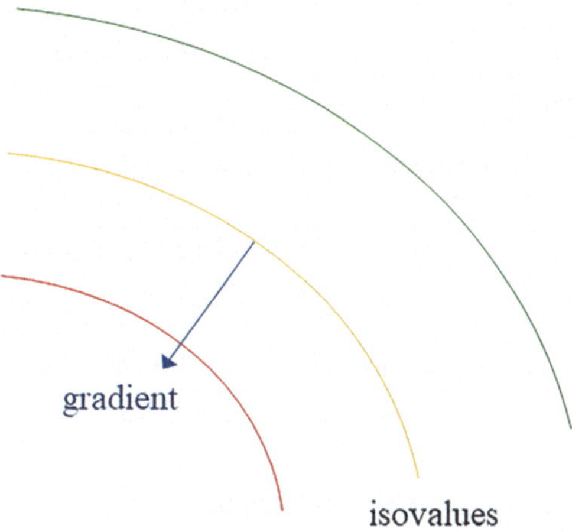

The gradient is a vector field that indicates at each point in which direction a scalar field is changing, as illustrated by Fig. A.2. The vectors of the gradient vector field are oriented towards the increasing values at the highest slope, the module of a gradient vector is equal to the slope. The vectors of the gradient field are perpendicular to isovalue curls (2D) or surfaces (3D).

A common notation for $\overrightarrow{\mathrm{grad}}f$ is ∇f.

A.2.2 Divergence

A.2.2.1 Definitions

We recall the Eq. A.1:

$$F(x, y, z) = P(x, y, z)i + Q(x, y, z)j + R(x, y, z)k$$

The divergence of a **vector** field produces a **scalar** field, defined by:

$$\mathrm{div}\,F = \frac{\partial}{\partial x}P(x, y, z) + \frac{\partial}{\partial y}Q(x, y, z) + \frac{\partial}{\partial z}R(x, y, z)$$

This scalar field gives the quantity of the vector field's source at each point. More technically, the divergence of the vector field F represents the volume density of the outward flux of F from an infinitesimal volume around a given point, being expressed by the Eq. A.3.

$$\text{div } \boldsymbol{F} = \lim_{d\tau \to 0} \left(\frac{1}{\tau} \iint_S \boldsymbol{F}.d\boldsymbol{S} \right) \tag{A.3}$$

Then, the Green-Ostrogradsky theorem:

$$\iiint_V \text{div } \boldsymbol{F} d\tau = \iint_S \boldsymbol{F}.d\boldsymbol{S} \tag{A.4}$$

A common notation for div $\boldsymbol{F}$ is $\nabla \boldsymbol{F}$.

A.2.3 Useful Formulas

Let $\boldsymbol{F}$ be a vector field, and $f(x, y, z)$, $g(x, y, z)$ scalar fields. We admit the well known following formulas :

$$\overrightarrow{\text{grad}}(f\,g) = f\,\overrightarrow{\text{grad}}g + g\,\overrightarrow{\text{grad}}f$$

$$\text{div } (f\,\boldsymbol{F}) = f\,\text{div } \boldsymbol{F} + (\overrightarrow{\text{grad}}f).\boldsymbol{F} \tag{A.5}$$

The formula A.5 finds an interesting application in neutronics. We recall the definitions of the chapter 2:

- $\nu(\boldsymbol{r}, \boldsymbol{\Omega}, E, t)$ is the energy and angular neutron density
- $\varphi(\boldsymbol{r}, \boldsymbol{\Omega}, E, t) = v(E)\nu(\boldsymbol{r}, \boldsymbol{\Omega}, E, t)$ is the angle dependent flux
- $\boldsymbol{i}(\boldsymbol{r}, \boldsymbol{\Omega}, E, t) = \boldsymbol{\Omega}\,\varphi(\boldsymbol{r}, \boldsymbol{\Omega}, E, t)$ is the angular neutron current

$\boldsymbol{i}$ is a vector field within the boundaries of the reactor, and ν, φ are scalar fields. $\boldsymbol{\Omega}$ is a vector and defines a constant vector field.

Equation A.5 enables to write:

$$\text{div } \boldsymbol{i}(\boldsymbol{r}, \boldsymbol{\Omega}, E, t) = \text{div } (\boldsymbol{\Omega}\,\varphi(\boldsymbol{r}, \boldsymbol{\Omega}, E, t)$$

$$= \boldsymbol{\Omega}.\overrightarrow{\text{grad}}\varphi(\boldsymbol{r}, \boldsymbol{\Omega}, E, t) + \underbrace{\varphi(\boldsymbol{r}, \boldsymbol{\Omega}, E, t)\,\text{div } \boldsymbol{\Omega}}_{\text{div } \boldsymbol{\Omega}=0 \text{ since } \boldsymbol{\Omega} \text{ is a constant vector field}}$$

And then :

$$\text{div } \boldsymbol{i}(\boldsymbol{r}, \boldsymbol{\Omega}, E, t) = \boldsymbol{\Omega}.\overrightarrow{\text{grad}}\varphi(\boldsymbol{r}, \boldsymbol{\Omega}, E, t) \tag{A.6}$$

Equation A.6 establishes at time t the rate of loss per unit time, by movements, of neutrons having the position $\boldsymbol{r}$, the energy E and moving in the direction $\boldsymbol{\Omega}$ per unit energy, per unit solid angle, per unit volume.

A.2.3.1 Divergence, Sources and Sinks

Concerning neutrons, since sinks nor sources exist for fluids, the equation A.17 established for fluids must be supplemented by additional terms taking account production of new neutrons (source, mainly by fission) and absorption of neutrons (sink).

$$\frac{\partial n}{\partial t} + \operatorname{div} \boldsymbol{J} = \underbrace{S}_{\substack{\text{Fissions and/or} \\ \text{neutrons source device}}} - \underbrace{\Sigma_a \Phi}_{\text{absorption}} \tag{A.7}$$

If divergence is negative for example, we can use the expression "concentration" and the two following cases are interesting:

- There is no source, nor sink. Then $\frac{\partial \rho}{\partial t} > 0$ (concentration)
- With a source, a steady state is possible and then div $\boldsymbol{J}$ = sink (absorption) < 0. The steady state is possible because the accumulation is "evacuated in the sink" and $\frac{\partial \rho}{\partial t} = 0$

If divergence is positive, we can use the expression "dispersion", or simply leakage out of the unit volume.

In a steady state, the divergence of a vector field at a point indicates if this point is a source or sink. This means that:

$$\operatorname{div} \boldsymbol{J} = \text{source (fission)} > 0 \text{ or sink (absorption)} < 0 \ (\text{n/cm}^3/\text{s}) \tag{A.8}$$

The Fig. A.3 illustrates different cases.

A.2.4 *Introduction to the Laplacian Operator (Buckling operator)*

The Laplacian is an operator which turns a scalar field into another scalar field.

A.2.4.1 Laplace Difference

Let's consider a graph with interconnected nodes. The Laplace difference is defined as the difference between the value n_i of a node and the mean value of its neighbors, for example: $n_i - \overline{(n_1 + n_2 + n_3)}$. It unveils the nodes in excess or in deficit, as compared to neighborhood.

Laplace difference has a quadratic behavior, that is easily highlighted in a one dimensional case. The Laplace difference divided par the distance squared has a limit that is equal to the Laplacian:

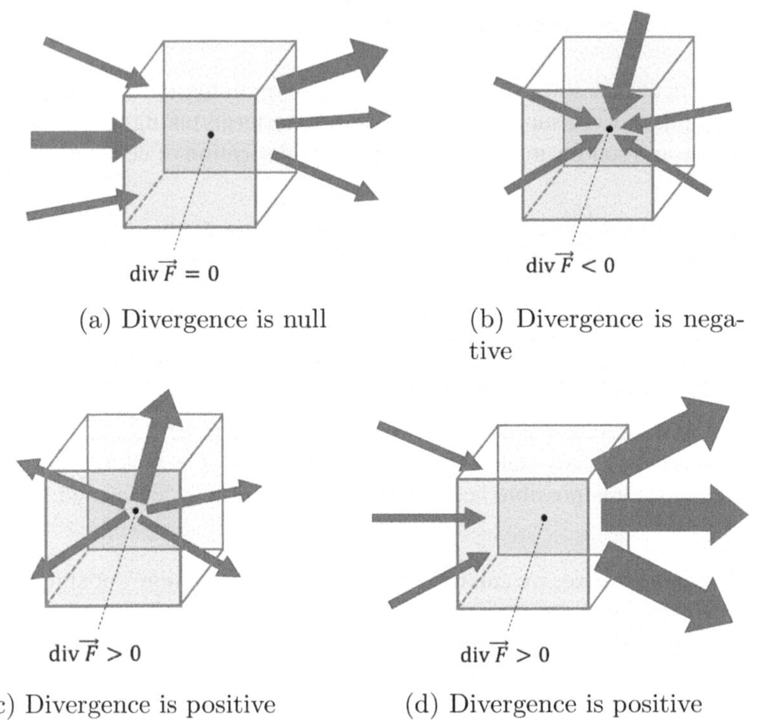

div $\vec{F} = 0$

(a) Divergence is null

div $\vec{F} < 0$

(b) Divergence is negative

div $\vec{F} > 0$

(c) Divergence is positive

div $\vec{F} > 0$

(d) Divergence is positive

Fig. A.3 Representation of null, negative, and positive divergences

$$\lim_{h \to 0} \frac{\frac{1}{2}(f(x_0 - h) + f(x_0 + h)) - f(x_0)}{h^2} = \Delta f(x_0) \qquad (A.9)$$

The Laplace difference sheds light on the Laplacian and helps to understand the Laplacian quantifies the curvature, or stress, of a scalar field. It tells you how much the value at a point of the field differs from the average value taken over the surrounding points.

This requires a measurable quantity (for example temperature at a point as a measure of thermal energy, and f(p) is the distribution of temperature), a physical relation describing how this quantity is exchanged (for example heat transfer between areas at different temperatures) and geometry. Neighborhood relationships are such that excess quantity is transferred to deficit adjacent areas. This is a diffusion mechanism. The harmony, or equilibrium, resultant from neighborhood relationships are reflected by Laplace differences which have a null value within the system.

At the boundary of the system, boundary conditions are imposed, for example the temperature at the external surface of an object.

A.2.4.2 Energy of Dirichlet

The energy of Dirichlet quantifies the amount of disparity within the system. It is not an energy strictly speaking, in $Joule$.

If we consider again a graph of connected nodes, and (i, j) are couples of two neighbor nodes:

$$E = \sum_{(i,j)} (n_i - n_j)^2$$

Energy of Dirichlet is a functional defined as half the integration over the domain of the gradients squared. For example, in the case of a surface:

$$E(f) = \frac{1}{2} \int_{surface} ||\overrightarrow{grad}f||^2 dS \tag{A.10}$$

A function for which the boundary conditions are satisfied, and the energy of Dirichlet minimal, is characterized by a null Laplacian within the domain.

A.2.4.3 Laplacian Mathematical Expression

The mathematical definition of the Laplacian of a scalar field is given by Eq. A.11:

$$\Delta f(x_0, y_0, z_0) = \mathrm{div}\ (\overrightarrow{grad}f(x_0, y_0, z_0))$$

$$= \frac{\partial^2}{\partial x^2} f(x_0, y_0, z_0) + \frac{\partial^2}{\partial y^2} f(x_0, y_0, z_0) + \frac{\partial^2}{\partial z^2} f(x_0, y_0, z_0)$$

$$\tag{A.11}$$

There are two notations for the Laplacian commonly used, $\nabla^2 f$ and Δf, the first one is mostly used by physicists, whereas mathematicians tend to prefer the latter one. The ∇^2 originates from:

$$\nabla^2 f = \mathrm{div}(\overrightarrow{grad}f) = \nabla \cdot (\nabla f)$$

A.2.5 Application of Laplacian in Thermal Science

The Fourier's law of Thermal Conduction quantifies heat transfer processes (Eq. A.12). This law states that the time rate of heat transfer through a material is proportional to the negative gradient in the temperature and to the area, at right angles to that gradient, through which the heat flows.

$$\overrightarrow{\varphi} = -\lambda \cdot \overrightarrow{grad}T \tag{A.12}$$

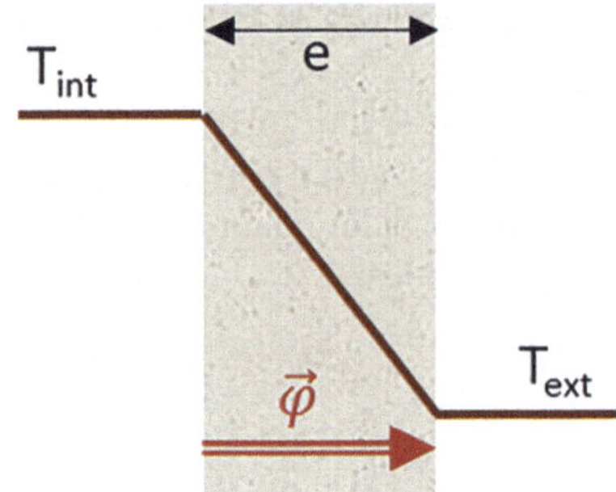

Fig. A.4 Temperature through an inert (no heat sources) wall

Where T is the temperature (K), λ the thermal conductivity of the material (W/m/K), $\vec{\varphi}$ is the local heat flux density (W/m^2).

The equation of continuity (Eq. A.13) for heat flux can be written as:

$$\text{div } \vec{\varphi} = q \tag{A.13}$$

By combining Eqs. A.12 and A.13, we can write:

$$\text{div } (-\lambda \cdot \overrightarrow{\text{grad}}T) = q$$

If the thermal conductivity is constant, then:

$$\lambda \cdot \Delta T + q = 0 \tag{A.14}$$

Equation A.14 enables to calculate the temperature profile in a heating rod (or plate), with the assumption that the thermal conductivity is constant.

If there are no heat sources in the rod, we can write the famous equation of Laplace:

$$\Delta T(r) = 0 \quad \forall r \tag{A.15}$$

For example, the solution of the Laplace equation applied to a wall or plate is linear, as illustrated on Fig. A.4.

Its equation is $\vec{\varphi} = -\frac{\lambda}{e}(T_{int} - T_{ext})\vec{x}$

A.2.6 *Laplacian and Sensemaking*

A simple and physically intuitive coordinate-free interpretation of the Laplacian of a scalar function f at a point P as a measure of the excess of a mean value of f through a small volume τ surrounding P over the value of the scalar function f at P.

As a consequence, the Laplacian of a scalar field reprents it curvature.

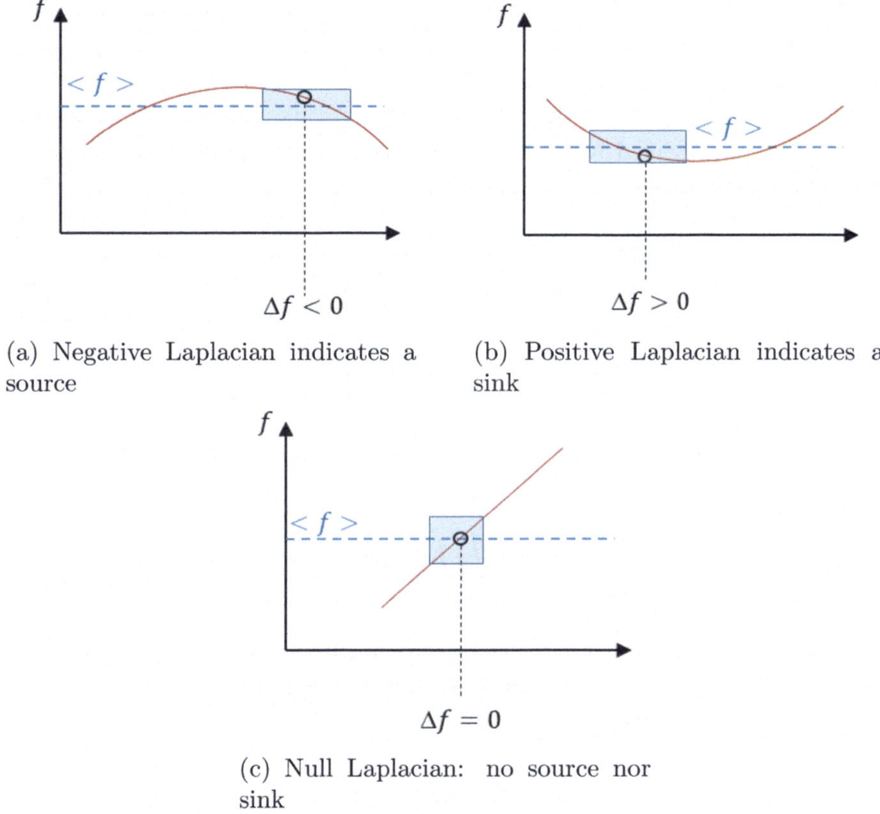

(a) Negative Laplacian indicates a source

(b) Positive Laplacian indicates a sink

(c) Null Laplacian: no source nor sink

Fig. A.5 Curvature of the scalar field and sign of the Laplacian

The Fig. A.5 enables us to compare the local value of a scalar field and the mean value of close neighborhood. Il the local value is greater than the mean value of close neighborhood (Fig. A.5a), then this indicates that there is a source at this point, and the Laplacian is negative (cf. Eq. A.9). For example, the temperature within a cylindric fuel rod is a paraboloid of revolution (the maximal temperature is in the fuel center), which has a convex curvature.

A.3 Equation of Continuity: Example of a Fluid

A fluid has a volumic mass ρ (g/cm^3). The local velocity of fluid particles is described by the vector field **V**. The flux of ρ **V** through the closed surface S delimiting the volume V establishes a **mass balance** per unit time:

$$\iint_S \rho \, \boldsymbol{V}.d\boldsymbol{S} \quad \text{(g/s), outflow is positive}$$

The mass balance within a volume V, delimited by a closed surface S crossed by a flow of fluid is represented by Eq. A.16. The imbalance between inlet flow and outlet flow shall be equal to the rate of change. The Eq. A.16 is the integral balance of mass within the volume D.

$$\underbrace{\iiint_V \frac{\partial \rho}{\partial t} d\tau}_{\substack{\text{Rate of change of total mass} \\ \text{within V}}} \quad + \quad \underbrace{\iint_S \rho \boldsymbol{V}.d\boldsymbol{S}}_{\substack{\text{Variation of mass within V due to} \\ \text{flow through boundary surface S}}} \quad = 0 \qquad (A.16)$$

The application of the Green-Ostrogradsky theorem (Eq. A.4) to the Eq. A.16 leads to Eq. A.17, which is a local mass balance.

$$\iiint_V \frac{\partial \rho}{\partial t} d\tau + \iiint_V \operatorname{div} \rho \boldsymbol{V} \cdot d\tau = 0$$

If $\operatorname{div} \rho \boldsymbol{V} \cdot d\tau < 0$, this means a that there is more fluid flowing into $d\tau$ through its border than flowing out: there is a **concentration** of fluid inside $d\tau$. If $\operatorname{div} \rho \boldsymbol{V} \cdot d\tau > 0$, then there is a **dispersion**.

$$\frac{\partial \rho}{\partial t} + \operatorname{div} \rho \boldsymbol{V} = 0 \qquad (A.17)$$

Reference

Edwin "Jed" Herman Gilbert Strang. *Calculus Volume 3*. OpenStax, 2016. URL https://openstax. org/books/calculus-volume-3/pages/1-introduction.

Appendix B
Special Functions

B.1 The Dirac's Delta Function

Dirac (1902–1984) introduced the delta function δ in the 1930s as a useful tool in his study of quantum mechanics. The delta function is sometimes called "Dirac's delta function" or the "impulse symbol".

As a generalized function, also called a "distribution", The Dirac delta function only makes sense under an integral (as any distribution).

B.1.1 The One Dimensional Dirac Delta Function

It can be viewed as the derivative of the Heaviside step function defined as:

$$H(x) = \begin{cases} 0, & x < 0 \\ \frac{1}{2}, & x = 0 \\ 1, & x > 0 \end{cases} \tag{B.1}$$

The delta function has the fundamental property that its integration over general intervals gives:

$$\int_a^b \delta(x)\,dx = \begin{cases} 1, & 0 \in [a, b] \\ 0, & 0 \notin [a, b] \end{cases} \tag{B.2}$$

This can be seen that the delta function is zero everywhere except at $x = 0$.

The delta function has the property that its convolution with any function f equals the value of f at zero:

© The Editor(s) (if applicable) and The Author(s), under exclusive license to Springer Nature Switzerland AG 2026
H. Grard, *Diffusion of Neutrons in Nuclear Reactors*,
https://doi.org/10.1007/978-3-032-05088-5

$$\int_{-\infty}^{+\infty} \delta(x) f(x)\, dx = f(0) \tag{B.3}$$

And more generaly, with a shift a :

$$\int_{-\infty}^{+\infty} \delta(x - a) f(x)\, dx = f(a)$$

And, in fact:

$$\int_{a-\epsilon}^{a+\epsilon} \delta(x - a) f(x)\, dx = f(a)$$

B.1.2 *The Dirac Delta Function on the Surface of the Disk*

We consider the Dirac function at the center of a disk. The integration of the Dirac function on the disk is equal to unity for any radius. Let's write it in polar coordinates:

$$\int_0^{\epsilon} r \int_0^{2\pi} \delta(r)\, d\theta\, dr = 1 \quad \forall \epsilon > 0$$

The delta function has the property that its convolution with any function $f(r, \theta)$ equals the value of f at $r = 0$:

$$\int_0^{+\infty} r \int_0^{2\pi} \delta(r) f(r, \theta)\, d\theta\, dr = f(0) \tag{B.4}$$

With a vectorial shift r_s from the center of the disk:

$$\int_0^{+\infty} r \int_0^{2\pi} \delta(r - r_s) f(r, \theta)\, d\theta\, dr = f(r_s) \tag{B.5}$$

B.1.3 *The Dirac Delta Function in a Sphere*

We consider the Dirac function at the center of a sphere. The integration of the Dirac function on the sphere volume is equal to unity for any radius:

$$\int_{\text{sphere vol.}} \delta(\mathbf{r})\, d\tau = 1 \quad \forall \text{ sphere radius}$$

B.2   Bessel Functions

B.2.1   Bessel's Differential Equation

Bessel functions are the standard form of the solutions to Bessel's differential equation:

$$x^2 \frac{\partial^2 y}{\partial x^2} + x \frac{\partial y}{\partial x} + (x^2 - \nu^2)y = 0 \tag{B.6}$$

Where n is the ordre of the Bessel equation.

The equation is named after the astronomer Friedrich Bessel (1784–1846). It arises in a number of settings, such as vibrations of a thin circular accoustic membrane (drum).

Bessel's differential equation is often obtained by the separation of the wave Eq. B.7 in cylindric or spherical coordinates.

$$\frac{\partial^2 u}{\partial t^2} = c^2 \Delta u \tag{B.7}$$

If $n = 0$, the equation takes the form:

$$\frac{\partial^2 y}{\partial x^2} + \frac{1}{x}\frac{\partial y}{\partial x} + y = 0 \tag{B.8}$$

This Eq. B.8 was discovered and solved by Daniel Bernoulli in 1732. He used the function of zero order as a solution to the problem of an oscillating chain suspended at one end.

B.2.2   Bessel's Modified Equation

$$x^2 \frac{\partial^2 y}{\partial x^2} + x \frac{\partial y}{\partial x} - (x^2 + \nu^2)y = 0 \tag{B.9}$$

B.2.3   Solutions of Bessel's Differential Equation

Since Bessel's differential equation is a second-order equation, there must be two linearly independent solutions. The solution to Bessel's equation yields Bessel functions of the first and second kind as follows:

$$y = A J_\nu(x) + B Y_\nu(x)$$

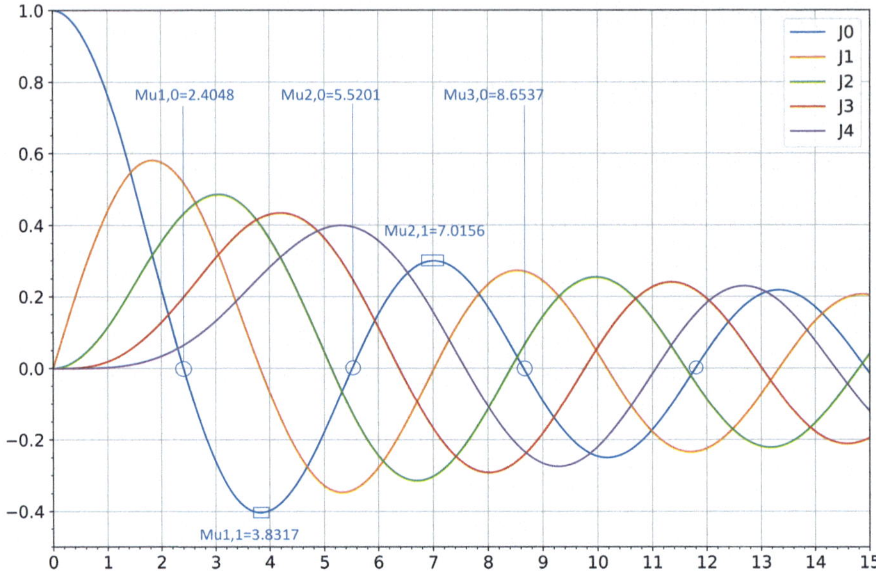

Fig. B.1 Bessel functions of the first kind, Integer order. The values of roots $\mu_{m,n}$ are given on Table E.1

Where A and B are arbitrary constants. While Bessel functions are often presented in text books and tables in the form of integer order, i.e. $\nu = 0, 1, 2\ldots$, in fact they are defined for all real values of ν.

B.2.3.1 Bessel Functions of the First Kind

$J_\nu(x)$ in the solution to Bessel's equation is referred to as a Bessel function of the first kind. These functions are finite at $x = 0$ for all real values of ν (Fig. B.1).

B.2.3.2 Bessel Functions of the Second Kind

$Y_\nu(x)$ in the solution to Bessel's equation is referred to as a Bessel function of the first kind, or sometimes the Neumann function. These functions are singular at $x = 0$ (Fig. B.2).

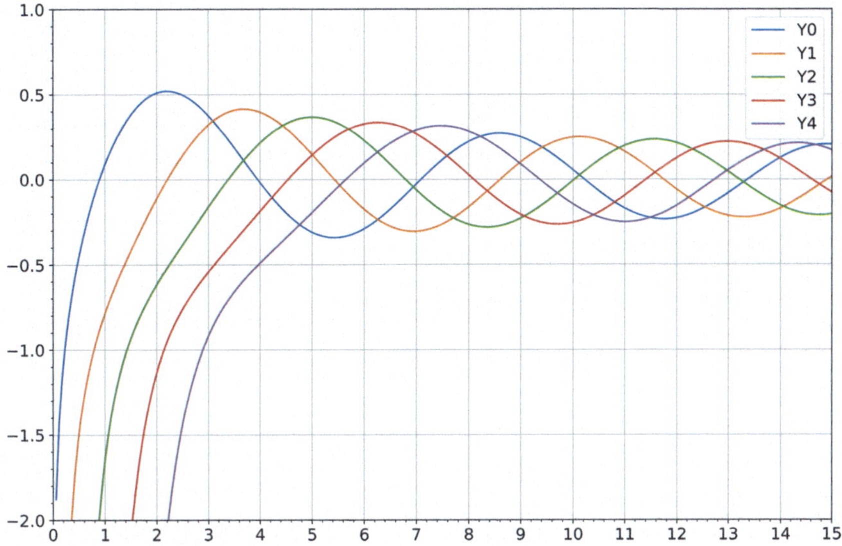

Fig. B.2 Bessel functions of the second kind, Integer order

B.2.4 Solutions of Bessel's Modified Equation

The solution to the modified Bessel equation yields modified Bessel functions of the first and second kind as follows:

$$y = C I_\nu(x) + D K_\nu(x)$$

B.2.4.1 Modified Bessel Functions of the First Kind

$I_\nu(x)$ in the solution to the modified Bessel's equation is referred to as a modified Bessel function of the first kind (Fig. B.3).

B.2.4.2 Modified Bessel Functions of the Second Kind

$K_\nu(x)$ in the solution to the modified Bessel's equation is referred to as a modified Bessel function of the first kind, or sometimes the Neumann function (Fig. B.4).

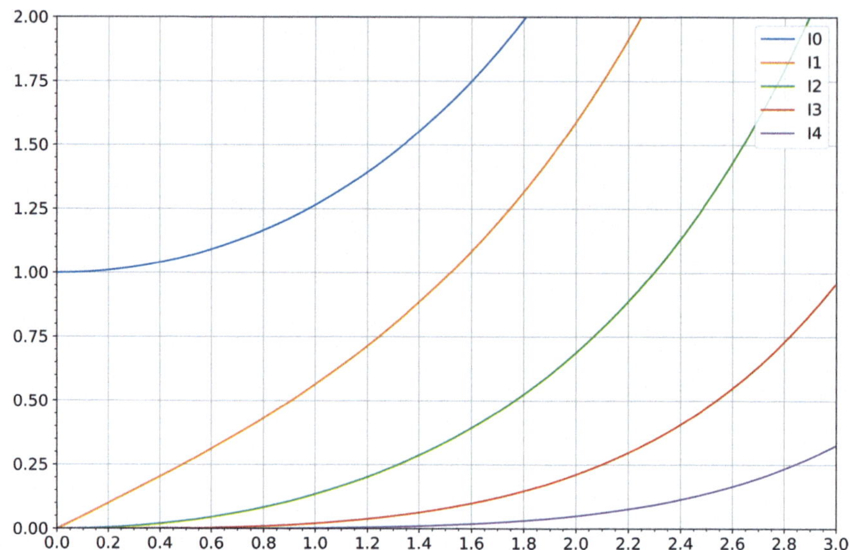

Fig. B.3 Modified bessel functions of the first kind, Integer order

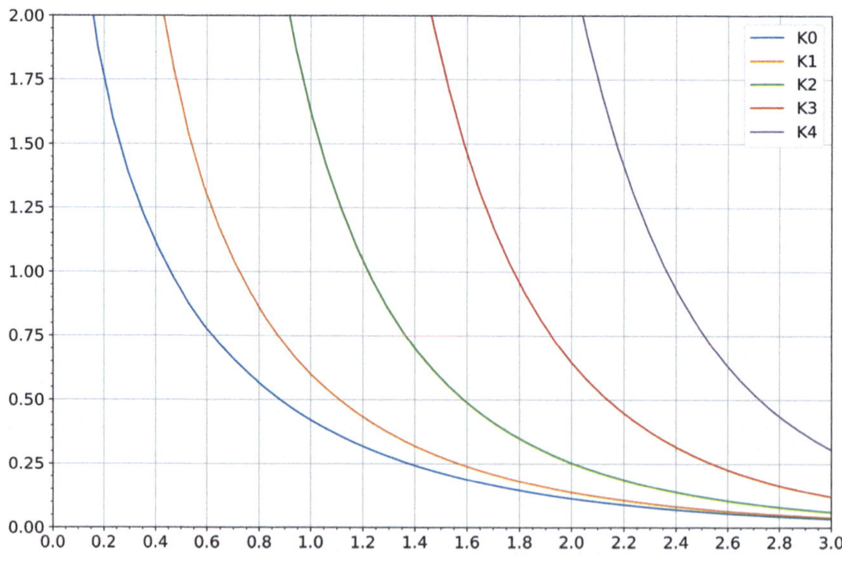

Fig. B.4 Modified bessel functions of the second kind, Integer order

B.2.5 First Order Derivatives of the Bessel's Functions

$$\frac{d}{dx}J_0(x) = -J_1(x)$$

$$\frac{d}{dx}Y_0(x) = -Y_1(x)$$

$$\frac{d}{dx}I_0(x) = I_1(x)$$

$$\frac{d}{dx}K_0(x) = -K_1(x)$$

B.3 The Gamma Function

The gamma function is defined to be an extension of the factorial to complex and real number arguments. It is related to the factorial by:

$$\Gamma(n) = (n-1)!$$

The gamma function can be defined as a definite integral for $\mathrm{Re}(z) > 0$ (Euler's integral form):

$$\Gamma(z) \overset{\text{def}}{=} \int_0^\infty t^{z-1}e^{-t}dt$$

Some special values of the Gamma function are:

$$\Gamma\left(\frac{1}{2}\right) = \sqrt{\pi}\,;\ \Gamma(1) = 1\,;\ \Gamma\left(\frac{3}{2}\right) = \frac{\sqrt{\pi}}{2}\,;\ \Gamma(2) = 1\,;\ \Gamma\left(\frac{5}{2}\right) = \frac{3\sqrt{\pi}}{4}$$

B.4 The Error Function

The integral from 0 to a finite upper limit a of the normal distribution, which is a normalized form of the Gaussian function, is the erf function:

$$\mathrm{erf}\,x = \frac{2}{\sqrt{\pi}}\int_0^x e^{-t^2}dt \tag{B.10}$$

Appendix C
Eigenvalues for a Few Geometrical Shapes

C.1 Calculation of the Laplacian Eigenvalues with Dirichlet Boundary Condition

The calculation of the eigen value can be done analytically, or with a numerical calculation with finite elemens method.

C.1.1 2D Geometrical Shapes

C.1.1.1 The Rectangle

The simplest and most widely known domain is the rectangle. Being the cartesian product of two intervals, its eigenstructure is expressible in terms of the corresponding one-dimensional eigenstructure which in turn is comprised of sines and cosines. The domain is a rectangle of dimension l_1 and l_2. The eignenvalues are givent by:

$$\lambda_{m,n} = \left(\frac{m\,\pi}{l_1}\right)^2 + \left(\frac{n\,\pi}{l_2}\right)^2$$

C.1.2 The Disk

The domain is a disk of radius R. The eigenvalues are:

$$\lambda = \left(\frac{\mu_{m,n}}{R}\right)^2$$

H. Grard, *Diffusion of Neutrons in Nuclear Reactors*,
https://doi.org/10.1007/978-3-032-05088-5

Where $\mu_{m,n}$ are the roots of the J_n Bessel functions, the values up to $m = 10$ and $n = 9$ are given in the Table E.1.

C.1.3   The L-Shape

The eigenvalues of the L-shape have to be calculated numerically.

In the case the L-shape is the assembly of three unit squares (width $= 1$), the first eigen values are:

$$9.6593, 15.199, 19.741, 29.529, 31.967, 41.530$$

C.1.4   The Equilateral Triangle

In the wake of Fourier, some fifty years later, the physician Gabriel Lamé, devoted to heating transmission, discovered analytical formula for the complete eigenstructure of the Laplacian on the equilateral triangle under either Dirichlet or Neumann boundary conditions.

The eigen values are:

$$\frac{4}{27} \left(\frac{\pi}{r}\right)^2 (m^2 + mn + n^2)$$

The eigenvalue of the fundamental mode is calculated with $m = n = 1$.

C.1.5   3D Geometrical Shapes

C.1.5.1   Rectangular Cuboid

The domain has six rectangular faces, its dimension are l_1, $l2$ and l_3. The eignenvalues are givent by:

$$\lambda_{m,n,p} = \left(\frac{m\,\pi}{l_1}\right)^2 + \left(\frac{n\,\pi}{l_2}\right)^2 + \left(\frac{p\,\pi}{l_3}\right)^2$$

C.1.5.2   The Sphere

The radius of the sphere is R. The eigen values are:

$$\left(\frac{n\,\pi}{R}\right)^2 \quad \text{with } n \geq 1$$

C.2 A Few Comparisons Between Eigenvalues of the Fundamental Mode

C.2.1 Disk and Square Having the Same Surface

The eigen value of the fundamental mode in a square of width π:

$$B_g^2 = \frac{1}{1} + \frac{1}{1} = 2$$

The eigen value of the fundamental mode in a disk of radius $\sqrt{\pi}$:

$$B_g^2 = \frac{2.4048^2}{\pi}$$

C.2.2 One Square, Two Squares and Three Squares in L-Shape

The considered squares are unit squares. The eigen value of the fundamental mode are:

- One square: $2\pi^2$
- Two squares making a rectangle: $\frac{\pi^2}{1} + \frac{\pi^2}{4} = \frac{5\pi^2}{4}$
- Three squares in L-shape: 9.6593

Appendix D
The Nordheim Equation

The appendix provides a method to calculate analytically the concentrations of neutrons and precursors, in the case the reactivity is constant.

D.1 Kinetics Without Independent Neutron Source

In the case there is no independant neutron source, and if the reactivity ρ is constant and not time dependent, the kinetics Eq. 14.15 constitute an homogeneous first order systems of differential equations with constant coefficients:

$$\begin{cases} \dfrac{dn(t)}{dt} = \dfrac{\rho - \sum_{j=1}^{g} \beta_j}{l} n(t) + \sum_{j=1}^{g} \lambda_j \, C_j(t) \\[2mm] \dfrac{dC_j(t)}{dt} = \dfrac{\beta_j}{l} n(t) - \lambda_j \, C_j(t) \end{cases}$$

Where g is the number of delayed neutron groups.

D.1.1 Eigenvalues of the Coefficient Matrix

We can define A as an 7×7 or 9×9 matrix with constants as entries, depending on the number of delayed neutron groups. And we define a vector $\boldsymbol{u}(t)$ whose components are $n(t)$, and then $C_i(t)$.

© The Editor(s) (if applicable) and The Author(s), under exclusive license to Springer Nature Switzerland AG 2026
H. Grard, *Diffusion of Neutrons in Nuclear Reactors*,
https://doi.org/10.1007/978-3-032-05088-5

Let's write the A matrix in the case of six neutron groups:

$$A = \begin{bmatrix} \frac{\rho-\beta}{l} & \lambda_1 & \lambda_2 & \lambda_3 & \lambda_4 & \lambda_5 & \lambda_6 \\ \frac{\beta_1}{l} & -\lambda_1 & 0 & \cdots & \cdots & \cdots & 0 \\ \frac{\beta_2}{l} & 0 & -\lambda_2 & \ddots & & & \vdots \\ \frac{\beta_3}{l} & \vdots & \ddots & -\lambda_3 & \ddots & & \vdots \\ \frac{\beta_4}{l} & \vdots & & \ddots & -\lambda_4 & \ddots & \vdots \\ \frac{\beta_5}{l} & \vdots & & & \ddots & -\lambda_5 & 0 \\ \frac{\beta_6}{l} & 0 & \cdots & \cdots & \cdots & 0 & -\lambda_6 \end{bmatrix} \qquad (D.1)$$

The system to solve is:

$$\frac{d}{dt}\,\boldsymbol{u}(t) = A\boldsymbol{u}(t) \qquad (D.2)$$

If ω is an eigen value of A with eigen vector $\boldsymbol{v}$, set $\boldsymbol{x} = e^{\omega t}\boldsymbol{v}$.

$$\frac{d}{dt}\,\boldsymbol{x} = \omega e^{\omega t}\boldsymbol{v}$$

$$A\boldsymbol{x} = e^{\omega t} A\boldsymbol{v} = e^{\omega t}\omega\boldsymbol{v} = \frac{d}{dt}\,\boldsymbol{x}$$

That is $e^{\omega t}\boldsymbol{v}$ is a solution of $\frac{d}{dt}\,\boldsymbol{u}(t) = A\boldsymbol{u}(t)$.

The eigenvalues ω are given by $det(A - \omega I) = 0$, where I is the identity matrix.

$$det(A - \omega I) = \begin{aligned} &\left(\frac{\rho-\beta}{l} - \omega\right) \times (\omega+\lambda_1)(\omega+\lambda_2)..(\omega+\lambda_6)(-1)^6 \\[2ex] &\qquad -\frac{\beta_1}{l} \times \lambda_1(\omega+\lambda_2)..(\omega+\lambda_6)(-1)^5 \\[2ex] &\qquad -\frac{\beta_2}{l} \times \lambda_2(\omega+\lambda_1)(\omega+\lambda_3)..(\omega+\lambda_6)(-1)^5 - (..) \\[2ex] &\qquad -\frac{\beta_6}{l} \times \lambda_6(\omega+\lambda_1)..(\omega+\lambda_5)(-1)^5 \\[2ex] &= \left(\frac{\rho-\beta}{l} - \omega\right) + \frac{1}{l}\left[\frac{\lambda_1\beta_1}{\omega+\lambda_1} + (..) + \frac{\lambda_6\beta_6}{\omega+\lambda_6}\right] \end{aligned}$$

The equation for eigenvalues becomes:

$$\frac{\rho - \beta}{l} + \frac{1}{l} \sum_{i=1}^{6} \frac{\lambda_i \beta_i}{\omega + \lambda_i} = \omega$$

$$\rho = l\omega + \beta - \sum_{i=1}^{6} \frac{\lambda_i \beta_i}{\omega + \lambda_i}$$

Regardless of reactivity value, there are 7 eigenvalues ω_i that are solutions to the equation. These solutions can be calculated numerically.

D.1.2 Eigenvectors Corresponding to Eigen Values

The eigenvectors v_i corresponding to the eigenvalues ω_i are the solutions of the equation:

$$(A - \omega_i I)v_i = 0$$

This can be easily solved with a Gaussian elimination.

Since the 7×7 matrix A has 7 linearly independent eigenvectors v_0, v_1, .. v_6 associated to 7 eigenvalue ω_i, we can define the fundamental matrix $M(t)$, for which every column solves the homogeneous system D.2. $M(t)$ is an invertible matrix; that is, the 7 columns of $M(t)$ are linearly independent.

$$M(t) = [e^{\omega_6 t} v_6 \quad e^{\omega_5 t} v_5 \quad (..) \quad e^{\omega_0 t} v_0]$$

The general solution of the homogeneous is:

$$u(t) = M(t)\,a$$

Where a is a column vector of constant values a_i, that have to be determined by initial conditions.

D.2 Solutions of the Coupled Kinetics Equations

The solution of the initial value problem is given by:

$$u(t) = M(t)\,M(t_0)^{-1}u(t_0)$$

Table D.1 Solutions of the Nordheim equation with $\rho = 100\,\text{pcm}$

ω_6	ω_5	ω_4	ω_3	ω_2	ω_1	ω_0
-268.17	-3.1307	-1.0670	-0.18861	-0.059954	-0.013647	0.023232

Table D.2 Coefficients a_i settled by the initial conditions

a_6	a_5	a_4	a_3	a_2	a_1	a_0
-0.23807	0.97794	-8.8561	99.512	445.75	-339.15	-2123.0

Thus the constants a_i are the solutions of:

$$\boldsymbol{u}(t_0) = \sum_{i=0}^{6} a_i \boldsymbol{v}_i$$

An analytical solution of $n(t)$ and C_j has been established, with numerically calculated values for ω_i, $\boldsymbol{v}_i$ and a_i. However, in the case of one delayed neutron group, analytical solutions can be written for ω_0 and ω_1.

D.2.1 *Illustration with an Example*

Our example is based on the kinetics parameters given in Table 14.5.

The transient is a reactivity step of $100\,\text{pcm}$ from a steady state at criticality, and performed at $t = 10\text{s}$. The steady state before the step is characterized by $n(0 \leq t < 10) = 1$. The matrix A is calculated with $\rho = 100\,\text{pcm} = 0.001$.

The eigenvalues ω_i are written in Table D.1

The associated eigen vectors are the colums of the matrix D.3. Each column is an eigen vector $\boldsymbol{v}_i$ associated to ω_i, with $i = 6$ to 0.

$$\begin{bmatrix}
8.5548 & -8.9973 \times 10^{-3} & 3.2125 \times 10^{-3} & -7.2460 \times 10^{-4} & -3.8449 \times 10^{-4} & 1.0166 \times 10^{-4} & -7.154 \times 10^{-4} \\
-0.030034 & 0.027165 & -0.028683 & 0.038757 & 0.076440 & -0.91372 & -0.18797 \\
-0.21514 & 0.19572 & -0.20906 & 0.30963 & 0.88934 & 0.39966 & -0.89287 \\
-0.19425 & 0.18172 & -0.20586 & 0.61760 & -0.40879 & 0.059752 & -0.31007 \\
-0.40119 & 0.40136 & -0.53647 & -0.72130 & -0.18952 & 0.042403 & -0.26586 \\
-0.14121 & 0.21035 & 0.79046 & -0.030191 & -0.014283 & 3.6344 \times 10^{-3} & -0.024834 \\
-0.048260 & -0.85001 & 0.021605 & -3.4924 \times 10^{-3} & -1.7793 \times 10^{-3} & 4.6379 \times 10^{-4} & -3.2276 \times 10^{-3}
\end{bmatrix}$$

$$(\text{D.3})$$

The coefficients a_i are calculated in order to equalize the sum of exponentials to the vector $\boldsymbol{u}(0 \leq t < 10)$, i.e. the values of $n(t)$ and $C_j(t)$ during the steady state prior to the step transient (Table D.2).

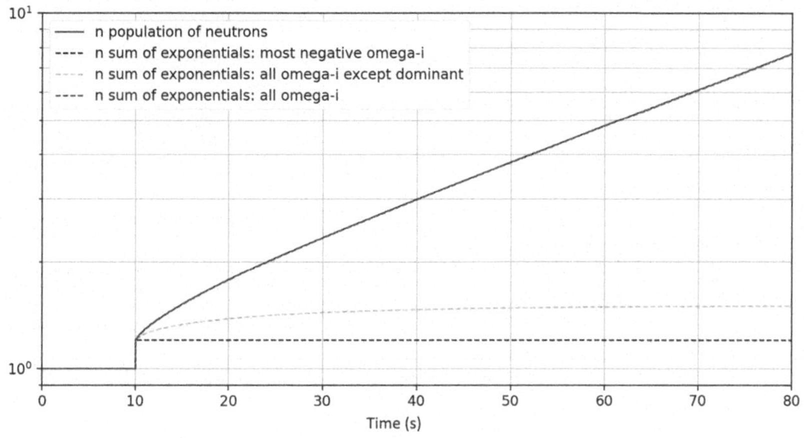

Fig. D.1 Contribution of the different exponentials to $n(t)$

The function $n(t)$ is:

$$n(t) = a_6\, v_6[0]\, e^{\omega_6(t-10)} + a_5\, v_5[0]\, e^{\omega_5(t-10)} + (..)a_0\, v_0[0]\, e^{\omega_0(t-10)}$$

Where $v_i[0]$ is the first component of the seven dimensional vector v_i.

$$n(t) = -0.23807 \times 0.85548 \times e^{-268.17(t-10)} + 0.9774 \times -8.9973 \times e^{-3.1307(t-10)} + (..)$$

The Fig. D.1 and enable to compare the numerical resolution[1] of the kinetics equations and the present semi-analytical sums of exponentials. The plotted functions are the following:

- $n(t)$ calculated with Crank-Nicolson
- $a_6\, v_6[0]\, e^{\omega_6(t-10)} + \sum_{i=0}^{5} a_i\, v_i[0]$
  Most negative exponential ω_6 and other constant terms.
- $\sum_{i=1}^{6} a_i\, v_i[0]\, e^{\omega_i(t-10)} + a_0\, v_0[0]$
  All the exponentials except the dominant one.

These figures show that the prompt-jump is determined by the most negative exponential. Quasi instantaneously, in one hundredth of a second, this exponential disappears. Then the other negative exponentials are gradually fading away during a couple of minutes, and $n(t)$ is finally determined by the dominant exponential $\omega_0 > 0$ since the reactor sur over critical (Fig. D.2).

---

[1] Crank-Nicholson implicit scheme in this case, that could be also an implicit Runge-Kutta.

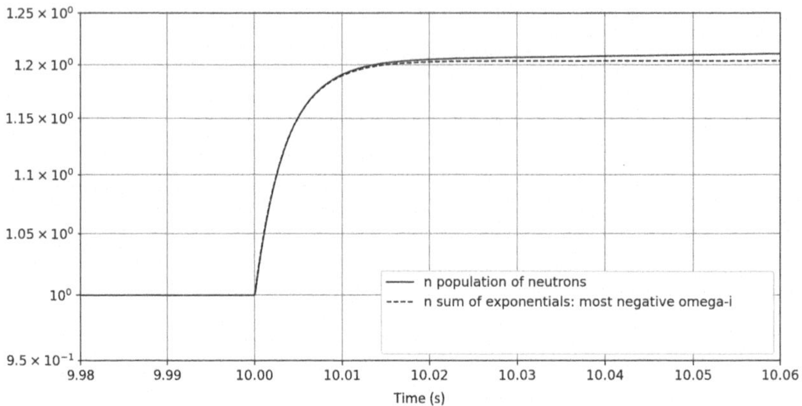

Fig. D.2 The most negative exponential and the prompt jump

D.3 Kinetics with an independant neutron source

With an independent neutron source, we get a non-homogeneous first order system.

$$\begin{cases} \dfrac{dn(t)}{dt} = \dfrac{\rho - \sum_{j=1}^{g} \beta_j}{l} n(t) + \sum_{j=1}^{g} \lambda_j \, C_j(t) + S(t) \\[2ex] \dfrac{dC_j(t)}{dt} = \dfrac{\beta_j}{l} n(t) - \lambda_j \, C_j(t) \end{cases} \tag{D.4}$$

We carry out the solution search in the case where the source $S(t)$ is time-dependent, although in practice, apart from the case of a source jerk, it is not time dependent.

A vector function $b(t)$ is defined; its first component is $S(t)$ and the other components are 0.

The system to solve is:

$$\frac{d}{dt}\, u(t) = A u(t) + b(t) \tag{D.5}$$

D.3.1 Finding the General Solution

Finding the general solution to D.5 splits into two steps: solving the reduced equation or associated homogeneous equation, that is D.2 and then finding one solution (called a particular solution) to D.5. The general solution to D.5 is the sum of the solution to the reduced equation and of the particular solution.

The homogeneous equation has already been solved in D.1, so we need now to find the particular solution.

D.3.2   Particular Solution

We will admit that the particular solution is given by the formula:

$$\boldsymbol{u}_p(t) = M(t) \int_{t_0}^{t} M(s)^{-1} \boldsymbol{b}(s)\, ds \tag{D.6}$$

D.3.3   Solution of the Initial Value Problem

The solution of the initial value problem is given by:

$$\boldsymbol{u}(t) = M(t)\, M(t_0)^{-1} \boldsymbol{u}(t_0) + M(t) \int_{t_0}^{t} M(s)^{-1} \boldsymbol{b}(s)\, ds \tag{D.7}$$

D.3.4   Matrix Exponentials

Exponentiation of a diagonal matrix can be performed simply by exponentiating each of the diagonal elements. We can define the diagonal matrix D with eigen values ω_i:

$$D = \begin{bmatrix} \omega_6 & & & 0 \\ & \ddots & & \\ & & \ddots & \\ 0 & & & \omega_0 \end{bmatrix} \tag{D.8}$$

The exponential of Dt is:

$$e^{Dt} = \begin{bmatrix} e^{\omega_6 t} & & & 0 \\ & \ddots & & \\ & & \ddots & \\ 0 & & & e^{\omega_0 t} \end{bmatrix} \tag{D.9}$$

Let's define a matrix P in which each column is a eigen vector $\boldsymbol{v}_i$. Since $\boldsymbol{v}_i$ are eigenvectors of A, and D is a diagonal matrix with the eigenvalues of A, we can write:

$$A\,D = P\,D$$

$$A = P\,D\,P^{-1}$$

For any integer n, we have:

$$A^n = P\, D^n\, P^{-1}$$

Given that the exponentional of a matrix is defined with a power series:

$$e^A = \sum_{n=0}^{\infty} \frac{A^n}{n!}$$

This enables us to write the exponential matrix of A:

$$e^A = P\, e^D\, P^{-1}$$

The matrix At has the same eigenvectors than A and its eigenvalues are multiplied by t, thus:

$$e^{At} = P\, e^{Dt}\, P^{-1}$$

The particular solution given in Eq. D.6 can be written as a function of the matrix P and D:

$$M(t) \int_{t_0}^{t} M(s)^{-1} = P\, e^{Dt}\, (P\, e^{Ds})^{-1}$$

$$= P\, e^{Dt}\, (e^{DS})^{-1}\, P^{-1}$$

$$= P\, e^{D(t-s)}\, P^{-1}$$

$$= e^{A(t-s)}$$

And then:

$$\boldsymbol{u}_p(t) = \int_{t_0}^{t} e^{A(t-s)}\, \boldsymbol{b}(s)\, ds \tag{D.10}$$

The solution to the reduced equation is:

$$\boldsymbol{u}_r(t) = M(t)\, \boldsymbol{a}$$

$$= P\, e^{Dt}\, \boldsymbol{a}$$

$$= P\, e^{Dt}\, P P^{-1}\, \boldsymbol{a}$$

$$= e^{At}\, \boldsymbol{c}$$

Where c is a column vector of constant values that have to be determined by initial conditions.

The general solution of the non-homogeneous system is $u_r(t) + u_p(t)$. From this we deduce the solution of the initial value problem:

$$u(t) = e^{A(t-t_0)} u(t_0) + \int_{t_0}^{t} e^{A(t-s)} b(s)ds \tag{D.11}$$

This Eq. D.11 is equivalent to Eq. D.7.

Appendix E
Stationnary Waves in Strings and Drums

Calculating the standing waves of a disk membrane with Dirichlet boundary conditions is similar to finding the fundamental mode and harmonics of a simplified reactor, defined as an homogeneous medium in an infinite cylinder. This appendix deals with the mathematical aspects. Physical comments on the potential energy of the vibrating membrane, and on the properties of harmonics, are likely to shed light on neutronics, by reasoning through analogies.

E.1 Stationnary Waves in Strings

E.1.1 The One Dimensionnal Wave Equation

A mass m (kg) suspended horizontally, or vertically, on a spring has one natural frequency at which it freely oscillates up and down:

$$\frac{1}{2\pi}\sqrt{\frac{k}{m}}$$

Where k is the spring stiffness (N/m). The oscillations correspond to energy exchanges between the potential energy (elasticity of the spring) and kinetic energy.

Whenever a wave (sound, heat, light, …) is confined to a finite region of space (string, pipe, cavity, …), the space fills up with a spectrum of vibrating patterns called "standing waves". Confining a wave quantizes the frequency.

We shall explore the physical characteristics of standing waves by considering waves on a taut string. A stretched string with fixed ends can oscillate up and down with a whole spectrum of frequencies and patterns of vibration.

Let's consider a string stretched between two fixed points, which we take to be at $x = 0$ and $x = L$, thus the displacement y is always zero at $x = 0$ and $x = L$: these are Dirichlet boundary conditions. The transverse displacement of the string is in the

H. Grard, *Diffusion of Neutrons in Nuclear Reactors*,
https://doi.org/10.1007/978-3-032-05088-5

y-direction $u(t, x)$ is equal to the deviation from the equilibrium of a transversally vibrating string at the point $x \in [0, L]$ at the time t. Transversal vibrations mean that each point of the string moves along the vertical line orthogonal to the equilibrium position. The function $u(t, x)$ satisfies the one-dimensional wave equation:

$$\frac{\partial^2}{\partial t^2} u(x, t) = c^2 \frac{d^2}{dx^2} u(x, t) \tag{E.1}$$

The term $\frac{\partial^2}{\partial t^2} u(x, t)$ is the vertical acceleration of a point on the string; the Laplacian $\frac{d^2}{dx^2}$ quantifies the concavity of the string at a given point, and c is the propagation speed of waves.

The constant c can be expressed in terms of the tension τ (N) of the string and the density ρ (the density is the mass of the string per unit length, in kg $\cdot$ m^{-1}):

$$c = \sqrt{\frac{\tau}{\rho}}$$

It is interesting to note that c is the speed of a pulse or wave on a string under tension.

Since the string is attached at both ends, these are Dirichlet boundary conditions and $u(t, 0) = u(t, L) = 0$.

E.1.2   Resolution of the Wave Equation

In order to find a solution of this equation, we use the Fourier method. The first step is to separate the variables and to look for a solution in the form:

$$u(t, x) = T(t)X(x)$$

From Eq. E.1, we get:

$$T''(t)\, X(x) = c^2\, T(t) X''(x)$$

And we obtain:

$$\frac{X''(x)}{X(x)} = \frac{T''(t)}{c^2\, T(t)} = -\lambda$$

Where λ is some constant. The first fraction does not depend on t, and the second one is independent of x, so both are equal to a constant. The functions $X(x)$ and $T(t)$ can be considered separately. The function $X(x)$ is the solution of the Laplacien eigen value problem:

$$X''(x) + \lambda X(x) = 0 \tag{E.2}$$

The eigen values and eigen functions are given by:

$$\lambda_m = \left(\frac{\pi m}{L}\right)^2 \quad X_m(x) = \sin\left(\frac{\pi m}{L}x\right) \quad \text{for } m \in \{1, 2, 3, ..\}$$

The $T_m(t)$ functions correspond to the functions $X_m(x)$ and we obtain:

$$T_m(t) = A_m \cos\left(\frac{c\pi m}{L}t\right) + B_m \sin\left(\frac{c\pi m}{L}t\right)$$

Where A_m and B_m are constants, depending on initial conditions.

These special modes of vibration of a string are called standing waves or normal modes. The expression standing wave comes from the fact that each normal mode has wave properties (wavelength λ, frequency f), but the wave pattern (sinusoidal shape) does not travel left or right through space: it stands still. All points on the string oscillate at the same frequency but with different amplitudes. Points that do not move (zero amplitude of oscillation) are called nodes. Points where the amplitude is maximum are called antinodes.

The functions $X_m(x)$ describes the shape of the stationnary wave, and the functions $T_m(t)$ describes the vertical oscillation as a function of time. The frequency of vibration is:

$$f_m = \frac{1}{2\pi}\left(\frac{c\pi m}{L}\right) = m\frac{c}{2L}$$

The fundamental mode of vibration, corresponding to $m = 1$, is the simplest standing wave on a string, with a single antinode at the center and a node at each end. This is also known as the first harmonicharmonic!first.

For the first harmonic, the length of the string is equal to one half of a wavelength of the standing wave, because the distance from one node to the next nearest node is one half a wavelength. The frequency of the first harmonic is called the fundamental frequency.

There are many other modes of vibration, each one corresponding to a value of m. Each mode is called a harmonic. Harmonics are multiples of the fundamental mode of vibration. The overtones are all the harmonics except the first harmonic.

The fundamental mode of a vibrating string, and the second and third harmonics can be seen in the Fig. 10.1, which shows the eigen functions of the Laplacian in a 1D problem.

E.2 Stationnary Waves in Drums

E.2.1 Wave Equation on a Disk

We're going to look at the simplest possible membrane model, which satisfies the following hypotheses:

- no thickness, the membrane is infinitely thin and uniform. Its mass density per unit area is σ (kg/m^2
- same surface tension T in all points and directions. The surface tension is a force per unit of length. If you draw a line segment of length dl on the membrane, then the force that the membrane on one side of the line segment exerts on the membrane on the other side is $T\,dl$
- no horizontal displacement
- no damping
- membrane immobile at edge

According to the previous hypothesis, membrane vertical displacement at the point r and time t can be represented by a function $u(r, t)$ defined on a disk (r refers to a point M of the disk such that $\overrightarrow{OM} = r$), and it can be demonstrated that the physical problem satisfies the following equations, called wave equation:

$$\frac{\partial^2}{\partial t^2} u(r, t) = c^2 \Delta u(r, t) \tag{E.3}$$

At the boundary R of the disk, $u = 0$.

In a similar way to the string example, c is a constant, the wave propagation speed at the surface of the membrane. In the case of a membrane, c can be expressed in terms of the surface tension T (N $\cdot$ m^{-1}) and the mass density σ (kg $\cdot$ m^{-2}):

$$c = \sqrt{\frac{T}{\sigma}}$$

The three-dimensionnal wave equation is also the same Eq. E.3. This equation models, for example, the small movements of a compressible fluid in an acoustic cavity. Solving the eigenvalue problem with appropriate boundary conditions boundary conditions corresponds to the calculation of the cavity's eigenfrequencies and the associated "harmonics". The set of eigenvalues is called the spectrum of the problem.

E.2.2 Resolution of the Wave Equation on a Disk

Similar to the one-dimensional problem, we separate the variables and search for the functions G as eigen functions of the Laplacian:

$$u(r, t) = F(t)\,G(r)$$

In polar coordinates, the vector r can be replaced by the distance from the center of the disk r and the angle θ:

$$u(r, \theta, t) = F(t)\,G(r, \theta)$$

Let's call $g(r, \theta)$ an eigen function of the Laplacian, and $\lambda > 0^2$ the associated eigen value:

$$\Delta g(r, \theta) + \lambda\, g(r, \theta) = 0$$

We separate the variables r and θ and we search eigen functions $G(r, \theta) = R(r)\Theta(\theta)$.

$$\Theta(\theta)\left(\frac{\partial^2}{\partial r^2}R(r) + \frac{1}{r}\frac{\partial}{\partial r}R(r)\right) + R(r)\frac{1}{r^2}\frac{\partial^2}{\partial\theta^2}\Theta(\theta) = -\lambda R(r)\Theta(\theta)$$

$$\frac{1}{R(r)}\left(\frac{\partial^2}{\partial r^2}R(r) + \frac{1}{r}\frac{\partial}{\partial r}R(r)\right) + \frac{1}{\Theta(\theta)}\frac{1}{r^2}\frac{\partial^2}{\partial\theta^2}\Theta(\theta) = -\lambda$$

$$\frac{1}{R(r)}\left(r^2\frac{\partial^2}{\partial r^2}R(r) + r\frac{\partial}{\partial r}R(r)\right) + \lambda r^2 = -\frac{1}{\Theta(\theta)}\frac{\partial^2}{\partial\theta^2}\Theta(\theta)$$

The left side depends only on r and the right side depends only on θ, so each one must be a constant. Θ is a periodic function with a 2π period, thus it can be put in the form of Fourrier series:

$$\Theta(\theta) = \sum_{n=0}^{\infty}\alpha_n\cos(n\theta) + \beta_n\sin(n\theta)$$

For the index n, we can write the equation:

$$\frac{1}{R_n(r)}\left(r^2\frac{\partial^2}{\partial r^2}R_n(r) + r\frac{\partial}{\partial r}R_n(r)\right) + \lambda r^2 = n^2$$

$$r^2\frac{\partial^2}{\partial r^2}R_n(r) + r\frac{\partial}{\partial r}R_n(r) + ((r\sqrt{\lambda})^2 - n^2)R_n(r) = 0$$

This equation is closely related to the Bessel Eq. B.9 (see appendix Special functions), by changing the variable and writing $y_n(r) = R_n(\frac{r}{\sqrt{\lambda}})$.

For a given integer n, Bessel's equation possesses, up to a multiplication constant, only one regular solution at $r = 0$, which is $J_n(r)$. It comes:

$$R_n(r) = J_n(r\sqrt{\lambda})$$

Taking into account the Dirichlet boundary condition:

$$R_n(R) = J_n(R\sqrt{\lambda}) = 0$$

---

[2] If $\lambda < 0$, the problem will reveal the modified Bessel's equation, and its solutions the modified Bessel's functions. See appendix Special functions.

We can conclude that the eigen values for the index n, .i.e. the possible values of λ, are given by the roots of J_n:

$$\lambda = \left(\frac{\mu_{m,n}}{R}\right)^2 \quad \text{with the index of the root } m \in \{1, 2, 3, ..\}$$

And the eigen function of the Laplacian for n is:

$$g_{m,n}(r, \theta) = \alpha_n J_n\left(r\frac{\mu_{m,n}}{R}\right)\cos(n\theta) + \beta_n J_n\left(r\frac{\mu_{m,n}}{R}\right)\sin(n\theta)$$

The coefficients α_n and β_n can be calculated to normalize the eigen functions. In that case, the functions $(g_{m,n})_{(m,n)\in\mathbb{N}^*\times\mathbb{R}}$ describe the standing waves, and are sometimes referred to as harmonics. They form an orthonormal basis.

Without normalization, we can just write the functions of an orthogonal basis:

$$g_{m,n}(r, \theta) = J_n\left(r\frac{\mu_{m,n}}{R}\right)\cos(n\theta) \tag{E.4}$$

Another basis can be defined with the equation:

$$g_{m,n}(r, \theta) = J_n\left(r\frac{\mu_{m,n}}{R}\right)(\cos(n\theta)\cos(\alpha) + \sin(n\theta)\sin(\alpha)) \tag{E.5}$$

This basis consists in a rotation by an angle α of the basis E.4. On the Fig. E.1, the angle α is equal to zero.

The orthogonality of the basis E.4 will be will be verified in E.2.3.

The time oscillation functions, times the harmonics functions, give the solutions to the wave equation:

$$u(r, t) = \sum_{m=1}^{\infty}\sum_{n=0}^{\infty} f_{m,n}(t)g_{m,n}(r, \theta)$$

$$\sum_{m=1}^{\infty}\sum_{n=0}^{\infty} g_{m,n}(r, \theta)\frac{\partial^2}{\partial t^2}f_{m,n}(t) = \sum_{m=1}^{\infty}\sum_{n=0}^{\infty} f_{m,n}(t)\underbrace{\Delta g_{m,n}(r, \theta)}_{=-\frac{\mu_{m,n}^2}{R^2}g_{m,n}(r,\theta)}$$

And given that $(g_{m,n})$ constitute a basis, it comes:

$$\frac{\partial^2}{\partial t^2}f_{m,n}(t) = -\frac{\mu_{m,n}^2}{R^2}g_{m,n}(r, \theta)$$

And then:

$$f_{m,n}(t) = A_{m,n}\cos(c\mu_{m,n}\,t) + B_{m,n}\sin(c\mu_{m,n}\,t)$$

Table E.1 Values of roots of J_n Bessel functions

	$m = 1$	2	3	4	5	6	7	8	9	10
$n = 0$	2.4048	5.5201	8.6537	11.7915	14.9309	18.0711	21.2116	24.3525	27.4935	30.6346
1	3.8317	7.0156	10.1735	13.3237	16.4706	19.6159	22.7601	25.9037	29.0468	32.1897
2	5.1356	8.4172	11.6198	14.7960	17.9598	21.1170	24.2701	27.4206	30.5692	33.7165
3	6.3802	9.7610	13.0152	16.2235	19.4094	22.5827	25.7482	28.9084	32.0649	35.2187
4	7.5883	11.0647	14.3725	17.6160	20.8269	24.0190	27.1991	30.3710	33.5371	36.6990
5	8.7715	12.3386	15.7002	18.9801	22.2178	25.4303	28.6266	31.8117	34.9888	38.1599
6	9.9361	13.5893	17.0038	20.3208	23.5861	26.8202	30.0337	33.2330	36.4220	39.6032
7	11.0864	14.8213	18.2876	21.6415	24.9349	28.1912	31.4228	34.6371	37.8387	41.0308
8	12.2251	16.0378	19.5545	22.9452	26.2668	29.5457	32.7958	36.0256	39.2404	42.4439
9	13.3543	17.2412	20.8070	24.2339	27.5837	30.8854	34.1544	37.4001	40.6286	43.8438

Where $A_{m,n}$ and $B_{m,n}$ are determined by the initial conditions. Each harmonic has its own frequency of vibration $\frac{1}{2\pi} c \mu_{m,n}$. The Table E.1 gives the first values of $\mu_{m,n}$.

E.2.3 Inner Product and Orthogonality of the Eigen Functions

In a disk and polar coordinates, the inner product is (see G.2):

$$\langle f, g \rangle = \int_0^R \int_0^{2\pi} f(r, \theta)\, g(r, \theta)\, r\, dr d\theta$$

The orthogonality of the basis $(g_{m,n})$ can be easily checked:

$$\langle g_{m,n}, g_{m',n'} \rangle = \begin{aligned} &\int_0^R \int_0^{2\pi} J_n\left(r \frac{\mu_{m,n}}{R}\right) \cos(n\theta) \\ &\quad \times J_n\left(r \frac{\mu_{m',n'}}{R}\right) \cos(n'\theta)\, r\, dr d\theta \\ &= \int_0^R J_n\left(r \frac{\mu_{m,n}}{R}\right) J_n\left(r \frac{\mu_{m',n'}}{R}\right) r\, dr \\ &\quad \times \int_0^{2\pi} \cos(n\theta) \cos(n'\theta)\, d\theta \end{aligned}$$

It is know that, if n and n' are integers:

$$\int_0^{2\pi} \cos(n\theta) \cos(n'\theta)\, d\theta = \pi\, \delta_{n,n'}$$

It follows that for $n \neq n'$ we have $\langle g_{m,n}, g_{m',n'} \rangle = 0$ and thus the eigen functions are orthogonal.

When $n = n'$, we can write:

$$
\begin{aligned}
\langle g_{m,n}, g_{m',n} \rangle &= \int_0^R \int_0^{2\pi} J_n\left(r \frac{\mu_{m,n}}{R}\right) \cos(n\theta) \\
&\quad \times J_n\left(r \frac{\mu_{m',n}}{R}\right) \cos(n\theta)\, r\, dr\, d\theta \\
&= \int_0^R J_n\left(r \frac{\mu_{m,n}}{R}\right) J_n\left(r \frac{\mu_{m',n}}{R}\right) r\, dr \times \pi
\end{aligned}
$$

Considering the well know orthogonality of Bessel functions:

$$
\int_0^1 x J_n(\mu_{m,n}x) J_n(\mu_{m',n}x)dx = 0 \quad \text{if } m \neq m'
$$

We can conclude that $\langle g_{m,n}, g_{m',n} \rangle = 0$ if $m \neq m'$.

Othogonality of the eigen functions ($g_{m,n}$) was to be expected since Laplace operator is both a linear and self ajoint operator, and it is known that if a linear operator is self adjoint, then eigenvectors with distinct eigenvalues are orthogonal (see Appendix G.5.1).

E.2.4 Potential and Kinetic Energy of the Membrane

We can set the potential energy of the disk membrane at zero in the case it is flat. Whenever it has a non flat shape, the membrane under tension has a potential energy associated to the deformation. It can be demonstrated that:

$$
Ke(t) = \frac{1}{2} \int_{\text{disk surf.}} \sigma \left(\frac{\partial u}{\partial t}\right)^2 dS
$$

$$
Kp(t) = \frac{1}{2} \int_{\text{disk surf.}} T \, (\nabla u)^2 dS
$$

The integral $\int_{\text{disk surf.}} (\nabla u)^2 dS$ is a surface. It is equal to zero in the case $u(r, t)$ is constant, which means that the membrane is still and flat. The integral can be interpreted as the increase of the area of the membrane due to deformation. An area increase times the surface tension has a strength times length unit, which is an energy. In the case of a string, the kinetic energy is:

$$
Kp(t) = \frac{1}{2} \int_{\text{segment length}} \tau \, (\nabla u)^2 dx
$$

In that case, the integral $\int_{\text{segment length}} (\nabla u)^2 dx$ is a length. can be interpreted as the increase of the length of the string due to deformation. The kinetic energy of the vibrating string has a formal analogy with the spring:

$$Kp = \frac{1}{2}k(x - x_0)(x - x_0)$$

where k is the spring stiffness (N/m), and x_0 the length at rest. $k(x - x_0)$ is the tension, analog to τ and $(x - x_0)$ is the length increase analog to the integral $\int_{\text{segment length}} (\nabla u)^2 dx$.

It should be noted that the energy of Dirichlet (presented in Appendix A.2.4.2) times the tension is equal to the potential energy.

The shape of each harmonic corresponds to a local minimum of the membrane's potential energy. This means that if the membrane shape deviates from the shape of an harmonic in any way, the potential energy increases.

E.2.5   Further Considerations

The book on the theory of sound (Strutt and Rayleigh, 1877) was the first comprehensive and systematic mathematical treatise on sound; this book opened the era of modern acoustics. The range of his interests covered almost all the subjects of physics including sound, light, electricity, heat, and gas. The Chap. 9 studies the vibrations of membranes.

> We have seen that the gravest tone of a membrane, whose boundary is approximatively circular, is nearly the same as that of a mechanically similar membrane in the form of a circle of the same mean radius or area. If the area of a membrane be given, there must evidently be some form of boundary for which the pitch (of the principal tone) is the gravest possible, and this form can be no other than the circle. In the case of approximate circularity an analytical demonstration may be given, of which the following is an outline. Strutt and Rayleigh (1877, p. 284)

This means that the disk minimizes the first eigenvalue (in ascending order) of the Laplacian with Dirichlet condition at fixed area. It was not until the 1920s, however, that a rigorous demonstration of this fact was established (Berger, 2015, p. 2).

The sounds emitted by a drum are made up of frequencies that are not overly damped, which are the drum's resonant frequencies. These frequencies depend directly on the drum's characteristics: shape and materials. The energy levels of the vibrating membrane, i.e. the potential energy of membrane deformation, are quantized.

Harmonics with the parameter $n = 0$ correspond to displacements in the vibrating drum problem where the drum is struck exactly in the center. This creates vibrational patterns that are radially symmetric. In this case, the zero-th order Bessel function of the first kind tells us the amplitude of the vibration of the skin at a given radius from the centre of the drum.

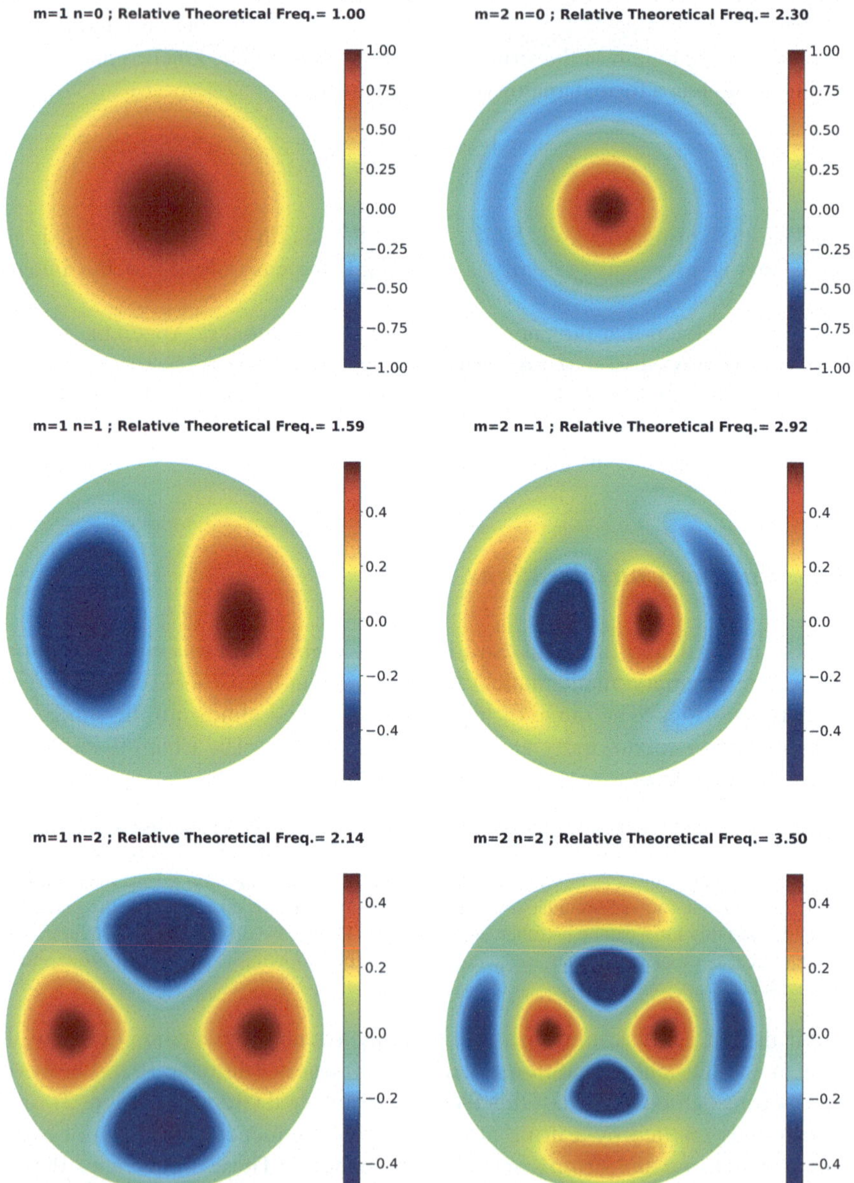

Fig. E.1 The more important normal modes of vibration $g_{m,n}(r, \theta)$ of a disk membrane, and the frequency referred to the gravest as unity

The Fig. E.1 shows that the area of the membrane is divided into segments by the nodal system, in such a manner that the sign of the vibration changes whenever a node is crossed (Strutt and Rayleigh, 1877, p. 206).

Calculations with high values of n and m show that (Grebenkov and Nguyen, 2013, Figs. 7.12 and 7.13, pp. 640, 641):

For a fixed m, an increase in the index n leads to stronger localization of eigenfunctions near the boundary of the disk. For a fixed n, an increase in the index m leads to stronger localization of eigenfunctions at the center of the disk.

References

Amandine Berger. *Optimisation du spectre du Laplacien avec conditions de Dirichlet et Neumann dans R^2 et R^3*. PhD thesis, Université Grenoble Alpes; Université de Neuchâtel (Suisse), 2015.

Denis S Grebenkov and B-T Nguyen. Geometrical structure of laplacian eigenfunctions. *siam REVIEW*, 55(4):601–667, 2013. URL https://pmc.polytechnique.fr/pagesperso/dg/publi/2013_06.pdf.

J.W. Strutt and J.W.S. Rayleigh. *The Theory of Sound*. Number vol. 1. Macmillan, 1877. URL https://gallica.bnf.fr/ark:/12148/bpt6k951307.image.

Appendix F
Numerical Resolution of the Diffusion Equations

This appendix gives an initial overview of the numerical resolution of the diffusion equations, in the simplified case of a single neutron group and a 1D geometry. The method presented here is the finite difference method associated with Power Iteration on Fission Source. Readers wishing to delve deeper into this subject and gain an overview of the steps leading to core calculation will study with profit Marguet (2018, Chaps. 14, 18) and Mohanakrishnan et al. (2021, Chap. 9).

F.1 Numerical Equation

The equation to solve, in the case of one goup is:

$$\nu\Sigma_f(r)\Phi(r) - \Sigma_a(r)\Phi(r) + \mathrm{div}(D(r) \cdot \overrightarrow{\mathrm{grad}}\Phi(r)) = 0$$

A numerical solution is achieved by replacing the continuous spatial dependence of the flux $\Phi(r)$ with the values of the flux at a number of discrete spatial locations, the solution for which will be the objective of the numerical technique [Stacey, 2007, page 78]. There are many ways to do this, and the most common one is to use the finite-difference method.

In a one-dimensional geometry, the numerical equation of diffusion is:

$$(-[D] - [A])\Phi = \frac{1}{k_{\mathrm{eff}}}[P]\Phi \tag{F.1}$$

Where $[P]$ is the production matrix:

The reactor is devided into cells numbered from 1 to I. Cell number 1 is the center cell (center of a sphere, center of a wall, or central axis of an infinite cylinder) where reflective boundary condition apply; cell number I is the last cell of the system where chosen boundary condition (Dirichlet, vacuum, etc.) apply.

© The Editor(s) (if applicable) and The Author(s), under exclusive license to Springer Nature Switzerland AG 2026

H. Grard, *Diffusion of Neutrons in Nuclear Reactors*,
https://doi.org/10.1007/978-3-032-05088-5

$$[P] = \begin{bmatrix} \nu\Sigma_{f,1} & 0 & \cdots\cdots\cdots\cdots\cdots\cdots\cdots & 0 \\ 0 & \nu\Sigma_{f,2} & & \vdots \\ \vdots & & \ddots & \vdots \\ & & & \nu\Sigma_{f,I} & 0 \\ 0 & \cdots\cdots\cdots\cdots\cdots\cdots & 0 & 0 \end{bmatrix}$$

$[P]$ matrix has $I + 1$ lines and columns, and the flux vector Φ has also $I + 1$ coordinates: the supplementary cell unables to take into account the boundary conditions. The matrices $[D]$ and $[A]$ operate diffusion and absorption. A detailed insight of the construction of these matrixes is given in McClarren [2017].

F.2 Resolution of the Numerical Equation

Let's write that $[L] = -[D] - [A]$

$$[L]\Phi = \frac{1}{k_{\text{eff}}}[P]\Phi$$

can also be written:

$$[L]^{-1}[P]\Phi = \lambda\Phi$$

The objective is to calculate the maximal value of λ and the associated eigen vector Φ. The maximal eigen value is k_{eff}. A comparison can be done with the system 10.4, reminded below:

$$\begin{cases} \Delta f(\boldsymbol{r}) = \lambda f(\boldsymbol{r}) \\ \\ f(\boldsymbol{r}) \text{ satisfies boundary conditions (BC)} \end{cases}$$

Numerical equation and Laplacian eigen value equations differ:

- In the numerical equation, the boundary conditions are embedded in the matrixes.
- The system 10.4 considers the eigen values of the Laplacian (that are negative); the numerical equation considers the eigen value of $[L]^{-1}[P]$ (that are positive)
- The fundamental mode of the Laplacian equation is associated to the smallest, in absolute value, of the negative eigen values. The fundamental mode of the numerial equation is associated to the highest positive eigen value.

F.2.1 Iterative Power Method

The iterative power methode is a classical numerical method, which is used to calculate the highest, in absolute value, eigen value and its associated eigen vector.

> The iterative method is an iterative technique used to determine the *dominant* eigenvalue of a matrix that is the eigen value with the largest magnitude. By modifying the method slightly it can also be used to determine other eigenvalues. One useful feature of the Power method is that it produces not only an eigen value, but an associated eigen vector. Faires and Burden (2003, p. 388)

For example, the matrix:

$$[A] = \begin{bmatrix} -4 & 14 & 0 \\ -5 & 13 & 0 \\ -1 & 0 & 2 \end{bmatrix}$$

has eigen values $\lambda_1 = 6, \lambda_2 = 3, \lambda_3 = 2$. And, indeed, the product of the eigen values is equal to $\det A = 36 = 6 \times 3 \times 2$.

So the Power method will converge. We define the vector $X^{(0)} = (1, 1, 1)$ The iterative method is based on the calculation of:

$$X^{(n+1)} = \frac{[A]X^{(n)}}{\| [A]X^{(n)} \|}$$

n	$X^{(n)}$	$\| [A]X^{(n)} \|$
0	(1, 1, 1)	12.845233
1	(0.778499, 0.622799, 0.077850)	7.034116
2	(0.796858, 0.597644, −0.088540)	6.488712
3	(0.798244, 0.583332, −0.150097)	6.232753
4	(0.797990, 0.576326, −0.176237)	6.113175
5	(0.797721, 0.572909, −0.188194)	6.055740
6	(0.797564, 0.571228, −0.193884)	6.027645
7	(0.797482, 0.570396, −0.196649)	6.013762
8	(0.797441, 0.569983, −0.198009)	6.006865
9	(0.797420, 0.56977, −0.198683)	6.003428
10	(0.797410, 0.569674, −0.199017)	6.001713
11	(0.797405, 0.569623, −0.199184)	6.000856

At each iteration, the powers of $[A]^n X^{(0)}$ are scaled to unit.

$[A]X^{(n)}$ tends to be colinear with $X^{(n)}$ (eigen vector) while $\| [A]X^{(n)} \|$ tends to the maximum eigen value.

F.2.2   Application to the Diffusion Equation in the Case of One Goup and One-Dimensional Geometry

In our example, the problem to solve is:

$$[C]\,\Phi = \lambda\,\Phi$$

With $[C] = [L]^{-1}\,[P]$. The iterative power method consists in initializing the vector with an initial guess $\phi^{(0)}$, and then to calculate the product $[C]\,\Phi^{(0)}$. It is established that if we set $\Phi^{(1)} = [C]\,\Phi^{(0)}$ and then repeat the process, the sucession of $\Phi^{(i)}$ vectors tend to be the eigen vector associated to the highest eigen value given by $\left\|\dfrac{\Phi^{(i+1)}}{\Phi^{(i)}}\right\|$.

However, at each iteration, the norm of the vectors increase, that's why at each iteration the vectors are normalized to unit: $\Phi^{(i+1)} = \frac{[C]\,\Phi^{(i)}}{\|[C]\,\Phi_{(i)}}\|$.

The iteration are done until $\|[C]\,\Phi^{(i)}\|$ converges towards a stable value, that gives the eigen value.

F.2.2.1   Power Iteration on Fission Source

The iterative power method with a prior calculation of $[C] = [L]^{-1}\,[P]$ and then iterative calculations of $[C]\,\Phi^{(i)}$ is not the used approach because it would require the computation of $[L]^{-1}$.

The numerical resolution avoids the computation of the inverse of $[L]$. A method called Power Iteration on Fission Source (Stacey, 2007, pp. 79–80) is commonly used. An initial guess of the flux vector ($I + 1$ values) $\Phi^{(0)}$ is arbitrarily constructed, which enables to calculate $[P]\Phi^{(0)}$: this quantity is coined fission source. The first iteration consists in:

- a numerical calculation of $\Phi^{(1)}$ based on equation $[P]\,\Phi^{(0)} = [L]\,\Phi^{(1)}$. This can be solved with a LU factorization.
- a normalization to unit by redefining $\Phi^{(1)}$ as $\frac{\Phi^{(1)}}{\|\Phi^{(1)}\|}$

The new flux vector $\Phi^{(1)}$ enables to calculate a new fission source, for the next iteration. The iteration process is continued until the eigen values obtained on two successive iterates differ by less than some convergence criterion (Stacey, 2007, p. 80).

References

J Douglas Faires and Richard L Burden. *Numerical methods.* Thomson, 2003.

S. Marguet. *The Physics of Nuclear Reactors.* Springer International Publishing, 2018. ISBN 9783319595603. URL https://books.google.fr/books?id=9DIODwAAQBAJ.

R. McClarren. *Computational Nuclear Engineering and Radiological Science Using Python.* Elsevier Science, 2017. ISBN 9780128123713. URL https://books.google.fr/books?id=YY-ZDgAAQBAJ.

P Mohanakrishnan, Om Pal Singh, and K Umasankari. *Physics of nuclear reactors.* Academic Press, 2021.

W.M. Stacey. *Nuclear Reactor Physics.* Wiley, 2007. ISBN 9783527406791. URL https://books.google.fr/books?id=y1UgcgVSXSkC.

Appendix G
Vector Space, Inner Product and Adjoint Operator

G.1 Vector Space

G.1.1 Definition

A vector space V over a field $\mathbb{F}$ is a nonempty set V of elements called vectors, under finite operations of vector addition $(+)$ and scalar multiplication $(\cdot)$, such that the following laws hold for all vectors $\boldsymbol{u}, \boldsymbol{v}, \boldsymbol{w} \in \mathbf{V}$ and scalars $a, b \in \mathbb{F}$:

1. Closure of vector addition: $\boldsymbol{u} + \boldsymbol{v} \in \mathbf{V}$
2. Commutativity of addition: $\boldsymbol{u} + \boldsymbol{v} = \boldsymbol{v} + \boldsymbol{u}$
3. Associativity of vector addition: $\boldsymbol{u} + (\boldsymbol{v} + \boldsymbol{w}) = (\boldsymbol{u} + \boldsymbol{v}) + \boldsymbol{w}$
4. Additive identity: There exists an element $\boldsymbol{0} \in \mathbf{V}$ such that $\boldsymbol{u} + \boldsymbol{0} = \boldsymbol{u} + \boldsymbol{0} = \boldsymbol{u}$
5. Additive inverse: There exists an element $-\boldsymbol{u} \in \mathbf{V}$ such that $\boldsymbol{u} + (-\boldsymbol{u}) = \boldsymbol{0} = (-\boldsymbol{u}) + \boldsymbol{u}$

And

1. Closure of scalar multiplication: $a \cdot \boldsymbol{u} \in \mathbf{V}$
2. Distributive property of scalar multiplication over vector addition: $a \cdot (\boldsymbol{u} + \boldsymbol{v}) = a \cdot \boldsymbol{u} + a \cdot \boldsymbol{v}$
3. Distributive property of scalar multiplication over scalar addition: $(a + b) \cdot \boldsymbol{u} = a \cdot \boldsymbol{u} + b \cdot \boldsymbol{u}$
4. Associativity of scalar multiplication: $(ab) \cdot \boldsymbol{u} = a \cdot (b \cdot \boldsymbol{u})$
5. Scalar multiplication identity: $1 \cdot \boldsymbol{u} = \boldsymbol{u}$

© The Editor(s) (if applicable) and The Author(s), under exclusive license to Springer Nature Switzerland AG 2026

H. Grard, *Diffusion of Neutrons in Nuclear Reactors*,

https://doi.org/10.1007/978-3-032-05088-5

G.1.2 Examples of Vector Spaces

G.1.2.1 Standard Vector Space of Dimension n Over the Reals

Given a positive integer n, we define the standard vector space of dimension n over the reals to be the set of vectors:

$$\mathbb{R}^n = \{(x_1, x_2, \ldots x_n) | x_1, x_2, \ldots x_n \in \mathbb{R}\}$$

together with the vector addition and scalar multiplication.

G.1.2.2 Vector Space of Real Functions

Let $C_{[a,b]}$ denote the set of real-valued functions that are continuous on the interval $[a, b]$ and use the standard function addition and scalar multiplication for these functions. That is, for f and g two functions of $C_{[a,b]}$ and real number c, we define the functions $f + g$ and $c \cdot f$ by:

$$(f + g)(x) = f(x) + g(x)$$

$$(c \cdot f)(x) = c \cdot (f(x))$$

It can be easily demonstrated that $C_{[a,b]}$ with the given operations is a vector space over $\mathbb{R}$.

G.1.2.3 Basis for a Vector Space V

Let $v_1, v_2, \ldots v_n$ be vectors in the vector space V. The span of these vectors, denoted by $\mathrm{span}\{v_1, v_2, \ldots v_n\}$, is the subset of V consisting of all linear combinations of these vectors:

$$\mathrm{span}\{v_1, v_2, \ldots v_n\} = \{c_1 v_1 + c_2 v_2 + \cdots + c_n v_n | c_1, c_2, \ldots c_n \text{ are scalars}\}$$

The vectors $v_1, v_2, \ldots v_n$ are said to be linearly dependant if there exists scalars $c_1, c_2, \ldots c_n$ not all zero, such that:

$$c_1 v_1 + c_2 v_2 + \cdots + c_n v_n = 0$$

Otherwise, the vectors are called linearly independant.

A basis of a vector space V is defined as a subset $\{v_1, v_2, \ldots v_n\}$ of vectors in V that are linearly independant and span V. Consequently, if $(v_1, v_2, \ldots v_n)$ is a list of vectors in V, then these vectors form a vector basis if and only if every $v \in V$ can be uniquely written as:

$$v = a_1 v_1 + a_2 v_2 + \cdots a_n v_n \,|\, a_1, a_2, \ldots a_n \text{ are scalars}$$

Every vector space has a basis. A vector space V has many different vector bases, but there are always the same number of basis vectors in each of them.

G.1.3 Inner Product

In a vector space, an inner product is a way to multiply vectors together, with the result of this multiplication being a scalar.

Theorem. Let V be a vector space over $\mathbb{R}$. An inner product on V is a map that assigns to each pair $(v, w) \in V^2$ a real number $\langle v, w \rangle$ such that, for all $u, v, w \in V$ and $\alpha \in \mathbb{R}$,

i $\langle v, v \rangle \geq 0$
ii $\langle v, v \rangle = 0 \Leftrightarrow v = 0$
iii $\langle u + v, w \rangle = \langle u, w \rangle + \langle v, w \rangle$
iv $\langle \alpha v, w \rangle = \alpha \langle v, w \rangle$

G.1.4 Inner Product Space

By definition, an inner product space is a vector space together with an inner product on it.

Examples of inner product spaces:

1. The vector space $\mathbb{R}^n$, where the inner product is given by:

$$\langle (x_1, x_2, \ldots x_n), (y_1, y_2, \ldots y_n) \rangle = x_1 y_1 + x_2 y_2 + \cdots + x_n y_n$$

2. The vector space of real functions whose domain is a closed interval $[a, b]$ with inner product:

$$\langle f, g \rangle = \int_a^b f(x) g(x) \, dx \tag{G.1}$$

3. The inner product of functions $f(r, \theta)$ and $g(r, \theta)$ defined in polar coordinates in a disk of radius R:

$$\langle f, g \rangle = \int_0^R \int_0^{2\pi} f(r, \theta) \, g(r, \theta) \, r \, dr d\theta \tag{G.2}$$

G.2   Linear Operators, Adoint of a Linear Operator

G.2.1   Linear Operator

A function L that maps elements of one vector space (domain) into another (target) over the **same field of scalars**, say $L : V \to W$ is called a **linear operator** if for all vectors $\boldsymbol{u}, \boldsymbol{v} \in V$ and scalars a, b, we have:

$$L(a\boldsymbol{u} + b\boldsymbol{v}) = aL(\boldsymbol{u}) + bL(\boldsymbol{v})$$

By taking $a = b = 1$ in the definition, we see that a linear function L is additive: $L(\boldsymbol{u} + \boldsymbol{v}) = L(\boldsymbol{u}) + L(\boldsymbol{v})$.

Also, by taking $b = 0$, we see that a linear function is outative, that is, $L(a\boldsymbol{u}) = aL(\boldsymbol{u})$

G.2.2   Linear Operator and Matrices

Theorem. Let $L : \mathbb{R}^n \to \mathbb{R}^m$ be a linear operator. Then there exists a unique matrix $[A]$ such that:

$$L(\boldsymbol{v}) = [A]\boldsymbol{v} \quad \forall \boldsymbol{v} \in \mathbb{R}^n$$

Every matrix multiplication is a linear transformation, and every linear transformation is a matrix multiplication. However, term linear transformation focuses on a property of the mapping, while the term matrix multiplication focuses on how such a mapping is implemented.

G.2.3   Case of Vector Space of Real Functions

We consider that the domain and the target are both a vector space of real functions, say V. An operator T of the vector space of real functions V associates a function $g = Tf$ to any function f of V.

G.3   Adjoint Operator

G.3.1   Definition

Let $L : V \to V$ be a linear operator on an inner product space V.

Definiton. We say that L has an adjoint on V if there exists a linear operator L^* on V satisfying for all $x, y \in V$:

$$\langle L(x), y \rangle = \langle x, L^*(y) \rangle$$

The existence of the adjoint is not guaranteed.

G.3.2 Existence and Uniqueness of the Adjoint Operator in a Finite-Dimensional Inner Product Space

Theorem. Let V be a finite dimensionnal inner product space, and let L be a linear operator on V.

There exists a unique function $L^* : V \to V$ such that $\langle L(x), y \rangle = \langle x, L^*(y) \rangle$ for all $x, y \in V$. Furthermore, L^* is linear.

It's important to note that the adjoint of L depends not only on L but on the inner product as well.

G.3.3 Properties of Adjoint Operators

Theorem. Let V be a finite dimensionnal inner product space over $\mathbb{R}$, and let L and M be linear operators on V, having adjoints. Then

  i $(L + M)^* = L^* + M^*$
 ii $(aL)^* = aL^*$ for any scalar $a \in \mathbb{R}$
iii $(LM)^* = M^*L^*$
iv If I is the identity operator, then $I^* = I$
 v $(L^*)^* = L$

G.3.4 Adjoint Matrix

Definition. Let $[A] = [a_{i,j}]$ be a $m \times n$ matrix with complex entries. The adjoint matrix of $[A]$ is the $n \times m$ matrix $[A]^* = [b_{i,j}]$ such that $b_{i,j} = \overline{a_{j,i}}$.

That is, $[A]^* = \overline{[A]^t}$.

In the case that $[A]$ has real entries, the adjoint matrix of $[A]$ is simply the transpose of $[A]$.

G.3.5 Properties of Ajoint Matrices

Theorem. Let $[A]$ and $[B]$ be matrices of order $m \times n$ over the field of complex scalars $\mathbb{C}$. Then

   i $([A] + [B])^* = [A]^* + [B]^*$
   ii $(a[A])^* = \overline{a}[A]^*$ for any scalar $a \in \mathbb{C}$
  iii $([A][B])^* = [B]^*[A]^*$
  iv If $[I]$ is the identity matrix, then $[I]^* = [I]$
   v $([A]^*)^* = [A]$

The properties of adjoint matrices are formally similar to the properties of adjoint operators.

G.3.6 How to Find Adjoint of a Linear Operator?

Theorem. Let V be a finite-dimensional inner product space, and let $\beta = \{\alpha_1, \alpha_2, \ldots \alpha_n\}$ be an ordered orthonormal basis for V. Let L be a linear operator on V and let $[A]$ be the matrix of L in the ordered basis β. Then the matrix of L^* is the conjugate transpose of the matrix of L. That is,

$$[L^*]_\beta = [L]_\beta^*$$

Through the use of an orthonormal basis, this adjoint operation on linear operators (passing from L to L^* is identified with the operation of forming the conjugate transpose of a matrix.

G.4 Operator, its Adjoint Operator and Eigen Values, Eigen Vectors

Theorem. Let V be a finite-dimensional inner product space over $\mathbb{C}$, and let L be a linear operator on V. λ is an eigen value of $L \Leftrightarrow \overline{\lambda}$ is an eigen value of L^*.

This means that the eigenvalues of L^* are the complex conjugate eigenvalues of L and reciprocally. If V is a finite-dimensional inner product space over $\mathbb{R}$, then a linear operator and its adjoint have the same eigenvalues. However, the eigen vectors might be different, for example:

$$\text{The eigenvector of the matrix } [A] = \begin{bmatrix} 0 & 1 \\ 0 & 0 \end{bmatrix} \text{ is } \begin{bmatrix} 1 \\ 0 \end{bmatrix}$$

$$\text{The eigen vector of its adjoint } [A]^* = \begin{bmatrix} 0 & 0 \\ 1 & 0 \end{bmatrix} \text{ is } \begin{bmatrix} 0 \\ 1 \end{bmatrix}$$

G.5   Self Adjoint Operator

G.5.1   Definition and Properties of Self Adjoint Operator

Definition. A linear operator or matrix is called self-adjoint or Hermitian if it is equal to its own adjoint. A symmetric matrix is self-adjoint.

Theorem. If the linear operator L is selfadjoint, then

  i Every eigenvalue of a self-adjoint operator is real.
 ii Eigenvectors with distinct eigenvalues are orthogonal.

G.5.2   Self Ajoint Operator in the Vector Space of Real Functions

We admit that the Laplace operator and multiplication by a function are self adjoint operators.

Index

© The Editor(s) (if applicable) and The Author(s), under exclusive license to Springer Nature Switzerland AG 2026
H. Grard, *Diffusion of Neutrons in Nuclear Reactors*,
https://doi.org/10.1007/978-3-032-05088-5

The manufacturer's authorised representative in the EU is Springer
Nature Customer Service Centre GmbH, Europaplatz 3, 69115 Heidelberg,
Germany. If you have any concerns regarding our products, please
contact ProductSafety@springernature.com

Printed and bound by CPI Group (UK) Ltd, Croydon, CR0 4YY
07/07/2026
02160901-0001